Adaptive Image Processing

Processing

A Computational Intelligence Perspective

SECOND EDITION

IMAGE PROCESSING SERIES

Series Editor: Phillip A. Laplante, Pennsylvania State University

Published Titles

Adaptive Image Processing: A Computational Intelligence Perspective, Second Edition
Kim-Hui Yap, Ling Guan, Stuart William Perry, and Hau-San Wong

Color Image Processing: Methods and Applications
Rastislav Lukac and Konstantinos N. Plataniotis

Image Acquisition and Processing with LabVIEW™
Christopher G. Relf

Image and Video Compression for Multimedia Engineering, Second Edition
Yun Q. Shi and Huiyang Sun

Multimedia Image and Video Processing
Ling Guan, S.Y. Kung, and Jan Larsen

Shape Classification and Analysis: Theory and Practice, Second Edition
Luciano da Fontoura Costa and Roberto Marcondes Cesar, Jr.

Single-Sensor Imaging: Methods and Applications for Digital Cameras
Rastislav Lukac

Software Engineering for Image Processing Systems
Phillip A. Laplante

Adaptive Image Processing

A Computational Intelligence Perspective

SECOND EDITION

Kim-Hui Yap

Ling Guan

Stuart William Perry

Hau-San Wong

CRC Press
Taylor & Francis Group
Boca Raton   London   New York

CRC Press is an imprint of the
Taylor & Francis Group, an **informa** business

CRC Press
Taylor & Francis Group
6000 Broken Sound Parkway NW, Suite 300
Boca Raton, FL 33487-2742

© 2010 by Taylor and Francis Group, LLC
CRC Press is an imprint of Taylor & Francis Group, an Informa business

No claim to original U.S. Government works

Printed in the United States of America on acid-free paper
10 9 8 7 6 5 4 3 2 1

International Standard Book Number: 978-1-4200-8435-1 (Hardback)

Library of Congress Cataloging-in-Publication Data

Adaptive image processing : a computational intelligence perspective / authors, Kim-Hui Yap ... [et al.]. -- 2nd ed.
    p. cm. -- (Image processing series)
    Rev. ed. of: Adaptive image processing : a computational intelligence perspective / Stuart William Perry, Hau-San Wong, Ling Guan.. 2002.
    "A CRC title."
    Includes bibliographical references and index.
    ISBN 978-1-4200-8435-1 (hardcover : alk. paper)
    1. Image processing--Data processing. 2. Image processing--Digital techniques. 3. Computational intelligence. I. Yap, Kim Hui. II. Perry, Stuart William. Adaptive image processing.

TA1637.P46 2010
621.36'7--dc22
                                                                 2009041602

Visit the Taylor & Francis Web site at
http://www.taylorandfrancis.com

and the CRC Press Web site at
http://www.crcpress.com

Contents

Preface

In this book, we consider the application of computational intelligence techniques to the problem of adaptive image processing. In adaptive image processing, it is usually required to identify each image pixel with a particular feature type (e.g., smooth regions, edges, textures, etc.) for separate processing, which constitutes a segmentation problem. We will then establish image models to describe the desired appearance of the respective feature types or, in other words, to characterize each feature type. Finally, we modify the pixel values in such a way that the appearance of the processed features conforms more closely with that specified by the feature models, where the degree of discrepancy is usually measured in terms of cost function. In other words, we are searching for a set of parameters that minimize this function, that is, an optimization problem.

To satisfy the above requirements, we consider the application of computational intelligence (CI) techniques to this class of problems. Here we will adopt a specific definition of CI, which includes neural network techniques (NN), fuzzy set theory (FS), and evolutionary computation (EC). A distinguishing characteristic of these algorithms is that they are either biologically inspired, as in the cases of NN and EC, or are attempts to mimic how human beings perceive everyday concepts, as in FS.

The choice of these algorithms is due to the direct correspondence between some of the above requirements with the particular capabilities of specific CI approaches. For example, segmentation can be performed by using NN. In addition, for the purpose of optimization, we can embed the image model parameters as adjustable network weights to be optimized through the network's dynamic action. In contrast, the main role of fuzzy set theory is to address the requirement of characterization, that is, the specification of human visual preferences, which are usually expressed in fuzzy languages, in the form of multiple fuzzy sets over the domain of pixel value configurations, and the role of EC is mainly in addressing difficult optimization problems.

In this book, the essential aspects of the adaptive image processing problems are illustrated through a number of applications organized in two parts. The first part of the book focuses on adaptive image restoration. The problem is representative of the general adaptive image processing paradigm in that the three requirements of segmentation, characterization, and optimization are present. The second part of the book centers on image analysis and retrieval. It examines the problems of edge detection and characterization, self-organization for pattern discovery, and content-based image categorization and retrieval. This section will demonstrate how CI techniques can be used to address various challenges in adaptive image processing including low-level image processing, visual content analysis, and feature representation.

This book consists of 11 chapters. The first chapter provides material of an introductory nature to describe the basic concepts and current state of the art in the field of computational intelligence for image restoration, edge detection, image analysis, and retrieval. Chapter 2 gives a mathematical description of the restoration problem from the Hopfield neural network perspective and describes current algorithms based on this method. Chapter 3 extends the algorithm presented in Chapter 2 to implement adaptive constraint restoration methods for both spatially invariant and spatially variant degradations. Chapter 4 utilizes a perceptually motivated image error measure to introduce novel restoration algorithms. Chapter 5 examines how model-based neural networks can be used to solve image-restoration problems. Chapter 6 examines image-restoration algorithms making use of the principles of evolutionary computation. Chapter 7 examines the difficult concept of image restoration when insufficient knowledge of the degrading function is available. Chapter 8 examines the subject of edge detection and characterization using model-based neural networks. Chapter 9 provides an in-depth coverage of the self-organizing tree map, and demonstrates its application in image analysis and retrieval. Chapter 10 examines content representation in compressed domain image classification using evolutionary algorithm. Finally, Chapter 11 explores the fuzzy user perception and small sample problem in content-based image retrieval and develops CI techniques to address these challenges.

Acknowledgments

We are grateful to our colleagues, especially Dr. Kui Wu in the Media Technology Lab of Nanyang Technological University, Singapore for their contributions and helpful comments during the preparation of this book. Our special thanks to Professor Terry Caelli for the many stimulating exchanges that eventually led to the work in Chapter 8. We would also like to thank Nora Konopka and Amber Donley of CRC Press for their advice and assistance. Finally, we are grateful to our families for their patience and support while we worked on the book.

1

Introduction

1.1 Importance of Vision

All life-forms require methods for sensing the environment. Being able to sense one's surroundings is of such vital importance for survival that there has been a constant race for life-forms to develop more sophisticated sensory methods through the process of evolution. As a consequence of this process, advanced life-forms have at their disposal an array of highly accurate senses. Some unusual sensory abilities are present in the natural world, such as the ability to detect magnetic and electric fields, or the use of ultrasound waves to determine the structure of surrounding obstacles. Despite this, one of the most prized and universal senses utilized in the natural world is vision.

Advanced animals living aboveground rely heavily on vision. Birds and lizards maximize their fields of view with eyes on each side of their skulls, while other animals direct their eyes forward to observe the world in three dimensions. Nocturnal animals often have large eyes to maximize light intake, while predators such as eagles have very high resolution eyesight to identify prey while flying. The natural world is full of animals of almost every color imaginable. Some animals blend in with surroundings to escape visual detection, while others are brightly colored to attract mates or warn aggressors. Everywhere in the natural world, animals make use of vision for their daily survival. The reason for the heavy reliance on eyesight in the animal world is due to the rich amount of information provided by the visual sense. To survive in the wild, animals must be able to move rapidly. Hearing and smell provide warning regarding the presence of other animals, yet only a small number of animals such as bats have developed these senses sufficiently to effectively utilize the limited amount of information provided by these senses to perform useful actions, such as to escape from predators or chase down prey. For the majority of animals, only vision provides sufficient information in order for them to infer the correct responses under a variety of circumstances.

Humans rely on vision to a much greater extent than most other animals. Unlike the majority of creatures we see in three dimensions with high resolution and color. In humans the senses of smell and hearing have taken second place to vision. Humans have more facial muscles than any other animal, because in our society facial expression is used by each of us as the primary indicator of the emotional states of other humans, rather than the scent signals used by many mammals. In other words, the human world revolves around visual stimuli and the importance of effective visual information processing is paramount for the human visual system.

To interact effectively with the world, the human vision system must be able to extract, process, and recognize a large variety of visual structures from the captured images. Specifically, before the transformation of a set of visual stimuli into a meaningful scene, the vision system is required to identify different visual structures such as edges and regions from the captured visual stimuli. Rather than adopting a uniform approach of processing these extracted structures, the vision system should be able to adaptively tune to the specificities of these different structures in order to extract the maximum amount of information for the subsequent recognition stage. For example, the system should selectively enhance the associated attributes of different regions such as color and textures in an adaptive manner such that for some regions, more importance is placed on the extraction and processing of the color attribute, while for other regions the emphasis is placed on the associated textural patterns. Similarly, the vision system should also process the edges in an adaptive manner such that those associated with an object of interest should be distinguished from those associated with the less important ones.

To mimic this adaptive aspect of biological vision and to incorporate this capability into machine vision systems have been the main motivations of image processing and computer vision research for many years. Analogous to the eyes, modern machine vision systems are equipped with one or more cameras to capture light signals, which are then usually stored in the form of digital images or video sequences for subsequent processing. In other words, to fully incorporate the adaptive capabilities of biological vision systems into machines necessitates the design of an effective *adaptive image processing* system. The difficulties of this task can already be foreseen since we are attempting to model a system that is the product of billions of years of evolution and is naturally highly complex. To give machines some of the remarkable capabilities that we take for granted is the subject of intensive ongoing research and the theme of this book.

1.2   Adaptive Image Processing

The need for adaptive image processing arises due to the need to incorporate the above adaptive aspects of biological vision into machine vision systems. For such systems the visual stimuli are usually captured through cameras and presented in the form of digital images that are essentially arrays of pixels, each of which is associated with a gray level value indicating the magnitude of the light signal captured at the corresponding position. To effectively characterize a large variety of image types in image processing, this array of numbers is usually modeled as a 2D discrete nonstationary random process. As opposed to stationary random processes where the statistical properties of the signal remain unchanged with respect to the 2D spatial index, the nonstationary process models the inhomogeneities of visual structures that are inherent in a meaningful visual scene. It is this inhomogeneity that conveys useful information of a scene, usually composed of a number of different objects, to the viewer. On the other hand, a stationary 2D random signal, when viewed as a gray level image, does not usually correspond to the appearances of real-world objects.

For a particular image-processing application (we interpret the term "image processing" in a wide sense such that applications in image analysis are also included), we usually assume the existence of an underlying *image model* [1–3], which is a mathematical description of a hypothetical process through which the current image is generated. If we suppose that an image is adequately described by a stationary random process, which, though not accurate in general, is often invoked as a simplifying assumption, it is apparent that only a single image model corresponding to this random process is required for further image processing. On the other hand, more sophisticated image-processing algorithms will account for the nonstationarity of real images by adopting *multiple* image models for more accurate representation. Individual regions in the image can usually be associated with a different image model, and the complete image can be fully characterized by a finite number of these local image models.

1.3 Three Main Image Feature Classes

The inhomogeneity in images implies the existence of more than one image feature type that convey independent forms of information to the viewer. Although variations among different images can be great, a large number of images can be characterized by a small number of feature types. These are usually summarized under the labels of smooth regions, textures, and edges (Figure 1.1). In the following, we will describe the essential characteristics of these three kinds of features, and the image models usually employed for their characterization.

1.3.1 Smooth Regions

Smooth regions usually comprise the largest proportion of areas in images, because surfaces of artificial or natural objects, when imaged from a distance, can usually be regarded as smooth. A simple model for a smooth region is the assignment of a constant gray-level value to a restricted domain of the image lattice, together with the addition of Gaussian noise of appropriate variance to model the sensor noise [2,4].

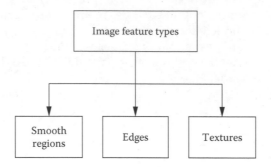

FIGURE 1.1
Three important classes of feature in images.

1.3.2 Edges

As opposed to smooth regions, edges comprise only a very small proportion of areas in images. Nevertheless, most of the information in an image is conveyed through these edges. This is easily seen when we look at the edge map of an image after edge detection: we can readily infer the original contents of the image through the edges alone. Since edges represent locations of abrupt transitions of gray-level values between adjacent regions, the simplest edge model is therefore a random variable of high variance, as opposed to the smooth region model that uses random variables with low variances. However, this simple model does not take into account the structural constraints in edges, which may then lead to their confusion with textured regions with equally high variances. More sophisticated edge models include the facet model [5], which approximates the different regions of constant gray level values around edges with separate piecewise continuous functions. There is also the edge-profile model, which describes the one-dimensional cross section of an edge in the direction of maximum gray level variation [6,7]. Attempts have been made to model this profile using a step function and various monotonically increasing functions. Whereas these models mainly characterize the *magnitude* of gray-level-value transition at the edge location, the edge diagram in terms of zero crossings of the second-order gray level derivatives, obtained through the process of Laplacian of Gaussian (LoG) filtering [8,9], characterizes the edge *positions* in an image. These three edge models are illustrated in Figure 1.2.

1.3.3 Textures

The appearance of textures is usually due to the presence of natural objects in an image. The textures usually have a noise-like appearance, although they are distinctly different from noise in that there usually exists certain discernible patterns within them. This is due to the correlations among the pixel values in specific directions. Due to this noise-like appearance, it is natural to model textures using a two-dimensional random field. The simplest approach

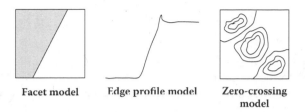

Facet model   **Edge profile model**   **Zero-crossing model**

FIGURE 1.2
Examples of edge models.

is to use i.i.d. (independent and identically distributed) random variables with appropriate variances, but this does not take into account the correlations among the pixels. A generalization of this approach is the adoption of Gauss–Markov random field (GMRF) [10–14] and Gibbs random field [15,16] which model these local correlational properties. Another characteristic of textures is their self-similarities: the patterns usually look similar when observed under different magnifications. This leads to their representation as fractal processes [17,18] that possess this very self-similar property.

1.4 Difficulties in Adaptive Image-Processing System Design

Given the very different properties of these three feature types, it is usually necessary to incorporate spatial adaptivity into image-processing systems for optimal results. For an image-processing system, a set of system parameters is usually defined to control the quality of the processed image. Assuming the adoption of spatial domain-processing algorithms, the gray-level value x_{i_1,i_2} at spatial index (i_1, i_2) is determined according to the following relationship.

$$x_{i_1,i_2} = f(\mathbf{y}; \mathbf{p}_{SA}(i_1, i_2)) \tag{1.1}$$

In this equation, the mapping f summarizes the operations performed by the image-processing system. The vector $\mathbf{y}$ denotes the gray-level values of the original image before processing, and $\mathbf{p}_{SA}$ denotes a vector of *spatially adaptive* parameters as a function of the spatial index (i_1, i_2). It is reasonable to expect that different parameter vectors are to be adopted at different positions (i_1, i_2), which usually correspond to different feature types. As a result, an important consideration in the design of this adaptive image-processing system is the proper determination of the parameter vector $\mathbf{p}_{SA}(i_1, i_2)$ as a function of the spatial index (i_1, i_2).

On the other hand, for nonadaptive image-processing systems, we can simply adopt a constant assignment for $\mathbf{p}_{SA}(i_1, i_2)$

$$\mathbf{p}_{SA}(i_1, i_2) \equiv \mathbf{p}_{NA} \tag{1.2}$$

where $\mathbf{p}_{NA}$ is a constant parameter vector.

We consider examples of $\mathbf{p}_{SA}(i_1, i_2)$ in a number of specific image-processing applications below:

- In image filtering, we can define $\mathbf{p}_{SA}(i_1, i_2)$ to be the set of filter coefficients in the convolution mask [2]. Adaptive filtering [19,20] thus corresponds to using a different mask at different spatial locations, while nonadaptive filtering adopts the same mask for the whole image.

- In image restoration [21–23], a *regularization parameter* [24–26] is defined that controls the degree of ill-conditioning of the restoration process, or equivalently, the overall smoothness of the restored image. The vector $\mathbf{p}_{SA}(i_1, i_2)$ in this case corresponds to the scalar regularization parameter. Adaptive regularization [27–29] involves selecting different parameters at different locations, and nonadaptive regularization adopts a single parameter for the whole image.

- In edge detection, the usual practice is to select a single *threshold* parameter on the gradient magnitude to distinguish between the edge and nonedge points of the image [2,4], which corresponds to the case of nonadaptive thresholding. This can be considered as a special case of adaptive thresholding, where a threshold value is defined at each spatial location.

Given the above description of adaptive image processing, we can see that the corresponding problem of adaptive parameterization, that of determining the parameter vector $\mathbf{p}_{SA}(i_1, i_2)$ as a function of (i_1, i_2), is particularly acute compared with the nonadaptive case. In the nonadaptive case, and in particular for the case of a parameter vector of low dimensionality, it is usually possible to determine the optimal parameters by interactively choosing different parameter vectors and evaluating the final processed results.

On the other hand, for adaptive image processing, it is almost always the case that a parameter vector of high dimensionality, which consists of the concatenation of all the local parameter vectors, will be involved. If we relax the previous requirement to allow the subdivision of an image into regions and the assignment of the same local parameter vector to each region, the dimension of the resulting concatenated parameter vector can still be large. In addition, the requirement to identify each image pixel with a particular feature type itself constitutes a nontrivial *segmentation* problem. As a result, it is usually not possible to estimate the parameter vector by trial and error. Instead, we should look for a parameter assignment algorithm that would automate the whole process.

To achieve this purpose, we will first have to establish image models that describe the desired local gray-level value configurations for the respective image feature types or, in other words, to *characterize* each feature type. Since the local gray-level configurations of the processed image are in general a function of the system parameters as specified in Equation (1.1), we can associate a *cost function* with each gray-level configuration that measures its degree of conformance to the corresponding model, with the local system parameters as arguments of the cost function. We can then search for those system parameter values that minimize the cost function for each feature type, that is, an *optimization* process. Naturally, we should adopt *different* image models in order to obtain different system parameters for each type of feature.

In view of these requirements, we can summarize the requirements for a successful design of an adaptive image-processing system as follows:

1.4.1 Segmentation

Segmentation requires a proper understanding of the difference between the corresponding structural and statistical properties of the various feature types, including those of edges, textures, and smooth regions, to allow partition of an image into these basic feature types.

1.4.2 Characterization

Characterization requires an understanding of the most desirable gray-level value configurations in terms of the characteristics of the human vision system (HVS) for each of the basic feature types, and the subsequent formulation of these criteria into cost functions in terms of the image model parameters, such that the minimization of these cost functions will result in an approximation to the desired gray-level configurations for each feature type.

1.4.3 Optimization

In anticipation of the fact that the above criteria will not necessarily lead to well-behaved cost functions, and that some of the functions will be nonlinear or even nondifferentiable, we should adopt powerful optimization techniques for the searching of the optimal parameter vector.

These three main requirements are summarized in Figure 1.3.

In this book, our main emphasis is on two specific adaptive image-processing systems and their associated algorithms: the adaptive image-restoration algorithm and the adaptive edge-characterization

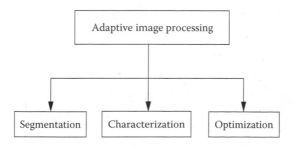

FIGURE 1.3
Three main requirements in adaptive image processing.

algorithm. For the former system, segmentation is first applied to partition the image into separate regions according to a local variance measure. Each region then undergoes characterization to establish whether it corresponds to a smooth, edge, or textured area. Optimization is then applied as a final step to determine the optimal regularization parameters for each of these regions. For the second system, a preliminary segmentation stage is applied to separate the edge pixels from nonedge pixels. These edge pixels then undergo the characterization process whereby the more salient ones among them (according to the users' preference) are identified. Optimization is finally applied to search for the optimal parameter values for a parametric model of this salient edge set.

1.5   Computational Intelligence Techniques

Considering the above stringent requirements for the satisfactory performance of an adaptive image-processing system, it will be natural to consider the class of algorithms commonly known as computational intelligence techniques. The term "computational intelligence" [30,31] has sometimes been used to refer to the general attempt to simulate human intelligence on computers, the so-called "artificial intelligence" (AI) approach [32]. However, in this book, we will adopt a more specific definition of computational intelligence techniques that are neural network techniques, fuzzy logic, and evolutionary computation (Figure 1.4). These are also referred to as the "numerical" AI approaches (or sometimes "soft computing" approach [33]) in contrast to the "symbolic" AI approaches as typified by the expression of human knowledge in terms of linguistic variables in expert systems [32].

A distinguishing characteristic of this class of algorithms is that they are usually biologically inspired: the design of neural networks [34,35], as the

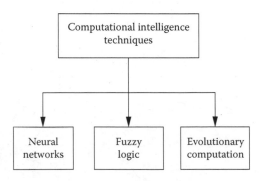

FIGURE 1.4
Three main classes of computational intelligence algorithms.

name implies, draws inspiration mainly from the structure of the human brain. Instead of adopting the serial processing architecture of the Von Neumann computer, a neural network consists of a large number of computational units or neurons (the use of this term again confirming the biological source of inspiration) that are massively interconnected with each other just as the real neurons in the human brain are interconnected with axons and dendrites. Each such connection between the artificial neurons is characterized by an adjustable *weight* that can be modified through a training process such that the overall behavior of the network is changed according to the nature of specific training examples provided, again reminding one of the human learning process.

On the other hand, fuzzy logic [36–38] is usually regarded as a formal way to describe how human beings perceive everyday concepts: whereas there is no exact height or speed corresponding to concepts like "tall" and "fast," respectively, there is usually a general consensus by humans as to approximately what levels of height and speed the terms are referring to. To mimic this aspect of human cognition on a machine, fuzzy logic avoids the arbitrary assignment of a particular numerical value to a single class. Instead, it defines each such class as a *fuzzy set* as opposed to a *crisp set*, and assigns a fuzzy set membership value within the interval [0,1] for each class that expresses the degree of membership of the particular numerical value in the class, thus generalizing the previous concept of crisp set membership values within the discrete set {0,1}.

For the third member of the class of computational intelligence algorithms, no concept is closer to biology than the concept of evolution, which is the incremental adaptation process by which living organisms increase their fitness to survive in a hostile environment through the processes of mutation and competition. Central to the process of evolution is the concept of a *population* in which the better adapted individuals gradually displace the not so well-adapted ones. Described within the context of an optimization algorithm, an *evolutionary computational algorithm* [39,40] mimics this aspect of evolution by generating a population of potential solutions to the optimization problem, instead of a sequence of single potential solutions, as in the case of gradient descent optimization or simulated annealing [16]. The potential solutions are allowed to compete against each other by comparing their respective cost function values associated with the optimization problem with each other. Solutions with high cost function values are displaced from the population while those with low cost values survive into the next generation. The displaced individuals in the population are replaced by generating new individuals from the survived solutions through the processes of mutation and recombination. In this way, many regions in the search space can be explored simultaneously, and the search process is not affected by local minima as no gradient evaluation is required for this algorithm.

We will now have a look at how the specific capabilities of these computational intelligence techniques can address the various problems encountered in the design and parameterization of an adaptive image-processing system.

1.5.1   Neural Networks

Artificial neural networks represent one of the first attempts to incorporate learning capabilities into computing machines. Corresponding to the biological neurons in human brain, we define artificial neurons that perform simple mathematical operations. These artificial neurons are connected with each other through *network weights* that specify the strength of the connection. Analogous to its biological counterpart, these network weights are adjustable through a learning process that enables the network to perform a variety of computational tasks. The neurons are usually arranged in *layers*, with the input layer accepting signals from the external environment, and the output layer emitting the result of the computations. Between these two layers are usually a number of *hidden* layers that perform the intermediate steps of computations. The architecture of a typical artificial neural network with one hidden layer is shown in Figure 1.5. In specific types of network, the hidden layers may be missing and only the input and output layers are present.

The adaptive capability of neural networks through the adjustment of the network weights will prove useful in addressing the requirements of segmentation, characterization, and optimization in adaptive image-processing system design. For segmentation, we can, for example, ask human users to specify which part of an image corresponds to edges, textures, and smooth regions, etc. We can then extract image features from the specified regions as training examples for a properly designed neural network such that the trained network will be capable of segmenting a previously unseen image into the primitive feature types. Previous works where a neural network is applied to the problem of image segmentation are detailed in References [41–43].

A neural network is also capable of performing characterization to a certain extent, especially in the process of *unsupervised competitive learning* [34,44], where both segmentation and characterization of training data are carried

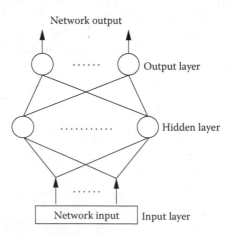

FIGURE 1.5
Architecture of a neural network with one hidden layer.

out: during the competitive learning process, individual neurons in the network, which represent distinct subclasses of training data, gradually build up *templates* of their associated subclasses in the form of *weight vectors*. These templates serve to characterize the individual subclasses.

In anticipation of the possible presence of nonlinearity in the cost functions for parameter estimation during the optimization process, a neural network is again an ideal candidate for accommodating such difficulties: the operation of a neural network is inherently nonlinear due to the presence of the sigmoid neuronal transfer function. We can also tailor the nonlinear neuronal transfer function specifically to a particular application. More generally, we can *map* a cost function onto a neural network by adopting an architecture such that the image model parameters will appear as adjustable weights in the network [45,46]. We can then search for the optimal image model parameters by minimizing the embedded cost function through the dynamic action of the neural network.

In addition, while the distributed nature of information storage in neural networks and the resulting fault-tolerance is usually regarded as an overriding factor in its adoption, we will, in this book, concentrate rather on the possibility of task localization in a neural network: we will subdivide the neurons into *neuron clusters*, with each cluster specialized for the performance of a certain task [47,48]. It is well known that similar localization of processing occurs in the human brain, as in the classification of the cerebral cortex into visual area, auditory area, speech area, and motor area, etc. [49,50]. In the context of adaptive image processing, we can, for example, subdivide the set of neurons in such a way that each cluster will process the three primitive feature types, namely, textures, edges, and smooth regions. The values of the connection weights in each subnetwork can be different, and we can even adopt different architectures and learning strategies for each subnetwork for optimal processing of its assigned feature type.

1.5.2 Fuzzy Logic

From the previous description of fuzzy techniques, it is obvious that its main application in adaptive image processing will be to address the requirement of characterization, that is, the specification of human visual preferences in terms of gray-level value configurations. Many concepts associated with image processing are inherently fuzzy, such as the description of a region as "dark" or "bright," and the incorporation of fuzzy set theory is usually required for satisfactory processing results [51–55]. The very use of the words "textures," "edges," and "smooth regions" to characterize the basic image feature types implies fuzziness: the difference between smooth regions and weak textures can be subtle, and the boundary between textures and edges is sometimes blurred if the textural patterns are strongly correlated in a certain direction so that we can regard the pattern as multiple edges. Since the image-processing system only recognizes gray-level configurations, it will be natural to define fuzzy sets with qualifying terms like "texture," "edge," and "smooth regions" over the set of corresponding gray-level configurations according to human preferences. However, one of the problems with this approach is that there is usually an extremely large number of possible gray-level configurations corresponding to each feature type, and human beings cannot usually relate what they perceive as a certain feature type to a particular configuration. In Chapter 5, a *scalar* measure has been established that characterizes the degree of resemblance of a gray-level configuration to either textures or edges. In addition, we can establish the exact interval of values of this measure where the configuration will more resemble textures than edges and vice versa. As a result, we can readily define fuzzy sets over this one-dimensional *universe of discourse* [37].

In addition, fuzzy set theory also plays an important role in the derivation of improved segmentation algorithms. A notable example is the *fuzzy c-means algorithm* [56–59], which is a generalization of the *k-means algorithm* [60] for data clustering. In the *k*-means algorithm, each data vector, which may contain feature values or gray-level values as individual components in image processing applications, is assumed to belong to one and only one class. This may result in inadequate characterization of certain data vectors that possess properties common to more than one class, but then get arbitrarily assigned to one of those classes. This is prevented in the fuzzy *c*-means algorithm, where each data vector is assumed to belong to every class to a different degree that is expressed by a numerical membership value in the interval [0,1]. This paradigm can now accommodate those data vectors that possess attributes common to more than one class, in the form of large membership values in several of these classes.

1.5.3   Evolutionary Computation

The often stated advantages of evolutionary computation include its implicit parallelism that allows simultaneous exploration of different regions of the search space [61], and its ability to avoid local minima [39,40]. However, in this book, we will emphasize its capability to search for the optimizer of a *nondifferentiable* cost function efficiently, that is, to satisfy the requirement of optimization. An example of a nondifferentiable cost function in image processing would be the metric that compares the probability density function (pdf) of a certain local attribute of the image (gray-level values, gradient magnitudes, etc.) with a desired pdf. We would, in general, like to adjust the parameters of the adaptive image-processing system in such a way that the distance between the pdf of the processed image is as close as possible to the desired pdf. In other words, we would like to minimize the distance as a function of the system parameters. In practice, we have to approximate the pdfs using histograms of the corresponding attributes, which involves the counting of discrete quantities. As a result, although the pdf of the processed image is a function of the system parameters, it is not differentiable with respect to these parameters. Although stochastic algorithms like simulated annealing can also be applied to minimize nondifferentiable cost functions,

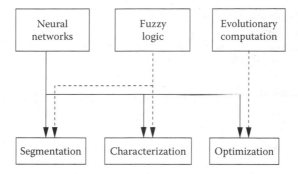

FIGURE 1.6
Relationships between the computational intelligence algorithms and the main requirements in adaptive image processing.

evolutionary computational algorithms represent a more efficient optimization approach due to the implicit parallelism of its population-based search strategy.

The relationship between the main classes of algorithms in computational intelligence and the major requirements in adaptive image processing is summarized in Figure 1.6.

1.6   Scope of the Book

In this book, as specific examples of adaptive image-processing systems, we consider the *adaptive regularization* problem in image restoration [27–29], the edge, characterization problem, the self-organization problem in image analysis, and the feature representation and fuzzy perception problem in image retrieval. We adopt computational intelligence techniques including neural networks, fuzzy methods, and evolutionary algorithms as the main approaches to address these problems due to their capabilities to satisfy all three requirements in adaptive image processing, as illustrated in Figure 1.6.

1.6.1   Image Restoration

The act of attempting to obtain the original image given the degraded image and some knowledge of the degrading factors is known as *image restoration*. The problem of restoring an original image, when given the degraded image, with or without knowledge of the degrading *point spread function* (PSF) or degree and type of noise present is an ill-posed problem [21,24,62,63] and can be approached in a number of ways such as those given in References [21,64,66]. For all useful cases a set of simultaneous equations is produced that is too large to be solved analytically. Common approaches to this problem can be divided into two categories: inverse filtering or transform-related techniques, and algebraic techniques. An excellent review of classical image-restoration techniques is given by Andrews and Hunt [21]. The following references also contain surveys of restoration techniques: Katsaggelos [23], Sondhi [67], Andrews [68], Hunt [69], and Frieden [70].

Image Degradations

There exist a large number of possible degradations that an image can suffer. Common degradations are blurring, motion, and noise. Blurring can be caused when an object in the image is outside the camera's depth of field some time during the exposure. Noise is generally a distortion due to the imaging system rather than the scene recorded. Noise results in random variations to pixel values in the image. This could be caused by the imaging system itself, or the recording or transmission medium. In this book, we consider only image degradations that may be described by a linear model. For these distortions, a suitable mathematical model is given in Chapter 2.

Adaptive Regularization

In regularized image restoration, the associated cost function consists of two terms: a data conformance term that is a function of the degraded image pixel values and the degradation mechanism, and the model conformance term that is usually specified as a continuity constraint on neighboring gray-level values to alleviate the problem of ill-conditioning characteristic of this kind of inverse problems. The *regularization parameter* [23,25] controls the relative contributions of the two terms toward the overall cost function. In general, if the regularization parameter is increased, the model conformance term is emphasized at the expense of the data conformance term, and the restored image becomes smoother while the edges and textured regions become blurred. On the contrary, if we decrease the parameter, the fidelity of the restored image is increased at the expense of decreased noise smoothing.

Perception-Based Error Measure for Image Restoration

The most common method to compare the similarity of two images is to compute their mean square error (MSE). However, the MSE relates to the power of the error signal and has little relationship to human visual perception. An important drawback to the MSE and any cost function that attempts to use the MSE to restore a degraded image is that the MSE treats the image as a stationary process. All pixels are given equal priority regardless of their relevance to human perception. This suggests that information is ignored. In view of the problems with classical error measures such as the MSE, Perry and Guan [71] and Perry [72] presented a different error measure, *local standard deviation mean square error* (LSMSE), which is based on the comparison of local standard deviations in the neighborhood of each pixel instead of their gray-level values. The LSMSE is calculated in the following way: Each pixel in the two images to be compared has its local standard deviation calculated over a small neighborhood centered on the pixel. The error between each pixel's local standard deviation in the first image and the corresponding pixel's local standard deviation in the second image is computed. The LSMSE is the mean squared error of these differences over all pixels in the image. The mean square error between the two standard deviations gives an indication of the degree of similarity between the two images. This error measure requires matching between the high- and low-variance regions of the image, which is more intuitive in terms of human visual perception. Generally throughout this book the size of the local neighborhoods used in the LSMSE calculation will be a 9-by-9 square centered on the pixel of interest. This alternative error measure will be heavily relied upon in Chapters 3 and 4. A mathematical description is given in Chapter 3.

Blind Deconvolution

In comparison with the determination of the regularization parameter for image restoration, the problem of *blind deconvolution* is considerably more difficult, since in this case the degradation mechanism, or equivalently the form of the point spread function, is unknown to the user. As a result, in addition to estimating the local regularization parameters, we have to estimate the coefficients of the point spread function itself. In Chapter 7, we describe an approach for blind deconvolution that is based on computational intelligence techniques. Specifically, the blind deconvolution problem is first formulated within the framework of evolutionary strategy where a pool of candidate PSFs is generated to form the population in evolutionary strategy (ES). A new cost function that incorporates the specific requirement of blind deconvolution in the form of a point spread function domain regularization term, which ensures the emergence of a valid PSF, in addition to the previous data fidelity measure and image regularization term is adopted as the fitness function in the evolutionary algorithm. This new blind deconvolution approach will be described in Chapter 7.

1.6.2   Edge Characterization and Detection

The characterization of important features in an image requires the detailed specification of those pixel configurations that human beings would regard as significant. In this work, we consider the problem of representing human preferences, especially with regard to image interpretation, again in the form of a model-based neural network with hierarchical architecture [48,73,74]. Since it is difficult to represent all aspects of human preferences in interpreting images using traditional mathematical models, we encode these preferences through a direct *learning* process, using image pixel configurations that humans usually regard as visually significant as training examples. As a first step, we consider the problem of edge characterization in such a network. This representation problem is important since its successful solution would allow computer vision systems to simulate to a certain extent the decision process of human beings when interpreting images.

Although the network can be considered as a particular implementation of the stages of segmentation and characterization in the overall adaptive image-processing scheme, it can also be regarded as a self-contained adaptive image-processing system on its own: the network is designed such that it automatically partitions the edges in an image into different classes depending on the gray-level values of the surrounding pixels of the edge, and applies different detection thresholds to each of the classes. This is in contrast to the usual approach where a single detection threshold is adopted across the whole image independent of the local context. More importantly, instead of providing *quantitative* values for the threshold as in the usual case, the users are asked to provide *qualitative* opinions on what they regard as edges by manually tracing their desired edges on an image. The gray-level configurations around the trace are then used as training examples for the model-based neural network to acquire an internal model of the edges, which is another example of the design of an adaptive image-processing system through the training process.

As seen above, we have proposed the use of a hierarchical model-based neural network for the solution of both these problems as a first attempt. It was observed later that, whereas the edge characterization problem can be satisfactorily represented by this framework, resulting in adequate characterization of those image edges that humans regard as significant, there are some inadequacies in using this framework exclusively for the solution of the adaptive regularization problem, especially in those cases where the images are more severely degraded. These inadequacies motivate our later adoption of fuzzy set theory and evolutionary computation techniques, in addition to the previous neural network techniques, for this problem.

1.6.3   Self-Organizing Tree Map for Knowledge Discovery

Computational technologies based on artificial neural networks have been the focus for much research into the problem of unsupervised learning and data clustering, where the goal is to formulate or discover significant patterns or features from within a given set of data without the guidance of a teacher. Input patterns are usually stored as a set of prototypes or clusters: representations or natural groupings of similar data. In forming a description of an unknown set of data, such techniques find application across a range of industrial tasks that warrant significant need for data mining, that is, bioinformatics research, high-dimensional image analysis and visualization, information retrieval, and computer vision. Inherently unsupervised in nature, neural-network architectures based on principles of self-organization appear to be a natural fit.

Such architectures are characterized by their adherence to four key properties [34]: synaptic self-amplification for mining correlated stimuli, competition over limited resources, cooperative encoding of information, and the implicit ability to encode pattern redundancy as knowledge. Such principles are, in many ways, a reflection of Turing's observations in 1952 [75]: "Global order can arise from Local interactions." Such mechanisms exhibit a basis in the process of associative memory, and, receiving much neurobiological support, are believed to be fundamental to the organization that takes place in the human brain.

Static architectures such as Kohonen's self-organizing feature map (SOFM) [76] represent one of the most fundamental realizations of such principles, and have been the foundation for much neural network-based research into knowledge discovery. Their popularity arises out of their ability to infer an ordered or topologically preserved mapping of the underlying data space. Thus, relationships between discovered prototypes are captured in an output layer that is connected by some predefined topology. Mappings onto this layer are such that order is maintained: thus patterns near to one another in the input space map to nearby locations in an output layer. Such mechanisms are useful for qualitative visualization [77,78] of high dimensional, multivariate data, where users are left to perceive possible clustered regions in a dimension that is more familiar to them (2D or 3D).

The hierarchical feature map (HFM) [79] extends such ideas to a pyramidal hierarchy of SOFMs, each progressively trained in a top-down manner, to achieve some semblance of hierarchical partitioning on the input space. At the other end of the self-organizing spectrum is neural gas (NG) [80], which completely abandons any imposed topology: instead relying on the consideration of k nearest neighbors for the refinement of prototypes.

One of the most challenging tasks in any unsupervised learning problem arises by virtue of the fact that an attempt is being made to quantitatively discover a set of dominant patterns (clusters or classes) in which to categorize underlying data without any knowledge of what an appropriate number of classes might be. There are essentially two approaches taken as a result: either attempt to perform a series of separate runs of a static clustering algorithm over a range of different class numbers and assess which yields a better result according to some independent index of quality, or maintain a purely dynamic architecture that attempts to progressively realize an appropriate number of classes throughout the course of parsing a data set. The latter of the two approaches is advantageous from a resource and time of execution point of view.

Many dynamic neural network-based architectures have been proposed, as they seem particularly suited to developing a model of an input space, one item of data at a time: they evolve internally, through progressive stimulation by individual samples. Such dynamic architectures are generally hierarchical or nonstationary in nature, and extend upon HFM/SOFM such as in the growing hierarchical SOM (GHSOM) [81,82], or extend upon NG as in the growing neural gas (GNG) algorithm [83] and its associated variants: growing grid (GG) [84] and growing cell structures (GCS) [85].

1.6.4   Content-Based Image Categorization and Retrieval

Content-based image categorization is the process of classifying a given image into one of the predefined categories based on content analysis. Content analysis of images refers to the extraction of features such as color, texture, shape, or spatial relationship from the images as the signatures, and from which the indexes of the images are built. Content analysis can be categorized into spatial domain analysis and compressed domain analysis. Spatial domain analysis performs feature extraction in the original image domain. On the other hand, compressed domain analysis extracts features in the compressed domain directly in order to reduce the computational time involved in the decompression of the images.

Content-based image retrieval (CBIR) has been developed as an alternative search technique that complements text-based image retrieval. It utilizes content analysis to retrieve images that are similar to the query from the database. Users can submit their query to the systems in the form of an example image, which is often known as query-by-example (QBE). The systems then perform image content analysis by extracting visual features from the query and compare them with the features of the images in the database. After similarity comparison, the systems display the retrieval results to the users.

Content Analysis

Previous approaches for content-based image classification mainly focus on spatial domain analysis. This, however, is often expensive in terms of computational and storage requirements as most digital images nowadays are stored in the compressed formats. Feature extraction performed in the spatial domain requires the decompression of the compressed images first, resulting in significant computational cost. To alleviate the computational load, some works have focused on performing content analysis in the compressed domain. Most of the approaches that adopt compressed domain features give more emphasis on computational efficiency than their effectiveness in content representation. It is worth noting that often the compressed domain features may not fully represent the actual image contents. To address this issue, evolutionary computation techniques have been applied to obtain proper transformation of the compressed domain features. In doing so, image classification accuracy can be improved using transformed features while retaining the efficiency of the compressed domain techniques. The detailed algorithms will be explained in Chapter 10.

Relevance Feedback in CBIR

Many CBIR systems have been developed over the years that include both commercial and research prototypes. However, a challenging issue that restricts the performance of the CBIR systems is the semantic gap between the low-level visual features and the high-level human perception. To bridge the semantic gap, *relevance feedback* has been introduced into the CBIR systems. The main idea is to incorporate human in the loop to enhance the retrieval accuracy. Users are allowed to provide their relevance judgement on the retrieved images. The feedbacks are then learned by the systems to discover user information needs. There have been a lot of studies on relevance feedback in CBIR in recent years with various algorithms developed. Although the incorporation of relevance feedback has been shown to boost the retrieval performance, there are still two important issues that need to be considered when developing an efficient and effective CBIR system: (1) imprecision of user perception on the relevance of the feedback images and (2) the small sample problem:

- Typically, in most interactive CBIR systems, a user is expected to provide a binary decision of either "fully relevant" or "totally irrelevant" on the feedback images. At times, this may not agree with the uncertainty embedded in user perception. For example, in a scenario application, a user intends to find pets, especially, dogs. If the retrieved results contain cats, the user would face a dilemma as to whether to classify the cats as either fully relevant or totally irrelevant. This is because these cat images only satisfy user information need up to a certain extent. Therefore, we need to take the potential imprecision of user perception into consideration when developing relevance feedback algorithm.

- In an interactive CBIR system with relevance feedback, it is tedious for users to label many images. This gives rise to the small sample problem where learning from a small number of training samples restricts the retrieval performance.

To address these two challenges, computational intelligence techniques, namely, neural networks, clustering, fuzzy reasoning, and SVM will be employed due to their effectiveness. We will describe the proposed approaches in more details in Chapter 11.

1.7   Contributions of the Current Work

With regard to the problems posed by the requirements of segmentation, characterization, and optimization in the design of an adaptive image-processing system, we have devised a system of interrelated solutions comprising the use of the main algorithm classes of computational intelligence techniques. The contributions of the work described in this book can be summarized as follows.

1.7.1   Application of Neural Networks for Image Restoration

Different neural network models, which will be described in Chapters 2, 3, 4, and 5, are adopted for the problem of image restoration. In particular, a model-based neural network with hierarchical architecture [48,73,74] is derived for the problem of adaptive regularization. The image is segmented into smooth regions and combined edge/textured regions, and we assign a single subnetwork to each of these regions for the estimation of the regional parameters. An important new concept arising from this work is our alternative viewpoint of the regularization parameters as model-based neuronal weights, which are then trainable through the supply of proper training examples. We derive the training examples through the application of *adaptive nonlinear filtering* [86] to individual pixel neighborhoods in the image for an *independent* estimate of the current pixel value.

1.7.2   Application of Neural Networks to Edge Characterization

A model-based neural network with hierarchical architecture is proposed for the problem of edge characterization and detection. Unlike previous edge-detection algorithms where various threshold parameters have to be specified [2,4], this parameterization task can be performed implicitly in a neural network by supplying training examples. The most important concept in this part of the work is to allow human users to communicate their preferences to the adaptive image-processing system through the provision of qualitative training examples in the form of edge tracings on an image, which is a more natural way of specifying preferences for humans, than the selection of *quantitative* values for a set of parameters. With the adoption of this network architecture and the associated training algorithm, it will be shown that the network can generalize from sparse examples of edges provided by human users to detect all significant edges in images not in the training set. More importantly, no retraining and alteration of architecture is required for applying the same network to noisy images, unlike conventional edge detectors that usually require threshold readjustment.

1.7.3   Application of Fuzzy Set Theory to Adaptive Regularization

For the adaptive regularization problem in image restoration, apart from the requirement of adopting different regularization parameters for smooth regions and regions with high gray-level variances, it is also desirable to further separate the latter regions into edge and textured regions. This is due to the different noise masking capabilities of these two feature types, which in turn requires different regularization parameter values. In our previous discussion of fuzzy set theory, we have described a possible solution to this problem, in the form of characterizing the gray-level configurations corresponding to the above two feature types, and then define fuzzy sets with qualifying terms like "texture" and "edge" over the respective sets of configurations. However, one of the problems with this approach is that there is usually an extremely large number of possible gray-level configurations corresponding to each feature type, and human beings cannot usually relate what they perceive as a certain feature type to a particular configuration. In Chapter 5, a *scalar* measure has been established that characterizes the degree of resemblance of a gray-level configuration to either textures or edges. In addition, we can establish the exact interval of values of this measure where the configuration will more resemble textures than edges and vice versa. As a result, we can readily define fuzzy sets over this one-dimensional *universe of discourse* [37].

1.7.4 Application of Evolutionary Programming to Adaptive Regularization and Blind Deconvolution

Apart from the neural network-based techniques, we have developed an alternative solution to the problem of adaptive regularization using evolutionary programming, which is a member of the class of evolutionary computational algorithms [39,40]. Returning again to the ETC measure, we have observed that the distribution of the values of this quantity assumes a typical form for a large class of images. In other words, the shape of the probability density function (pdf) of this measure is similar across a broad class of images and can be modeled using piecewise continuous functions. On the other hand, this pdf will be different for blurred images or incorrectly regularized images. As a result, the model pdf of the ETC measure serves as a kind of *signature* for correctly regularized images, and we should minimize the difference between the corresponding pdf of the image being restored and the model pdf using some kind of distance measure. The requirement to approximate this pdf using a histogram, which involves the counting of discrete quantities, and the resulting nondifferentiability of the distance measure with respect to the various regularization parameters, necessitates the use of evolutionary computational algorithms for optimization. We have adopted evolutionary programming that, unlike the genetic algorithm which is another widely applied member of this class of algorithms, operates directly on real-valued vectors instead of binary-coded strings and is therefore more suited to the adaptation of the regularization parameters. In this algorithm, we have derived a parametric representation that expresses the regularization parameter value as a function of the local image variance. Generating a population of these *regularization strategies* that are vectors of the above hyperparameters, we apply the processes of mutation, competition, and selection to the members of the population to obtain the optimal regularization strategy. This approach is then further extended to solve the problem of blind deconvolution by including the point spread function coefficients in the set of hyperparameters associated with each individual in the population.

1.7.5 Application of Self-Organization to Image Analysis and Retrieval

A recent approach known as the self-organizing tree map (SOTM) [87] inherently incorporates hierarchical properties by virtue of its growth, in a manner that is far more flexible in terms of revealing the underlying data space without being constrained by an imposed topological framework. As such, the SOTM exhibits many desirable properties over traditional SOFM-based strategies.

Chapter 9 of the book will provide an in-depth coverage of this architecture. Due to the adaptive nature, this family of unsupervised methods exhibits a number of desirable properties over the SOFM and its early derivatives such as (1) better topological preservation to ensure the ability to adapt to different datasets; (2) consistent topological descriptions of the underlying datasets; (3) robust and succinct allocation of cluster prototypes; (4) built-in awareness of topological information, local density, and variance indicators for optimal selection of cluster prototypes at runtime; and (5) a true automatic mode to deduce simultaneously, optimal number of clusters, their prototypes, and an appropriate topological mapping associating them.

Chapter 9 will then cover a series of pertinent real-world applications with regards to the processing of image and video data—from its role in more generic image-processing techniques such as the automated modeling and removal of impulse noise in digital images, to problems in digital asset management including the modeling of image and video content, indexing, and intelligent retrieval.

1.7.6   Application of Evolutionary Computation to Image Categorization

To address the issue of accuracy in content representation that is crucial for compressed domain image classification, we propose to perform transformation on the compressed-domain features. These feature values are modeled as realizations of random variables. The transformation on the random variable is then carried out by the merging and removal of histogram bin counts. To search for the optimal transformation on the random variable, genetic algorithm (GA) has been employed to perform the task. The approach has been further extended by adopting individually optimized transformations for different image classes, where a set of separate classification modules is associated with each of these transformations.

1.7.7   Application of Computational Intelligence to Content-Based Image Retrieval

In order to address the imprecision of user perception in relevance feedback of CBIR systems, a fuzzy labeling scheme that integrates the user's uncertain perception of image similarity is proposed. In addition to the "relevant" and "irrelevant" choices, the proposed scheme provides a third "fuzzy" option to the user. The user can provide a feedback as "fuzzy" if the image only satisfies his or her partial information needs. Under this scheme, the soft relevance of the fuzzy images has to be estimated. An a posteriori probability estimator is developed to achieve this. With the combined relevant, irrelevant, and fuzzy images, a recursive fuzzy radial basis function network (RFRBFN) has been developed to learn the user information needs. To address the small sample problem in relevance feedback, a predictive-label fuzzy support vector machine (PLFSVM) framework has been developed. Under this framework, a clustering algorithm together with consideration of the correlation between labeled and unlabeled images has been used to select the predictive-labeled images. The selected images are assigned predictive-labels of either relevant or irrelevant using a proposed measure. A fuzzy membership function is then designed to estimate the significance of the predictive-labeled images. A fuzzy support vector machine (FSVM), which can deal with the class uncertainty of training images is employed by using the hybrid of both labeled and predictive-labeled images.

1.8 Overview of This Book

This book consists of 11 chapters. The first chapter provides material of an introductory nature to describe the basic concepts and current state of the art in the field of computational intelligence for image restoration, edge detection, image analysis, and retrieval. Chapter 2 gives a mathematical description of the restoration problem from the Hopfield neural network perspective, and describes current algorithms based on this method. Chapter 3 extends the algorithm presented in Chapter 2 to implement adaptive constraint restoration methods for both spatially invariant and spatially variant degradations. Chapter 4 utilizes a perceptually motivated image error measure to introduce novel restoration algorithms. Chapter 5 examines how model-based neural networks [73] can be used to solve image-restoration problems. Chapter 6 examines image restoration algorithms making use of the principles of evolutionary computation. Chapter 7 examines the difficult concept of image restoration when insufficient knowledge of the degrading function is available. Chapter 8 examines the subject of edge detection and characterization using model-based neural networks. Chapter 9 provides an in-depth coverage of the self-organizing tree map, and demonstrates its application in image analysis and retrieval. Chapter 10 examines content representation in compressed domain image classification using evolutionary algorithm. Finally, Chapter 11 explores the fuzzy user perception and small sample problem in content-based image retrieval, and develops computational intelligence techniques to address these challenges.

2

Fundamentals of CI-Inspired Adaptive Image Restoration

2.1 Neural Networks as a CI Architecture

In the next few chapters, we examine CI (computational intelligence) concepts by looking at a specific problem: image restoration. Image restoration is the problem of taking a blurred, noisy image and producing the best estimate of the original image possible. From a CI perspective, we should not simply treat the image as a single instantiation of a random signal, but rather consider its content and adapt our approach accordingly. As the content of the image varies, we need to vary the processing that is applied. An excellent framework to allow this variation of processing is the Hopfield neural network approach to image restoration. The representation of each pixel as a neuron, and the point spread function (PSF) as the strength of the neuronal weights gives us the flexibility we need to treat the image restoration problem from a CI perspective. In this chapter, we will define the problem of image restoration and introduce the Hopfield neural network approach. We will investigate connections between the neural network approach and other ways of thinking about this problem. In later chapters, we will show how the neural network approach to image restoration allows considerable freedom to implement CI approaches.

2.2 Image Distortions

Images are often recorded under a wide variety of circumstances. As imaging technology is rapidly advancing, our interest in recording unusual or irreproducible phenomena is increasing as well. We often push imaging technology to its very limits. For this reason we will always have to handle images suffering from some form of degradation.

Since our imaging technology is not perfect, every recorded image is a degraded version of the scene in some sense. Every imaging system has a limit to its available resolution and the speed at which images can be recorded. Often the problems of finite resolution and speed are not crucial to the applications of the images produced, but there are always cases where this is not so. There exists a large number of possible degradations that an image can suffer. Common degradations are blurring, motion, and noise. Blurring can be caused when an object in the image is outside the camera's depth of field some time during the exposure. For example, a foreground tree might be blurred when we have set up a camera with a telephoto lens to take a photograph of a distant mountain. A blurred object loses some small-scale detail and the blurring process can be modeled as if high-frequency components have been attenuated in some manner in the image [4,21]. If an imaging system internally attenuates the high-frequency components in the image, the result will again appear blurry, despite the fact that all objects in the image were in the camera's field of view. Another commonly encountered image degradation is motion blur. Motion blur can be caused when an object moves relative to the camera during an exposure, such as a car driving along a highway in an image. In the resultant image, the object appears to be smeared in one direction. Motion blur can also result when the camera moves during the exposure. Noise is generally a distortion due to the imaging system rather than the scene recorded. Noise results in random variations to pixel values in the image. This could be caused by the imaging system itself, or the recording or transmission medium. Sometimes the definitions are not clear as in the case where an image is distorted by atmospheric turbulence, such as heat haze. In this case, the image appears blurry because the atmospheric distortion has caused sections of the object to be imaged to move about randomly. This distortion could be described as random motion blur, but can often be modeled as a standard blurring process. Some types of image distortions, such as certain types of atmospheric degradations [88–92], can be best described as distortions in the phase of the signal. Whatever the degrading process, image distortions may be placed into two categories [4,21]:

- Some distortions may be described as spatially invariant or space invariant. In a space-invariant distortion, the parameters of the distortion function are kept unchanged for all regions of the image and all pixels suffer the same form of distortion. This is generally caused by problems with the imaging system such as distortions in the optical system, global lack of focus, or camera motion.

- General distortions are what is called spatially variant or space variant. In a space-variant distortion, the degradation suffered by a pixel in the image depends upon its location in the image. This can be caused by internal factors, such as distortions in the optical system, or by external factors, such as object motion.

In addition, image degradations can be described as linear or nonlinear [21]. In this book, we consider only those distortions that may be described by a linear model.

All linear image degradations can be described by their impulse response. A two-dimensional impulse response is often called a point spread function. It is a two-dimensional function that smears a pixel at its center with some of the pixel's neighbors. The size and shape of the neighborhood used by the PSF is called the PSF's *region of support*. Unless explicitly stated, we will from now on consider PSFs with square-shaped neighborhoods. The larger the neighborhood, the more smearing occurs and the worse the degradation to the image. Here is an example of a 3 by 3 discrete PSF.

$$\frac{1}{5} \begin{bmatrix} 0.5 & 0.5 & 0.5 \\ 0.5 & 1.0 & 0.5 \\ 0.5 & 0.5 & 0.5 \end{bmatrix}$$

where the factor $1/5$ ensures energy conservation.

The final value of the pixel acted upon by this PSF is the sum of the values of each pixel under the PSF mask, each multiplied by the matching entry in the PSF mask.

Consider a PSF of size P by P acting on an image of size N by M. Let $L = MN$. In the case of a two-dimensional image, the PSF may be written as $h(x, y; \alpha, \beta)$. The four sets of indices indicate that the PSF may be spatially variant, hence the PSF will be a different function for pixels in different locations of an image. When noise is also present in the degraded image, as is often the case in real-world applications, the image degradation model in the discrete case becomes [4]

$$g(x, y) = \sum_{\alpha}^{N} \sum_{\beta}^{M} f(\alpha, \beta) h(x, y; \alpha, \beta) + n(x, y) \qquad (2.1)$$

where $f(x, y)$ and $g(x, y)$ are the original and degraded images, respectively, and $n(x, y)$ is the additive noise component of the degraded image. If $h(x, y; \alpha, \beta)$ is a linear function then Equation (2.1) may be restated by lexicographically ordering $g(x, y)$, $f(x, y)$, and $n(x, y)$ into column vectors of size NM. To lexicographically order an image, we simply scan each pixel in the image row by row and stack them one after another to form a single-column vector. Alternately, we may scan the image column by column to form the vector. For example, assume the image $f(x, y)$ looks like

$$f(x, y) = \begin{bmatrix} 11 & 12 & 13 & 14 \\ 21 & 22 & 23 & 24 \\ 31 & 32 & 33 & 34 \\ 41 & 42 & 43 & 44 \end{bmatrix}$$

After lexicographic ordering the following column vector results

$$\mathbf{f} = [11\ 12\ 13\ 14\ 21\ 22\ 23\ 24\ 31\ 32\ 33\ 34\ 41\ 42\ 43\ 44]^T$$

If we are consistent and order $g(x, y)$, $f(x, y)$, and $n(x, y)$ and in the same way, we may restate Equation (2.1) as a matrix operation [4,21]:

$$\mathbf{g} = \mathbf{Hf} + \mathbf{n} \tag{2.2}$$

where $\mathbf{g}$ and $\mathbf{f}$ are the lexicographically organized degraded and original image vectors, $\mathbf{n}$ is the additive noise component, and $\mathbf{H}$ is a matrix operator whose elements are an arrangement of the elements of $h(x, y; \alpha, \beta)$ such that the matrix multiplication of $\mathbf{f}$ with $\mathbf{H}$ performs the same operation as convolving $f(x, y)$ with $h(x, y; \alpha, \beta)$. In general, $\mathbf{H}$ may take any form. However, if $h(x, y; \alpha, \beta)$ is spatially invariant with $P \ll \min(N, M)$ then $h(x, y; \alpha, \beta)$ becomes $h(x-\alpha, y-\beta)$ in Equation (2.1) and $\mathbf{H}$ takes the form of a block-Toeplitz matrix.

A Toeplitz matrix [2] is a matrix where every element lying on the same diagonal line has the same value. Here is an example of a Toeplitz matrix:

$$\begin{bmatrix} 1 & 2 & 3 & 4 & 5 \\ 2 & 1 & 2 & 3 & 4 \\ 3 & 2 & 1 & 2 & 3 \\ 4 & 3 & 2 & 1 & 2 \\ 5 & 4 & 3 & 2 & 1 \end{bmatrix}$$

A block-Toeplitz matrix is a matrix that can be divided into a number of equal-sized blocks. Each block is a Toeplitz matrix, and blocks lying on the same *block diagonal* are identical. Here is an example of a 6 by 6 block-Toeplitz matrix:

$$\begin{bmatrix} 1 & 2 & 3 & 4 & 5 & 6 \\ 2 & 1 & 4 & 3 & 6 & 5 \\ 3 & 4 & 1 & 2 & 3 & 4 \\ 4 & 3 & 2 & 1 & 4 & 3 \\ 5 & 6 & 3 & 4 & 1 & 2 \\ 6 & 5 & 4 & 3 & 2 & 1 \end{bmatrix} = \begin{bmatrix} \mathbf{H}_{11} & \mathbf{H}_{22} & \mathbf{H}_{33} \\ \mathbf{H}_{22} & \mathbf{H}_{11} & \mathbf{H}_{22} \\ \mathbf{H}_{33} & \mathbf{H}_{22} & \mathbf{H}_{11} \end{bmatrix}$$

where

$$\mathbf{H}_{11} = \begin{bmatrix} 1 & 2 \\ 2 & 1 \end{bmatrix}, \quad \mathbf{H}_{22} = \begin{bmatrix} 3 & 4 \\ 4 & 3 \end{bmatrix}, \quad \mathbf{H}_{33} = \begin{bmatrix} 5 & 6 \\ 6 & 5 \end{bmatrix}$$

Notice that a Toeplitz matrix is also a block-Toeplitz matrix with a block size of 1 by 1, but a block Toeplitz matrix is usually not Toeplitz. The block-Toeplitz structure of $\mathbf{H}$ comes about due to the block structure of $\mathbf{f}$, $\mathbf{g}$, and $\mathbf{n}$ created by the lexicographic ordering. If $h(x, y; \alpha, \beta)$ has a simple form of space variance then $\mathbf{H}$ may have a simple form, resembling a block-Toeplitz matrix.

2.3 Image Restoration

When an image is recorded suffering some type of degradation, such as mentioned above, it may not always be possible to take another, undistorted, image of the interesting phenomenon or object. The situation may not recur, like the image of a planet taken by a space probe, or the image of a crime in progress. On the other hand, the imaging system used may introduce inherent distortions to the image which cannot be avoided, for example, a magnetic resonance imaging system. To restore an image degraded by a linear distortion, a restoration cost function can be developed. The cost function is created using knowledge about the degraded image and an estimate of the degradation, and possibly noise, suffered by the original image to produce the degraded image. The free variable in the cost function is an image, that we will denote by $\hat{\mathbf{f}}$, and the cost function is designed such that the $\hat{\mathbf{f}}$ that minimizes the cost function is an estimate of the original image. A common class of cost functions is based on the mean square error (MSE) between the original image and the estimate image. Cost functions based on the MSE often have a quadratic nature.

2.4 Constrained Least Square Error

In this work, the cost function we consider minimizing starts with the constrained least square error measure [4]:

$$E = \frac{1}{2}\|\mathbf{g} - \mathbf{H}\hat{\mathbf{f}}\|^2 + \frac{1}{2}\lambda\|\mathbf{D}\hat{\mathbf{f}}\|^2 \qquad (2.3)$$

where $\hat{\mathbf{f}}$ is the restored image estimate, λ is a constant, and $\mathbf{D}$ is a smoothness constraint operator. Since $\mathbf{H}$ is often a low-pass distortion, $\mathbf{D}$ will be chosen to be a high-pass filter. The second term in Equation (2.3) is the regularization term. The more noise that exists in an image, the greater the second term in Equation (2.3) should be, hence minimizing the second term will involve reducing the noise in the image at the expense of restoring sharpness.

Choosing λ becomes an important consideration when restoring an image. Too great a value of λ will oversmooth the restored image, whereas too small a value of λ will not properly suppress noise. At their essence, neural networks minimize cost functions such as that above. It is not unexpected that there exist neural network models to restore degraded imagery, and such models will be explored in the next section. However it is always enlightening to make links between different restoration models and philosophies to gain new insights. The constrained least square error measure can be viewed from different perspectives, and let's explore two such viewpoints before going on to the neural network models.

2.4.1 A Bayesian Perspective

The restoration problem may be considered in terms of probability distributions and Bayesian theory.

Consider the image degradation model given by Equation (2.2)

$$g = Hf + n$$

Consider the case where n is a normally distributed random variable with mean 0 and a covariance matrix given by R. This is often written as

$$n : N(0, R)$$

Written out more formally, the probability density function of n is [93]

$$p(n) = \frac{1}{2\pi^{N/2}(\det R)^{1/2}} \exp\left\{-\frac{1}{2}n^T R^{-1} n\right\} \qquad (2.4)$$

where $\det R$ is the determinate of R, which is assumed to be invertible. Since n is a random variable, then the observed image g will also be a random variable. Each realization of n in Equation (2.2) will produce a different g. It can be shown that given a fixed f, g has the following distribution:

$$g|f : N(Hf, R)$$

This is also called the *conditional probability density* of g given f and is denoted as $p(g|f)$. When written out formally, the conditional probability density function of g given f is [93]

$$p(g|f) = \frac{1}{2\pi^{N/2}(\det R)^{1/2}} \exp\left\{-\frac{1}{2}(g - Hf)^T R^{-1}(g - Hf)\right\} \qquad (2.5)$$

If we find the value of f that maximizes Equation (2.5), then we have found the *maximum likelihood* solution for f. The maximum likelihood solution is the value of f that maximizes the probability of obtaining our observed g. When finding the maximum of Equation (2.5), we note that the denominator is not a function of f, so we can ignore it. We can also take the logarithm of Equation (2.5) without affecting the solution, since the logarithm operator is a monotonic function and hence $\max_u(\ln(f(u))) = \max_u(f(u))$. The problem then reduces to finding the argument f that minimizes

$$E(f) = (g - Hf)^T R^{-1}(g - Hf) \qquad (2.6)$$

A common assumption to make is that each of the components n_i of the additive noise is independent of the other components and identically distributed. This means that

$$R = \sigma^2 I \qquad (2.7)$$

where I is the identity matrix and σ^2 is the variance of each of the n_i components of n. Substituting Equation (2.7) into Equation (2.6) and noting that σ^2

is not a function of **f** and can hence be ignored, we obtain

$$E(\mathbf{f}) = (\mathbf{g} - \mathbf{Hf})^T (\mathbf{g} - \mathbf{Hf}) \tag{2.8}$$
$$= \|(\mathbf{g} - \mathbf{Hf})\|^2$$

Comparing this with the constrained least squares penalty function [Equation (2.3)], we see that the maximum likelihood solution [Equation (2.8)] in this case is just the image fidelity term in Equation (2.3). We can now see that setting λ to zero and minimizing the constrained least squares penalty function [Equation (2.3)] gives us the maximum likelihood estimate in the case of i.i.d. (independent and identically distributed) zero mean noise.

It is rare that we have absolutely no additional knowledge of **f** outside of that supplied by our observation **g** and our knowledge of the degrading PSF **H**. Assume that we know that $\mathbf{f} : N(\mathbf{0}, \mathbf{Q})$ for some invertible covariance matrix **Q**. This knowledge can also be written as

$$p(\mathbf{f}) = \frac{1}{2\pi^{N/2} (\det \mathbf{Q})^{1/2}} \exp\left\{-\frac{1}{2}\mathbf{f}^T \mathbf{Q}^{-1}\mathbf{f}\right\} \tag{2.9}$$

Equation (2.9) is called the *prior distribution* or simply the *prior* of **f**. According to the Bayes rule [93]

$$p(\mathbf{f}|\mathbf{g}) = \frac{p(\mathbf{g}|\mathbf{f})p(\mathbf{f})}{p(\mathbf{g})} \tag{2.10}$$

where $p(\mathbf{g})$ is the marginal distribution of **g** and is computed by integrating out **f** from Equation (2.5)

$$p(\mathbf{g}) = \int p(\mathbf{g}|\mathbf{f})p(\mathbf{f})d\mathbf{f} \tag{2.11}$$

Since $p(\mathbf{g})$ has **f** integrated out, it is not a function of **f**. The left-hand side of Equation (2.10) is called the *posterior* probability density of **f**. This is the conditional probability of **f** given a known observation **g**. The posterior density of **f** takes into account our prior knowledge about **f**. The **f** that maximizes the posterior density is the **f** that is most probable given the observation and the prior density. This is called the maximum a posteriori estimate and is obtained by finding the **f** that maximizes Equation (2.10). When we ignore $p(\mathbf{g})$ and other elements that are not a function of **f**, the posterior density of **f** is

$$p(\mathbf{f}|\mathbf{g}) \propto \exp\left\{-\frac{1}{2}(\mathbf{g} - \mathbf{Hf})^T \mathbf{R}^{-1}(\mathbf{g} - \mathbf{Hf}) - \frac{1}{2}\mathbf{f}^T \mathbf{Q}^{-1}\mathbf{f}\right\} \tag{2.12}$$

If we take the natural logarithm of Equation (2.12), then we get the following function to be minimized:

$$E(\mathbf{f}) = \frac{1}{2}(\mathbf{g} - \mathbf{Hf})^T \mathbf{R}^{-1}(\mathbf{g} - \mathbf{Hf}) + \frac{1}{2}\mathbf{f}^T \mathbf{Q}^{-1}\mathbf{f} \tag{2.13}$$

If we make the assumption that $\mathbf{n}$ is i.i.d. as we did above in the case of the maximum likelihood estimate, we obtain the following result:

$$E(\mathbf{f}) = \frac{1}{2}(\mathbf{g} - \mathbf{Hf})^T(\mathbf{g} - \mathbf{Hf}) + \frac{1}{2}\mathbf{f}^T\mathbf{Q}^{-1}\mathbf{f} \qquad (2.14)$$

If we assume that the inverse of $\mathbf{Q}$ can be decomposed as $\mathbf{Q}^{-1} = \lambda\mathbf{D}^T\mathbf{D}$ for some matrix $\mathbf{D}$, then we obtain our final result:

$$E(\mathbf{f}) = \frac{1}{2}(\mathbf{g} - \mathbf{Hf})^T(\mathbf{g} - \mathbf{Hf}) + \frac{1}{2}\lambda\mathbf{f}^T\mathbf{D}^T\mathbf{Df} \qquad (2.15)$$

$$= \frac{1}{2}\|\mathbf{g} - \mathbf{Hf}\|^2 + \frac{1}{2}\lambda\|\mathbf{Df}\|^2$$

This is simply the constrained least square error penalty function given by Equation (2.3). We can see that constrained least square image restoration computes the maximum a posteriori estimate of the original image given the degraded image, knowledge of the blur function, and the assumptions we have made above regarding the form of the additive noise in the system and the prior distribution of the original image.

2.4.2 A Lagrangian Perspective

The restoration problem may also be considered in terms of the theory of Lagrange [94]. We will see in this section the consequences of this analysis for constrained least squares-based restoration. We will expand the restoration problem in terms of the theory of Karush–Kuhn–Tucker (a generalization of the theory of Lagrange) and then show what solutions result.

Problem Formulation

Consider the image degradation model given by Equation (2.2)

$$\mathbf{g} = \mathbf{Hf} + \mathbf{n}$$

where $\mathbf{g}$ and $\mathbf{f}$ are L-dimensional lexicographically organized degraded and original images, respectively, $\mathbf{n}$ is an L-dimensional additive noise vector and $\mathbf{H}$ is a convolutional operator. The problem we are considering is that given $\mathbf{g}$ and $\mathbf{H}$, we wish to recover $\mathbf{f}$.

The restoration problem is ill-posed due to the fact that $\mathbf{g}$ is usually of equal size or smaller than $\mathbf{f}$. If $\mathbf{g}$ is smaller than $\mathbf{f}$ then the problem in general has an infinite number of solutions. If $\mathbf{g}$ is of equal size as $\mathbf{f}$, then there is no additional information about the imaging system or $\mathbf{f}$ contained in $\mathbf{g}$ to help us to disambiguate the contributions of $\mathbf{Hf}$ and $\mathbf{n}$ on $\mathbf{g}$. Depending on the form of $\mathbf{H}$, there may be an infinite number of solutions, or a single solution that is extremely sensitive to the exact noise realization $\mathbf{n}$ present in the system.

For this reason, extra constraints will be required. Consider a matrix $\mathbf{D} \in \Re^{L \times L}$ designed to act as a high-pass operator. $\mathbf{Df}$ is therefore the high-pass component of the estimated image. To reduce the effects of noise, we can constrain $\|\mathbf{Df}\|^2$ to be equal to a preset constant. If the constant is chosen too low, the image will be blurred, and if the constant is chosen too high, the image will contain unacceptable levels of noise.

This problem can then be stated as

$$\mathbf{f}^* = \arg\min_f \|\mathbf{g} - \mathbf{Hf}\|^2 \text{ subject to } \mathbf{h}(\mathbf{f}^*) = 0 \text{ and } \mathbf{p}(\mathbf{f}^*) \le 0$$

where $h(\mathbf{f}) \in \Re$, $h(\mathbf{f}) = \|\mathbf{Df}\|^2 - \sigma^2$ and $\mathbf{p}(\mathbf{f}) \in \Re^L$, $\mathbf{p}(\mathbf{f}) = -\mathbf{f}$, $p_j(\mathbf{f}) = -f_j$, $1 \le j \le L$.

Note that $\mathbf{p}$ is a set of positivity constraints, while $\mathbf{h}$ comprises the constraint on the norm of the high-frequency components in the image.

Let us look now at how we can solve this equation. The following definition and theory will be useful, and we will be making use of these in subsequent chapters.

Definition 2.1
Let $\mathbf{f}^*$ satisfy $\mathbf{h}(\mathbf{f}^*) = 0$, $\mathbf{p}(\mathbf{f}^*) \le 0$ and let $J\{\mathbf{f}^*\}$ be the index set of active inequality constraints, that is, $J\{\mathbf{f}^*\} = \{j; \ p_j(\mathbf{f}^*) = 0\}$.

Then $\mathbf{f}^*$ is a regular point if $\nabla h(\mathbf{f}^*), \nabla p_j(\mathbf{f}^*)$ are linearly independent $\forall j \in J\{\mathbf{f}^*\}$ [94].

Theorem of Karush–Kuhn–Tucker:
Let $E, \mathbf{h}, \mathbf{p} \in C^1$. Let $\mathbf{f}^*$ be a regular point and a local minimizer for the problem of minimizing E subject to $\mathbf{h}(\mathbf{f}^*) = 0$ and $\mathbf{p}(\mathbf{f}^*) \le 0$.

Then there exists $\lambda \in \Re$, $\mu \in \Re^L$ such that

1. $\mu \ge 0$
2. $\nabla E(\mathbf{f}^*) + \lambda \nabla h(\mathbf{f}^*) + \sum_{j=1}^{L} \mu_j \nabla p_j(\mathbf{f}^*) = 0$
3. $\mu^T \mathbf{p}(\mathbf{f}^*) = 0$

Problem Solution

The theorem of Karush–Kuhn–Tucker (KKT) can be used to solve this problem. Note the following equations:

$$h(\mathbf{f}^*) = \|\mathbf{Df}^*\|^2 - \sigma^2 = \mathbf{f}^{*T}\mathbf{D}^T\mathbf{Df}^* - \sigma^2 \tag{2.16}$$

$$\nabla h(\mathbf{f}^*) = 2\mathbf{D}^T\mathbf{Df}^* \tag{2.17}$$

$$p_j(\mathbf{f}^*) = -f_j^* \tag{2.18}$$

$$\nabla p_j(\mathbf{f}^*) = -\mathbf{e}_j \tag{2.19}$$

where

$$\mathbf{e}_j \in \Re^L; [\mathbf{e}_j]_i = \begin{cases} 1, & i = j \\ 0, & i \ne j \end{cases} \tag{2.20}$$

Note that $\nabla h(\mathbf{f}^*)$ and $\nabla p_j(\mathbf{f}^*) \in \Re^L$ for each $j \in J\{\mathbf{f}^*\}$.

If we assume $\mathbf{f}^*$ is a regular point, the solution to the problem can be obtained as follows:

$$E(\mathbf{f}^*) = \|\mathbf{g} - \mathbf{Hf}^*\|^2 = (\mathbf{g} - \mathbf{Hf}^*)^T (\mathbf{g} - \mathbf{Hf}^*)$$
$$= \mathbf{g}^T \mathbf{g} - 2\mathbf{g}^T \mathbf{Hf}^* + \mathbf{f}^{*T} \mathbf{H}^T \mathbf{Hf}^* \tag{2.21}$$

Note that we have used the fact that $\mathbf{f}^{*T} \mathbf{H}^T \mathbf{g} = (\mathbf{g}^T \mathbf{Hf}^*)^T = \mathbf{g}^T \mathbf{Hf}^*$ since this quantity is a scalar.

$$\nabla E(\mathbf{f}^*) = 2\mathbf{H}^T \mathbf{Hf}^* - 2\mathbf{H}^T \mathbf{g} \tag{2.22}$$

By KKT condition 2 we have

$$\nabla E(\mathbf{f}^*) + \lambda \nabla h(\mathbf{f}^*) + \sum_{j=1}^{L} \mu_j \nabla p_j(\mathbf{f}^*) = 0$$

$$2\mathbf{H}^T \mathbf{Hf}^* - 2\mathbf{H}^T \mathbf{g} + 2\lambda \mathbf{D}^T \mathbf{Df}^* - \sum_{j=1}^{L} \mu_j \mathbf{e}_j = 0$$

$$2\left(\mathbf{H}^T \mathbf{H} + \lambda \mathbf{D}^T \mathbf{D}\right) \mathbf{f}^* = 2\mathbf{H}^T \mathbf{g} + \sum_{j=1}^{L} \mu_j \mathbf{e}_j$$

Therefore,

$$\mathbf{f}^* = \left[\mathbf{H}^T \mathbf{H} + \lambda \mathbf{D}^T \mathbf{D}\right]^{-1} \left(\mathbf{H}^T \mathbf{g} + \frac{1}{2} \sum_{j=1}^{L} \mu_j \mathbf{e}_j\right) \tag{2.23}$$

Notice that in the solution given by Equation (2.23), the actual constant σ^2 in Equation (2.16) controlling the level of high-frequency components allowed in the solution has disappeared. The effect of this constant has been absorbed into λ. If we ignore the positivity constraint term in Equation (2.23), then Equation (2.23) is in fact the solution to the constrained least squares cost function we presented in Equation (2.3) above. This shows the equivalence of the constrained least squares penalty function approach and the Lagrangian approach.

KKT condition 3 states that $\mu^T \mathbf{p}(\mathbf{f}^*) = -\sum_{j=1}^{L} \mu_j f_j^* = 0$. If $f_j^* \neq 0$ then $j \notin J\{\mathbf{f}^*\}$ and $\mu_j = 0$. If $f_j^* = 0$, then μ_j can be any value and satisfy KKT condition 3. In fact, by examining Equation (2.23) we discover that the nonzero values of μ are the solutions to

$$\frac{1}{2} \sum_{p \in J(\mathbf{f}^*)} \left[\left(\mathbf{H}^T \mathbf{H} + \lambda \mathbf{D}^T \mathbf{D}\right)^{-1}\right]_{jp} \mu_p = -\left[\left(\mathbf{H}^T \mathbf{H} + \lambda \mathbf{D}^T \mathbf{D}\right)^{-1} \mathbf{H}^T \mathbf{g}\right]_j \tag{2.24}$$

$$\forall j; \, j \in J(\mathbf{f}^*)$$

The complexity of this equation should not be surprising. A single pixel constrained to be positive by an active constraint will change the solution point of the problem and affect all the other pixels' values. In this way, it affects the degree to which other pixels need to be constrained in order to be positive. Hence the values of all the elements in μ are closely interrelated.

The equality constraint $h(\mathbf{f}^*) = 0$ gives us the extra equation required to solve for the value of λ. Substituting Equation (2.23) into Equation (2.16) we obtain λ as the solution to

$$\mathbf{X}^T \mathbf{D}^T \mathbf{D} \mathbf{X} = \sigma^2$$

where

$$\mathbf{X} = \left[\mathbf{H}^T \mathbf{H} + \lambda \mathbf{D}^T \mathbf{D}\right]^{-1} \left(\mathbf{H}^T \mathbf{g} + \frac{1}{2} \sum_{j \in J\,\{\mathbf{f}^*\}} \mu_j \mathbf{e}_j\right) \tag{2.25}$$

In theory, Equation (2.24) and Equation (2.25) form a set of $1 + |J(\mathbf{f}^*)|$ simultaneous equations that may be solved to determine the elements of λ and μ (where $|J(\mathbf{f}^*)|$ is the number of active inequality constraints). Once the values of λ and μ are found, the solution to the problem can be found using Equation (2.23). At least in theory. In practice, solving Equations (2.25) and (2.24) can be extremely difficult. It can be seen that the value of λ and the elements in μ are closely related. This is not surprising since the constraint on the high-frequency components can reasonably be expected to influence the degree to which pixels need to be constrained to positive. In turn, pixels constrained to be positive will influence the high-frequency components of the image, and hence λ. In practice the problem is not so complex. The value of λ is set mainly by the constant σ. If the σ is not known beforehand (as is often the case), then the value of λ may be set in a trial-and-error fashion without using Equation (2.25). Once the optimal value of the λ is found, an iterative algorithm which operates on one pixel at a time and constrains each pixel to be positive during each iteration will implicitly calculate the μ_j.

2.5 Neural Network Restoration

Neural network restoration approaches are designed to minimize a quadratic programming problem [46,95–98]. The generalized Hopfield network can be applied to this case [35]. The general form of a quadratic programming problem can be stated as follows:

Minimize the energy function associated with a neural network given by

$$E = -\frac{1}{2} \hat{\mathbf{f}}^T \mathbf{W} \hat{\mathbf{f}} - \mathbf{b}^T \hat{\mathbf{f}} + c \tag{2.26}$$

Comparing this with Equation (2.3), $\mathbf{W}$, $\mathbf{b}$ and c are functions of $\mathbf{H}$, $\mathbf{D}$, λ, and $\mathbf{n}$, and other problem related constraints. In terms of a neural network energy function, the (i, j)th element of $\mathbf{W}$ corresponds to the interconnection strength between neurons (pixels) i and j in the network. Similarly, vector **b** corresponds to the bias input to each neuron.

Equating the formula for the energy of a neural network with Equation (2.3), the bias inputs and interconnection strengths can be found such that as the neural network minimizes its energy function, the image will be restored.

Expanding Equation (2.3) we get

$$
\begin{aligned}
E &= \frac{1}{2}\sum_{p=1}^{L}\left(g_p - \sum_{i=1}^{L} h_{pi}\hat{f}_i\right)^2 + \frac{1}{2}\lambda\sum_{p=1}^{L}\left(\sum_{i=1}^{L} d_{pi}\hat{f}_i\right)^2 \\
&= \frac{1}{2}\sum_{p=1}^{L}\left(g_p - \sum_{i=1}^{L} h_{pi}\hat{f}_i\right)\left(g_p - \sum_{j=1}^{L} h_{pj}\hat{f}_j\right) + \frac{1}{2}\lambda\sum_{p=1}^{L}\left(\sum_{i=1}^{L} d_{pi}\hat{f}_i\right)\left(\sum_{j=1}^{L} d_{pj}\hat{f}_j\right) \\
&= \frac{1}{2}\sum_{p=1}^{L}\left((g_p)^2 - 2g_p\sum_{i=1}^{L} h_{pi}\hat{f}_i + \sum_{i=1}^{L} h_{pi}\hat{f}_i\sum_{j=1}^{L} h_{pj}\hat{f}_j\right) \\
&\quad + \frac{1}{2}\lambda\sum_{p=1}^{L}\left(\sum_{i=1}^{L} d_{pi}\hat{f}_i\right)\left(\sum_{j=1}^{L} d_{pj}\hat{f}_j\right) \\
&= \frac{1}{2}\sum_{p=1}^{L}(g_p)^2 - \sum_{p=1}^{L}\sum_{i=1}^{L} g_p h_{pi}\hat{f}_i + \frac{1}{2}\sum_{p=1}^{L}\sum_{i=1}^{L} h_{pi}\hat{f}_i\sum_{j=1}^{L} h_{pj}\hat{f}_j \\
&\quad + \frac{1}{2}\lambda\sum_{p=1}^{L}\sum_{i=1}^{L} d_{pi}\hat{f}_i\sum_{j=1}^{L} d_{pj}\hat{f}_j \\
&= \frac{1}{2}\sum_{p=1}^{L}\sum_{i=1}^{L}\sum_{j=1}^{L} h_{pj}\hat{f}_j h_{pi}\hat{f}_i + \frac{1}{2}\lambda\sum_{p=1}^{L}\sum_{i=1}^{L}\sum_{j=1}^{L} d_{pj}\hat{f}_j d_{pi}\hat{f}_i \\
&\quad - \sum_{p=1}^{L}\sum_{i=1}^{L} g_p h_{pi}\hat{f}_i + \frac{1}{2}\sum_{p=1}^{L}(g_p)^2
\end{aligned}
$$

Hence,

$$
E = \frac{1}{2}\sum_{i=1}^{L}\sum_{j=1}^{L}\left(\sum_{p=1}^{L} h_{pj}h_{pi} + \lambda\sum_{p=1}^{L} d_{pj}d_{pi}\right)\hat{f}_i\hat{f}_j - \sum_{i=1}^{L}\sum_{p=1}^{L} g_p h_{pi}\hat{f}_i + \frac{1}{2}\sum_{p=1}^{L}(g_p)^2
$$

$$(2.27)$$

Expanding (2.26) we get

$$
E = -\frac{1}{2}\sum_{i=1}^{L}\sum_{j=1}^{L} w_{ij}\hat{f}_i\hat{f}_j - \sum_{i=1}^{L} b_i\hat{f}_i + c \tag{2.28}
$$

By equating the terms in Equations (2.27) and (2.28), we find that the neural network model can be matched to the constrained least square error cost function by ignoring the constant, c, and setting

$$w_{ij} = -\sum_{p=1}^{L} h_{pi} h_{pj} - \lambda \sum_{p=1}^{L} d_{pi} d_{pj} \tag{2.29}$$

$$= -\mathbf{H}^T \mathbf{H} - \lambda \mathbf{D}^T \mathbf{D} \tag{2.30}$$

and

$$b_i = \sum_{p=1}^{L} g_p h_{pi} \tag{2.31}$$

$$= \mathbf{H}^T \mathbf{g} \tag{2.32}$$

where w_{ij} is the interconnection strength between pixels i and j, and b_i is the bias input to neuron (pixel) i. In addition, h_{ij} is the (i, j)th element of matrix $\mathbf{H}$ from Equation (2.3) and d_{ij} is the (i, j)th element of matrix $\mathbf{D}$.

Now let's look at some neural networks in the literature to solve this problem.

2.6 Neural Network Restoration Algorithms in the Literature

In the network described by Zhou et al. [46] for an image with $S + 1$ gray levels, each pixel is represented by $S + 1$ neurons. Each neuron can have a value of 0 or 1. The value of the ith pixel is then given by

$$\hat{f}_i = \sum_{k=0}^{S} v_{i,k} \tag{2.33}$$

where $v_{i,k}$ is the state of the kth neuron of the ith pixel. Each neuron is visited sequentially and has its input calculated according to

$$u_i = b_i + \sum_{j=1}^{L} w_{ij} \hat{f}_j \tag{2.34}$$

where u_i is the input to neuron i, and $\hat{f}_j$ is the state of the jth neuron. Based on u_i, the neuron's state is updated according to the following rule:

$$\Delta \hat{f}_i = G(u_i)$$

where

$$G(u) = \begin{cases} 1, & u > 0 \\ 0, & u = 0 \\ -1, & u < 0 \end{cases} \tag{2.35}$$

The change in energy resulting from a change in neuron state of $\Delta \hat{f}_i$ is given by

$$\Delta E = -\frac{1}{2} w_{ii}(\Delta \hat{f}_i)^2 - u_i \Delta \hat{f}_i \qquad (2.36)$$

If $\Delta E < 0$, then the neuron's state is updated. This algorithm may be summarized as

Algorithm 2.1
repeat
{
  For $i = 1, \ldots, L$ do
  {
    For $k = 0, \ldots, S$ do
    {
      $u_i = b_i + \sum_{j=1}^{L} w_{ij} \hat{f}_j$
      $\Delta \hat{f}_i = G(u_i)$

      where $G(u) = \begin{cases} 1, & u > 0 \\ 0, & u = 0 \\ -1, & u < 0 \end{cases}$

      $\Delta E = -\frac{1}{2} w_{ii}(\Delta \hat{f}_i)^2 - u_i \Delta \hat{f}_i$
      If $\Delta E < 0$, then $v_{i,k} = v_{i,k} + \Delta \hat{f}_i$
      $\hat{f}_i = \sum_{k=0}^{S} v_{i,k}$
    }
  }
  $t = t + 1$
}
until $\hat{f}_i(t) = \hat{f}_i(t-1) \forall i = 1, \ldots, L)$

In the paper by Paik and Katsaggelos, Algorithm 2.1 was enhanced to remove the step where the energy reduction is checked following the calculation of $\Delta \hat{f}_i$ [95]. Paik and Katsaggelos presented an algorithm which made use of a more complicated neuron. In their model, each pixel was represented by a single neuron that takes discrete values between 0 and S, and is capable of updating its value by ± 1, or keeping the same value during a single step. A new method for calculating $\Delta \hat{f}_i$ was also presented:

$$\Delta \hat{f}_i = \acute{G}_i(u_i)$$

where

$$\acute{G}_i = \begin{cases} -1, & u < -\theta_i \\ 0, & -\theta_i \le u \le \theta_i \\ 1, & u > \theta_i \end{cases} \qquad (2.37)$$

where $\theta_i = -\frac{1}{2} w_{ii} > 0$.

This algorithm may be presented as

Algorithm 2.2
repeat
{
  For $i = 1, \ldots, L$ do
  {

$$u_i = b_i + \sum_{j=1}^{L} w_{ij} \hat{f}_j$$

$$\Delta \hat{f}_i = \acute{G}_i(u_i)$$

$$\text{where } \acute{G}_i(u) = \begin{cases} -1, & u < -\theta_i \\ 0, & -\theta_i \leq u \leq \theta_i \\ 1, & u > \theta_i \end{cases}$$

$$\text{where } \theta_i = -\tfrac{1}{2} w_{ii} > 0$$

$$\hat{f}_i(t+1) = K(\hat{f}_i(t) + \Delta \hat{f}_i)$$

$$\text{where } K(u) = \begin{cases} 0, & u < 0 \\ u, & 0 \leq u \leq S \\ S, & u \geq S \end{cases}$$

  }
  $t = t + 1$
}
until $\hat{f}_i(t) = \hat{f}_i(t-1) \forall i = 1, \ldots, L)$

Algorithm 2.2 makes no specific check that energy has decreased during each iteration and so in [95] they proved that Algorithm 2.2 would result in a decrease of the energy function at each iteration. Note that in Algorithm 2.2, each pixel only changes its value by ± 1 during an iteration. In Algorithm 2.1, the pixel's value would change by any amount between 0 and S during an iteration since each pixel was represented by $S + 1$ neurons. Although Algorithm 2.2 is much more efficient in terms of the number of neurons used, it may take many more iterations than Algorithm 2.1 to converge to a solution (although the time taken may still be faster than Algorithm 2.1). If we consider that the value of each pixel represents a dimension of the L dimensional energy function to be minimized, then we can see that Algorithms 2.1 and 2.2 have slightly different approaches to finding a local minimum. In Algorithm 2.1, the energy function is minimized along each dimension in turn. The current image estimate can be considered to represent a single point in the solution space. In Algorithm 2.1, this point moves to the function minimum along each of the L axes of the solution space until it eventually reaches a local minimum of the energy function. In Algorithm 2.2, for each pixel, the point takes a unit step in a direction that reduces the network energy along that dimension. If the weight matrix is negative definite ($-\mathbf{W}$ is positive definite), however, regardless of how these algorithms work, the end results must be similar (if each algorithm ends at a minimum). The reason for this is that when the weight matrix is negative definite, there is only the global minimum. That is, the function has only one minimum. In this case the matrix $\mathbf{W}$ is invertible and taking Equation (2.26) we see that

$$\frac{\delta E}{\delta \hat{\mathbf{f}}} = -\mathbf{W}\hat{\mathbf{f}} - \mathbf{b} \tag{2.38}$$

Hence the solution is given by

$$\hat{\mathbf{f}}^* = -\mathbf{W}^{-1}\mathbf{b} \tag{2.39}$$

(assuming that $\mathbf{W}^{-1}$ exists).

The $\hat{\mathbf{f}}^*$ is the only minimum and the only stationary point of this cost function, so we can state that if $\mathbf{W}$ is negative definite and Algorithm 2.1 and Algorithm 2.2 both terminate at a local minimum, the resultant image must be close to $\hat{\mathbf{f}}^*$ for both algorithms. Algorithm 2.1 approaches the minimum in a zigzag fashion, whereas Algorithm 2.2 approaches the minimum along a smooth curve. If $\mathbf{W}$ is not negative definite, then a local minimum may exist and Algorithms 2.1 and 2.2 may not produce the same results. If Algorithm 2.2 is altered so that instead of changing each neuron's value by ±1 before going to the next neuron, the current neuron is iterated until the input to that neuron is zero, then Algorithms 2.1 and 2.2 will produce identical results. Each algorithm will terminate in the same local minimum.

2.7 Improved Algorithm

Although Algorithm 2.2 is an improvement on Algorithm 2.1, it is not optimal. From iteration to iteration, neurons often oscillate about their final value, and during the initial iterations of Algorithm 2.1, a neuron may require 100 or more state changes in order to minimize its energy contribution. A faster method to minimize the energy contribution of each neuron being considered is suggested by examination of the mathematics involved. For an image where each pixel is able to take on any discrete integer intensity between 0 and S, we assign each pixel in the image to a single neuron able to take any discrete value between 0 and S. Since the formula for the energy reduction resulting from a change in neuron state $\Delta \hat{f}_i$ is a simple quadratic, it is possible to solve for the $\Delta \hat{f}_i$ which produces the maximum energy reduction. Theorem 2.1 states that this approach will result in the same energy minimum as Algorithm 2.1 and hence the same final state of each neuron after it is updated.

Theorem 2.1

For each neuron i in the network during each iteration, there exists a state change $\Delta \hat{f}_i^*$ such that the energy contribution of neuron i is minimized.

Proof

Let u_i be the input to neuron i which is calculated by

$$u_i = b_i + \sum_{j=1}^{L} w_{ij} \hat{f}_j$$

Let ΔE be the resulting energy change due to $\Delta \hat{f}_i$.

$$\Delta E = -\frac{1}{2} w_{ii} (\Delta \hat{f}_i)^2 - u_i \Delta \hat{f}_i \qquad (2.40)$$

Differentiating ΔE with respect to $\Delta \hat{f}_i$ gives us

$$\frac{\delta \Delta E}{\delta \hat{f}_i} = -w_{ii} \Delta \hat{f}_i - u_i$$

The value of $\Delta \hat{f}_i$ which minimizes Equation (2.40) is given by

$$0 = -w_{ii} \Delta \hat{f}_i^* - u_i$$

Therefore,

$$\Delta \hat{f}_i^* = \frac{-u_i}{w_{ii}} \qquad (2.41)$$

QED. ■

Based on Theorem 2.1, an improved algorithm is presented below.

Algorithm 2.3

repeat
{
　For $i = 1, \ldots, L$ do
　{
　　$u_i = b_i + \sum_{j=1}^{L} w_{ij} \hat{f}_j$
　　$\Delta \hat{f}_i = G(u_i)$

　　where $G(u) = \begin{cases} -1, & u < 0 \\ 0, & u = 0 \\ 1, & u > 0 \end{cases}$

$$\Delta E_{ss} = -\frac{1}{2} w_{ii} (\Delta \hat{f}_i)^2 - u_i \Delta \hat{f}_i \qquad (2.42)$$

　If $\Delta E_{ss} < 0$ then $\Delta \hat{f}_i^* = \frac{-u_i}{w_{ii}}$

　$\hat{f}_i(t+1) = K(\hat{f}_i(t) + \Delta \hat{f}_i^*)$

$$\text{where } K(u) = \begin{cases} 0, & u < 0 \\ u, & 0 \le u \le S \\ S, & u \ge S \end{cases}$$

$\}$

$t = t + 1$

$\}$

until $\hat{f}_i(t) = \hat{f}_i(t-1) \forall i = 1, \ldots, L)$

Each neuron is visited sequentially and has its input calculated. Using the input value, the state change needed to minimize the neuron's energy contribution to the network is calculated. Note that since $\Delta \hat{f}_i \in \{-1, 0, 1\}$ and $\Delta \hat{f}_i$ and $\Delta \hat{f}_i^*$ must be the same sign as u_i, [Equation (2.42)] is equivalent to checking that at least a unit step can be taken which will reduce the energy of the network. If $\Delta E_{ss} < 0$, then

$$-\frac{1}{2} w_{ii} - u_i \Delta \hat{f}_i < 0$$
$$-\frac{1}{2} w_{ii} - |u_i| < 0$$
$$-w_{ii} < 2|u_i|$$

Substituting this result into the formula for $\Delta \hat{f}_i^*$ we get

$$\Delta \hat{f}_i^* = \frac{-u_i}{w_{ii}} > \frac{u_i}{2|u_i|} = \frac{1}{2} \Delta \hat{f}_i$$

Since $\Delta \hat{f}_i^*$ and $\Delta \hat{f}_i$ have the same sign and $\Delta \hat{f}_i = \pm 1$ we obtain

$$|\Delta \hat{f}_i^*| > \frac{1}{2} \tag{2.43}$$

In this way, $\Delta \hat{f}_i^*$ will always be large enough to alter the neuron's discrete value.

Algorithm 2.3 makes use of concepts from both Algorithms 2.1 and 2.2. Like Algorithm 2.1, the energy function is minimized in solution space along each dimension in turn until a local minimum is reached. In addition, the efficient use of space by Algorithm 2.2 is utilized. Note that the above algorithm is much faster than either Algorithm 2.1 or 2.2 due to the fact that this algorithm minimizes the current neuron's energy contribution in one step rather than through numerous iterations as did Algorithms 2.1 and 2.2.

2.8   Analysis

In the paper by Paik and Katsaggelos, it was shown that Algorithm 2.2 would converge to a fixed point after a finite number of iterations and that the fixed point would be a local minimum of E in Equation (2.3) in the case of a sequential algorithm [95]. Here we will show that Algorithm 2.3 will also converge to a fixed point which is a local minimum of E in (2.3).

Algorithm 2.2 makes no specific check that energy has decreased during each iteration and so in [95] they proved that Algorithm 2.2 would result in a decrease of the energy function at each iteration. Algorithm 2.3, however, changes the current neuron's state if and only if an energy reduction will occur and $|\Delta \hat{f}_i| = 1$. For this reason Algorithm 2.3 can only reduce the energy function and never increase it. From this we can observe that each iteration of Algorithm 2.3 brings the network closer to a local minimum of the function. The next question is: *Does Algorithm 2.3 ever reach a local minimum and terminate?* Note that the gradient of the function is given by

$$\frac{\delta E}{\delta \hat{\mathbf{f}}} = -\mathbf{W}\hat{\mathbf{f}} - \mathbf{b} = -\mathbf{u} \tag{2.44}$$

where $\mathbf{u}$ is a vector whose ith element contains the current input to neuron i. Note that during any iteration, $\mathbf{u}$ will always point in a direction that reduces the energy function. If $\hat{\mathbf{f}} \neq \hat{\mathbf{f}}^*$ then for at least one neuron a change in state must be possible that would reduce the energy function. For this neuron, $u_i \neq 0$. The algorithm will then compute the change in state for this neuron to move closer to the solution. If $|\Delta \hat{f}_i^*| > 1/2$ the neuron's state will be changed. In this case, we assume that no boundary conditions have been activated to stop neuron i from changing value. Due to the discrete nature of the neuron states, we see that the step size taken by the network is never less than 1.

To restate the facts obtained so far,

- During each iteration Algorithm 2.3 will reduce the energy of the network.

- A reduction in the energy of the network implies that the network has moved closer to a local minimum of the energy function.

- There is a lower bound to the step size taken by the network and a finite range of neuron states. Since the network is restricted to changing state only when an energy reduction is possible, the network cannot iterate forever.

From these observations, we can conclude that the network reaches a local minimum in a finite number of iterations, and that the solution given by Algorithm 2.3 will be close to the solution given by Algorithm 2.1 for the same problem. The reason Algorithms 2.1 and 2.3 must approach the same local minimum is the fact that they operate on the pixel in an identical manner. In Algorithm 2.1, each of the $S + 1$ neurons associated with pixel i is adjusted to reduce its contribution to the energy function. The sum of the contributions of the $S + 1$ neurons associated with pixel i in Algorithm 2.1 equals the final grayscale value of that neuron. Hence during any iteration of Algorithm 2.1 the current pixel can change to any allowable value. There are $S + 1$ possible output values of pixel i and only one of these values results when the algorithm minimizes the contribution of that pixel. Hence whether the pixel is represented by $S + 1$ neurons or just a single neuron, the output grayscale value that occurs when the energy contribution of that pixel is minimized during the current iteration remains the same. Algorithms 2.1 and 2.3 both minimize the current pixel's energy contribution; hence they must both produce the same results. In practice the authors have found that all three algorithms generally produce identical results, which suggests that for reasonable values of the parameter λ, only a single global minimum is present.

Note that in the discussion so far, we have not made any assumptions regarding the nature of the weighting matrix, $\mathbf{W}$, or the bias vector, $\mathbf{b}$. $\mathbf{W}$ and $\mathbf{b}$ determine where the solution is in the solution space, but as long as they are constant during the restoration procedure the algorithm will still terminate after a finite number of iterations. This is an important result, and implies that even if the degradation suffered by the image is space variant or if we assign a different value of λ to each pixel in the image, the algorithm will still converge to a result. Even if $\mathbf{W}$ and $\mathbf{b}$ are such that the solution lies outside of the bounds on the values of the neurons, we would still expect that there exists a point or points which minimize E within the bounds. In practice we would not expect the solution to lie entirely out of the range of neuron values. If we assume that Algorithm 2.3 has terminated at a position where no boundary conditions have been activated, then the condition

$$\left| \Delta \hat{f}_i^* \right| = \left| \frac{-u_i}{w_{ii}} \right| < \frac{1}{2}, \forall i \in \{0, 1, \ldots, L\}$$

must have been met. This implies that

$$|u_i| < \frac{1}{2} w_{ii}, \forall i \in \{0, 1, \ldots, L\} \tag{2.45}$$

Paik and Katsaggelos [95] noticed this feature as well, since the same termination conditions apply to Algorithm 2.2. The self-connection weight, w_{ii}, controls how close to the ideal solution the algorithm will approach before terminating. Since increasing the value of λ increases the value of w_{ii}, we would expect also that the algorithm would terminate more quickly and yet be less accurate for larger values of λ. This is found to occur in practice. When λ is increased, the number of iterations before termination drops rapidly.

2.9 Implementation Considerations

Despite the increase in efficiency and speed of Algorithm 2.3 when compared to Algorithms 2.1 and 2.2, there are still a number of ways that the algorithm can be made more efficient. The ith row of the weighting matrix describes the interconnection strengths between neuron i and every other neuron in the network from the *viewpoint* of neuron i. The weighting matrix is NM by NM, which is clearly a prohibitively large amount of data that requires storage. However, the mathematical discussion in the previous sections implies a shortcut.

By examining Equation (2.29) we observe that in the case of $P \ll \min(M, N)$, where P is the size of the degrading PSF, it can be seen that when calculating the input to each neuron, only pixels within a certain rectangular neighborhood of the current pixel contribute nonzero components to the neuron input. In addition, it can be seen that the nonzero interconnection strengths between any given pixel and a neighboring pixel depend only on the position of the pixels relative to each other in the case of spatially invariant distortion. Using the above observations, the input to any neuron (pixel) in the image can be calculated by applying a mask to the image centered on the pixel being examined. For a P by P distortion, each weighting mask contains only $(2P - 1)^2$ terms. A 5 by 5 degrading PSF acting on a 250 by 250 image requires a weight matrix containing 3.9×10^9 elements, yet a weighting mask of only 81 elements. In addition, by considering the finite regions of support of the degrading and filtering impulse functions represented by **H** and **D**, the weighting masks and bias inputs to each neuron may be calculated without storing matrices **H** and **D** at all. They may be calculated using only the impulse responses of the above matrices.

2.10 Numerical Study of the Algorithms

To compare the three algorithms, let us look at an example. In the example, the efficiency of Algorithms 2.1, 2.2, and 2.3 will be compared to one another. In later chapters of this book, practical examples of the use of this method will be given.

2.10.1 Setup

In both these examples, the images were blurred using a Gaussian PSF with the impulse response

$$h(x, y) = \frac{1}{2\pi \sigma_x \sigma_y} \exp\left[-\left(\frac{x^2}{2\sigma_x^2} + \frac{y^2}{2\sigma_y^2} \right) \right] \tag{2.46}$$

where σ_x and σ_y are the standard deviations of the PSF in the x and y directions, respectively.

2.10.2 Efficiency

The time taken to restore an image was compared among Algorithms 2.1, 2.2, and 2.3. A degraded image was created by blurring a 256 by 256 image with a Gaussian blur of size 5 by 5 and standard deviation 2.0. Noise of variance 4.22 was added to the blurred image. Each algorithm was run until at least 85% of the pixels in the image no longer changed value or many iterations had passed with the same number of pixels changing value during each iteration. Algorithm 2.1 was stopped after the sixth iteration when no further improvement was possible, and took 6067 seconds to run on a SUN Ultra SPARC 1. Algorithm 2.2 was stopped after the 30th iteration with 89% of pixels having converged to their stable states and took 126 seconds to run. Algorithm 2.3 was stopped after the 18th iteration with 90% of pixels stable and took 78 seconds to run. Algorithm 2.3 is much faster than Algorithms 2.1 and 2.2, despite the fact that Algorithms 2.1 and 2.3 approach the same local minimum and hence give the same results. The computation time of Algorithm 2.3 can be expected to increase linearly with the number of pixels in the image, as can the computation times of Algorithms 2.1 and 2.2. The single step neuron energy minimization technique of Algorithm 2.3 provides its superior speed and this trend holds for any size image. Various types of distortions and noise would not be expected to change the speed relationship between Algorithms 2.1, 2.2, and 2.3 for any given image. This is because each algorithm was shown to converge to similar points in solution space and just use different methods to reach this point.

2.11 Summary

In this chapter, we have examined the basic neural network optimization algorithm. We first looked at the problem of restoring an image acted upon by a degrading PSF in the presence of noise. The problem of reversing the degrading process and approximating the original image can be formulated as a minimization process on a constrained least square error measure function. We explored the relationship between the constrained least square error measure function and Bayesian and Lagrangian approaches to the image restoration problem. This error measure can be minimized by relating the terms of the error measure with the formula for the energy of a Hopfield-style neural network. By selecting suitable interconnection strengths (weights) and bias inputs for the neurons, we can equate the neural network energy function with the constrained least square error cost function. The problem then reduces to selecting the optimal algorithm to minimize the network energy function. We examined two such algorithms presented previously in the literature and considered the advantages and disadvantages of both. We then presented a new algorithm which brings together features from both the previous algorithms. The new algorithm was shown to converge to a solution in a finite number of steps with much faster speed than either of the previous algorithms. We looked at some interesting properties of the algorithm, including the fact that the convergence of the algorithm was irrespective of whether the degradation was spatially invariant or spatially variant. We examined implementation considerations involved with using the algorithm in practice and studied an example of the algorithm's use in practice. The example showed that the proposed algorithm is much faster than previous algorithms.

3

Spatially Adaptive Image Restoration

3.1 Introduction

In the previous chapter, it was seen that as long as the weights of the network remain constant from iteration to iteration, Algorithm 2.3 will converge. This result holds regardless of whether the weights of the network vary from pixel to pixel across the image to be restored. We will refer to this type of variation of the weights from region to region in the image as *spatial variation*.

The analysis in the previous chapter focused on the case when the weights are not spatially variant. But there are situations when we may want to solve problems that involve spatially variant weights. The first case occurs when the type and/or strength of the degrading function varies from region to region in the image. For example, imagine taking a picture of a distant mountain when a nearby tree is in the picture. If the image of the mountain is in focus, then the image of the tree will most likely be out of focus. The pixels belonging to the image of the tree have suffered a more severe distortion than the pixels belonging to the image of the mountain. We will call this type of effect *spatially variant distortion*. In spatially variant distortion, the columns of the degradation matrix, **H**, that correspond to pixels in the image of the tree will be different from the rest of the degradation matrix. In turn, this will cause the weights assigned to the neurons associated with these pixels to also be different from the rest of the neurons in the network. Spatially variant distortion may also be caused by moving objects in the picture or local problems with the imaging system such as impurities in the camera lens. In fact, spatially variant distortion is the most common type of degradation occurring in optical imagery. The problem of varying the restoration weights to handle spatially variant distortion will be examined in this chapter and examples given.

The second case of when we may want to vary the weights of the network spatially is more fundamental than the case of spatially variant distortion. The restoration model we developed in Chapter 2 has a single parameter to control high-frequency components in the image and balance the desire to restore the image against the desire to reduce noise in the image. But this approach makes the assumption that in all regions of the image, the balance is approximately the same. That is, all parts of the image have the same statistical properties of signal and noise. Real images almost always violate this assumption. Most images of interesting objects or phenomena contain many different regions with different statistical properties. For example, real images contain edges, smooth areas, textures of different types, and fine and coarse details. Each of these areas has its own optimal balance between restoration and noise suppression, which may be different from other areas. It seems that the best way to handle this problem would be to vary the restoration parameter from pixel to pixel or region to region.

In *adaptive constraint restoration*, the weights are varied to implement different values of regularization parameter for different regions of the image. By doing this, we can adjust restoration parameters to suppress noise more greatly in regions where it is most noticeable, and less so in regions where image sharpness is the dominant consideration. There are two ways to look at this problem.

One way to look at this problem is to think of the constrained least square error formula in Equation (2.3) as containing a restoration term and a penalty term. The natural approach is then to assign each neuron in the network its own regularization parameter rather than letting all neurons have the same value. Then the problem becomes how to select the best value of regularization parameter for each pixel. This may be done using energy considerations with a gradient descent algorithm, or the statistical properties of the local region of the neuron. In this chapter we will analyze the penalty function concept and show that the method for selecting the regularization parameter based on gradient descent can find only suboptimal solutions. We also mathematically verify a fact, which has been observed in practice, that this method would generally use a small value of regularization parameter for textured regions, and a large value for smooth regions. Using this observation, we then introduce a regional processing approach based on local statistics to select the optimal regularization parameter for each pixel. This approach also has some relationship to biological vision systems in that it emphasizes edges. The algorithms described in the previous chapter are expanded to vary the neural weights to take account of a spatially variant point spread function (PSF) and to combine the adaptive constraint concept developed. We then look at how adaptive constraint restoration can be used to compensate for incomplete knowledge of the degrading PSF or incomplete knowledge of the spatial variance of the degradation.

The second way to look at this problem is to consider the constrained least square error formula in Equation (2.3) as being a Lagrangian formulation of the restoration problem as was explored to a certain extent in Chapter 2. In this chapter we will revisit this concept and show that the natural extension to adaptive constraint restoration is to divide the image into regions, with a value of regularization parameter assigned to each region rather than each pixel.

This chapter is divided into a number of sections. Section 3.2 examines the problem of handling spatially variant distortions, while Section 3.3 introduces the adaptive constraint algorithms derived from the penalty function concept. Section 3.4 describes how adaptive constraint restoration can be merged with methods to handle spatially variant distortions, while Section 3.5 considers the problem of semiblind restoration. This is restoration when only incomplete knowledge of the degrading function is available. Section 3.6 details some implementation considerations, while Section 3.7 presents examples to showcase the concepts presented in the previous sections, and Section 3.8 briefly describes the concept behind the Lagrange approach to adaptive constraint restoration. Section 3.9 summarizes this chapter.

3.2 Dealing with Spatially Variant Distortion

In the analysis in Chapter 2, no conditions were placed upon the form of the matrix **H**. In fact the neural network cost function only assumes that the weighting matrix, **W**, is symmetric. In the case of a space invariant degradation, the matrix **H** will indeed be symmetric. However, in the real world this is not often the case. As discussed in Chapter 2, the general form of an image degradation will be space variant. However, many of these distortions will still be linear and so the model given by Equation (2.2) still holds. In the case of a linear space variant distortion, **H** will not be symmetric, but the weighting matrix, **W**, will still be symmetric. This can be clearly shown by examining an equation describing the neural weights:

$$w_{ij} = -\sum_{p=1}^{l} h_{pi} h_{pj} - \lambda \sum_{p=1}^{L} d_{pi} d_{pj} \tag{3.1}$$

By converting Equation (3.1) into matrix notation we see that the weighting matrix is given by

$$\mathbf{W} = -\mathbf{H}^T \mathbf{H} - \lambda \mathbf{D}^T \mathbf{D} \tag{3.2}$$

Assume that **H** is not symmetric, such that $\mathbf{H} \neq \mathbf{H}^T$; then

$$
\begin{aligned}
\mathbf{W}^T &= -\left(\mathbf{H}^T \mathbf{H} + \lambda \mathbf{D}^T \mathbf{D}\right)^T \\
&= -\left(\mathbf{H}^T \mathbf{H}\right)^T - \lambda \left(\mathbf{D}^T \mathbf{D}\right)^T \\
&= -\mathbf{H}^T (\mathbf{H}^T)^T - \lambda \mathbf{D}^T (\mathbf{D}^T)^T \\
&= -\mathbf{H}^T \mathbf{H} - \lambda \mathbf{D}^T \mathbf{D} = \mathbf{W} \tag{3.3}
\end{aligned}
$$

Hence the weighting matrix, **W**, is symmetric. This is a powerful feature of the neural network approach. If the image suffers a known form of space variance, then the weights of the neurons in the network can be adjusted to restore the image exactly with very little additional computational overhead.

If the degrading PSFs follow some regular or cyclic pattern then further optimizations can be made to the algorithm. Instead of developing the full nonsymmetric version of **H** for the entire image, certain regular patterns of space variance allow us to compute a set number of simpler alternative versions of the **H** matrix that describe each different type of degrading PSF the image has suffered. From this we can compute a number of alternative versions of the **W** matrix. Hence we have a number of different sets of weights to select from when restoring the image. In the worst case, completely random spatial variance, there are $L = NM$ unique sets of weights to restore the image. However, if patterns of spatial variance can be discovered, the number of unique sets of weights can be significantly reduced. Let us analyze the cyclic spatially variant case in detail. Consider a cyclic spatially variant distortion obtained by the use of V PSFs $h_0(x, y), \ldots, h_{V-1}(x, y)$. The pixels in any one row of the image are acted upon by the same PSF; however, the PSF applied to each row is varied cyclically through the sequence

$$S_H = \{h_0(x, y), h_1(x, y), \ldots, h_{V-1}(x, y), h_{V-2}(x, y), \ldots, h_1(x, y)\} \qquad (3.4)$$

The sequence S_H has a period of $2V - 2$, and hence $2V - 2$ unique sets of weights are required to restore the image. This type of distortion is similar to the degradations involved with side-scan sonar images. The method employed in References [98,99] to handle spatially variant distortion was to precompute a number of sets of weights to handle the various PSFs degrading the image. Since the spatial variation of the PSFs was known, the correct set of weights could be chosen to restore the image accurately. Let us look now at an example of restoring an image degraded by a cyclic distortion.

An image was created using a cyclic variation of 7 by 7 Gaussian PSFs. Each PSF had the impulse response

$$h(x, y) = \frac{1}{2\pi \sigma_x \sigma_y} \exp \left[-\left(\frac{x^2}{2\sigma_x^2} + \frac{y^2}{2\sigma_y^2} \right) \right] \qquad (3.5)$$

where σ_x and σ_y are the standard deviations of the PSF in the x and y directions, respectively.

Using the analysis in this section, V was set to be 4. Table 3.1 details the degrading PSFs used to blur the image as per Equation (3.4).

The original image is shown in Figure 3.1a and the degraded image is shown in Figure 3.1b. This image was restored using two techniques. The first technique was a spatially invariant approach. The spatially variant distortion was approximated as a spatially invariant distortion by using a 7 by 7 Gaussian PSF of standard deviation 2.55. That is, all pixels were acted upon by

TABLE 3.1

Degrading PSFs

Standard Deviation	1.5	2.0	3.0	4.0
PSF	$h_0(x, y)$	$h_1(x, y)$	$h_2(x, y)$	$h_3(x, y)$

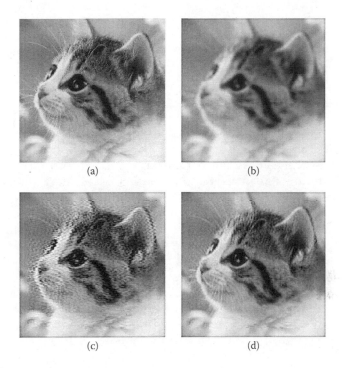

FIGURE 3.1
Original, degraded, and restored images suffering spatially variant distortion.

one weighting mask, whose components were calculated by approximating the space variant distortion as a space invariant distortion. Since the space variant distortion was very severe, the approximation caused instability in the restored image. The time to restore this image using the spatially invariant approximation was 230 seconds on a SUN Ultra SPARC 1. This image is shown in Figure 3.1c. The second technique was the proposed spatially variant approach with the six correctly calculated weighting masks. This restoration not only is better than Figure 3.1c, but recovers almost all the fine details lost in distortion. The time to restore this image using the spatially variant method was 314 seconds on a SUN Ultra SPARC 1. This image is shown in Figure 3.1d. Note that using Algorithm 2.3, the image could be restored using the correct spatially variant weights with only an extra 36% time penalty when compared to the spatially invariant approximation.

3.3 Adaptive Constraint Extension of the Penalty Function Model

Now we will consider the adaptation of the constraint factor to take account of the nonstationary nature of the image. We will analyze the penalty function model of the restoration problem and develop algorithms based on that model.

3.3.1 Motivation

The basic nonadaptive restoration cost function is given by

$$E = \frac{1}{2}\|\mathbf{g} - \mathbf{H}\hat{\mathbf{f}}\|^2 + \frac{1}{2}\lambda\|\mathbf{D}\hat{\mathbf{f}}\|^2 \tag{3.6}$$

In the penalty function model of this problem, the first term in Equation (3.6) attempts to control the fidelity of the image restoration, while the second term penalizes image estimates that contain too much noise. In Equation (3.6), a single value of the constraint parameter, λ, is used for every pixel in the image. This formula assumes that the statistics of the image do not change from region to region. Such a process is called *stationary*.

Images are, by nature, ensembles of nonstationary processes. Hence they do not fit the model given by Equation (3.6). For this reason, solutions based on this model can only produce a suboptimal restoration. An optimal restoration may be achieved by treating statistically dissimilar regions of an image with different restoration strategies or parameters. One method to achieve this is by using an adaptive regularization scheme. When implementing an adaptive regularization parameter scheme, an important consideration is on what basis should the regularization value be determined. In this section, we study the space invariant case. The result will be generalized to the space variant case in the next section.

To substantiate this issue, we first generalize the quadratic model in Equation (3.6) to

$$E = \frac{1}{2}\|\mathbf{g} - \mathbf{H}\hat{\mathbf{f}}\|^2 + \frac{1}{2}\|\sqrt{\Lambda}\mathbf{D}\hat{\mathbf{f}}\|^2 \tag{3.7}$$

where

$$\sqrt{\Lambda} = \begin{bmatrix} \sqrt{\lambda_1} & 0 & 0 \\ 0 & \ldots & 0 \\ 0 & 0 & \sqrt{\lambda_{NM}} \end{bmatrix} \tag{3.8}$$

is a diagonal matrix to reflect the adaptive processing nature. When $\lambda_1 = \lambda_2 = \ldots = \lambda_{NM}$, Equation (3.7) is reduced to the conventional constrained least squares formula given in Equation (3.6).

By relating Equation (3.7) to the formula for the energy of a neural network, the neural weight between neurons i and j in the network is given by

$$w_{ij} = -\sum_{p=1}^{L} h_{pi}h_{pj} - \sum_{p=1}^{L} \lambda_p d_{pi}d_{pj} \tag{3.9}$$

Equation (3.9) is overly complicated and can be approximated as

$$w_{ij} = -\sum_{p=1}^{L} h_{pi}h_{pj} - \lambda_i \sum_{p=1}^{L} d_{pi}d_{pj} \tag{3.10}$$

Note that the matrix $\mathbf{W}$, whose (i, j)th element is given by Equation (3.10), is not symmetric. As long as a check is made on whether the energy will be decreased before updating a pixel value, the lack of symmetry will not cause the algorithm to fail to converge. This is because $\mathbf{W}$ may be replaced by a symmetric equivalent version, $\mathbf{W}^*$, without changing the energy function.

Define $\mathbf{W}^* = \frac{(\mathbf{W}+\mathbf{W}^T)}{2}$, where $\mathbf{W}$ is the nonsymmetric weighting matrix whose (i, j)th element is given by Equation (3.10). Note $\mathbf{W}^*$ is symmetric since

$$(\mathbf{W}^\star)^T = \frac{(\mathbf{W}+\mathbf{W}^T)^T}{2} = \frac{(\mathbf{W}^T+\mathbf{W})}{2} = \mathbf{W}^\star$$

Taking the formula for the energy of the neural network, and noting that $a^T = a$ when a is a scalar, then we obtain

$$E = -\frac{1}{2}\hat{\mathbf{f}}^T\mathbf{W}^\star\hat{\mathbf{f}} - \mathbf{b}^T\hat{\mathbf{f}} + c$$

$$= -\frac{1}{2}\hat{\mathbf{f}}^T\frac{(\mathbf{W}+\mathbf{W}^T)}{2}\hat{\mathbf{f}} - \mathbf{b}^T\hat{\mathbf{f}} + c$$

$$= -\frac{1}{4}\hat{\mathbf{f}}^T\mathbf{W}\hat{\mathbf{f}} - \frac{1}{4}\hat{\mathbf{f}}^T\mathbf{W}^T\hat{\mathbf{f}} - \mathbf{b}^T\hat{\mathbf{f}} + c$$

$$E = -\frac{1}{4}\hat{\mathbf{f}}^T\mathbf{W}\hat{\mathbf{f}} - \left(\frac{1}{4}\hat{\mathbf{f}}^T\mathbf{W}\hat{\mathbf{f}}\right)^T - \mathbf{b}^T\hat{\mathbf{f}} + c = -\frac{1}{2}\hat{\mathbf{f}}^T\mathbf{W}\hat{\mathbf{f}} - \mathbf{b}^T\hat{\mathbf{f}} + c \qquad (3.11)$$

From Equation (3.11) we can see that the form of the energy functions given by using $\mathbf{W}$ and $\mathbf{W}^*$ are identical. This means that if we assign neuron weights according to Equation (3.10), the algorithm will converge.

The elements of $\mathbf{W}^*$ can be considered an alternative approximation of Equation (3.9) and are given by

$$w_{ij}^\star = -\sum_{p=1}^{L}h_{pi}h_{pj} - \left(\frac{\lambda_i + \lambda_j}{2}\right)\sum_{p=1}^{L}d_{pi}d_{pj} \qquad (3.12)$$

However, since we have shown that both approximations produce identical energy functions, we will hence use the approximation given by Equation (3.10). Next we need to show that approximating Equation (3.9) as Equation (3.10) does not seriously affect the quality of the restored images.

Equation (3.10) is much easier to calculate for each neuron since it only relies on the value of λ assigned to neuron i, and does not depend on the values of λ assigned to other neurons in the neighborhood of neuron i. If we have a finite set of available values of λ, then by using Equation (3.10) we will require only one weighing mask for each λ value in the set. If we instead use Equation (3.9) we will have to recalculate the weights for each neuron in the network. This will, for obvious reasons, greatly degrade the performance of the network in terms of computational efficiency. The next question we must ask is how well Equation (3.10) approximates Equation (3.9) and when we will expect the approximation to break down.

Note that the concept of a *weighting mask* described in Chapter 2 used the fact that a neuron only has nonzero connections to other neurons within a certain neighborhood centered on the neuron. If a neuron lies in a region where all its neighbors were assigned the same value of λ, then Equation (3.9) would be the same as the weighting mask produced by using a constant value of λ for all pixels in the image. This means that a value of λ can be found such that Equation (3.9) can be accurately approximated by Equation (3.10) and $w_{ij} = w_{ji}$ in the vicinity of this neuron.

If some of the neurons in the neighborhood of neuron i have different values of λ, but the difference between λ_i and λ_j is not great, then Equation (3.10) is still a good approximation although $w_{ij} \neq w_{ji}$. However, we can expect that $w_{ij} \approx w_{ji}$.

If some of the neurons in the neighborhood of neuron i have very different values of λ, then the approximation will not hold. We cannot expect that $w_{ij} \approx w_{ji}$. This is not a big problem, however, since regions where neighboring neurons have greatly different values of λ must be edges and highly textured regions. The reasons for this will become apparent later. In such regions, noise is not readily noticeable and so any errors made by approximating Equation (3.9) as Equation (3.10) will be masked. Hence the approximation provides increased ease of use and only breaks down in those regions where the effects are least noticeable.

Since approximation Equation (3.10) is much easier to implement than Equation (3.9), this approximation will be used in the subsequent sections. The next question is how to determine the values of λ_i in Equation (3.10) for each pixel in the image.

Let us investigate two methods for determining the regularization parameter. The first is an extension of the gradient-based method considered in the last chapter. In effect, this method is motivated by the fact that since we used gradient descent to restore the image in Algorithm 2.3, then it seems natural to use gradient descent to find the optimal values of λ for each pixel. The second method is based on local statistics in the framework of a neural network.

3.3.2   Gradient-Based Method

Since Algorithm 2.3 in the previous chapter used the principles of gradient descent to solve this problem, let us first look at how gradient descent might be used to select unique values of λ for each pixel.

Equations (2.7) and (2.8) indicate that differences in the regularization parameter affect only the weights of the neural network, not the initial values or the bias inputs.

Equation (2.12) gives the energy change resulting from a change of neuron state $\Delta \hat{f}_i$. Substituting Equation (2.10) into Equation (2.12) yields

$$\Delta E_i = -\left[\sum_{j=1}^{L} w_{ij} \hat{f}_i + b_i \right] \Delta \hat{f}_i - \frac{1}{2} w_{ii} (\Delta \hat{f}_i)^2 \qquad (3.13)$$

(a)   (b)

FIGURE 3.2
Original images.

It should be noted that $\sum_{j=1}^{L} w_{ij} \hat{f}_i + b_i$ is the input to each neuron and has the same sign as $\Delta \hat{f}_i$ if ΔE is to be negative. It is important to note that we are only considering the cases where ΔE is negative since when $\Delta E \geq 0$, there will be no pixel value update at pixel $\Delta \hat{f}_i$.

Definition 3.1
The greatest energy minimization (GEM) method for selecting the constraint value for each pixel is defined as choosing for each pixel in each iteration the value of λ from a range of allowable λ values, $\lambda_a \leq \lambda \leq \lambda_h, \langle \lambda_a, \lambda_b \geq 0 \rangle \in \Re$, which best minimizes Equation (3.13). [71,100,101].

Let us have a look at an example of applying this algorithm to real imagery. In this example, the images in Figure 3.2 were blurred using Equation (3.5). The number of choices of regularization parameter R was set to 5. The impulse response of the constraint matrix, $d(x, y)$, was as follows:

$$d(x, y) = \frac{1}{6} \begin{bmatrix} 1.0 & 1.0 & 1.0 & 1.0 & 1.0 \\ 1.0 & 4.0 & -3.0 & 4.0 & 1.0 \\ 1.0 & -3.0 & -20.0 & -3.0 & 1.0 \\ 1.0 & 4.0 & -3.0 & 4.0 & 1.0 \\ 1.0 & 1.0 & 1.0 & 1.0 & 1.0 \end{bmatrix}$$

Each image was degraded by the same PSF. For this experiment, a PSF of size $P = 5$ was used with $\sigma_x = \sigma_y = 2.0$. Two levels of white noise, with variances approximately equal to 4 and 18, were added to each degraded image.

Table 3.2 shows the signal to noise ratio (SNR) and local standard deviation mean square error (LSMSE) between the original and the degraded images, and images restored with a constant value of λ, and the greatest energy reduction technique for selecting λ, and the adaptive constraint technique of

TABLE 3.2

Statistics of Degraded and Restored Images for
Various Levels of Noise

		SNR(dB)		
Noise	DI	NA	GER	KK
		Cat	*Image*	
4.24	12.61	13.30	12.55	11.38
18.30	12.25	12.81	11.02	10.83
		Flower	*Image*	
4.22	15.67	16.65	15.13	15.74
18.52	14.94	15.93	12.61	14.32
		LSMSE		
	DI	NA	GER	KK
		Cat	*Image*	
4.24	42.98	14.58	14.98	30.57
18.30	38.65	17.45	30.43	30.46
		Flower	*Image*	
4.22	20.53	4.35	9.74	6.56
18.52	19.35	5.94	33.54	10.88

Legend: DI = degraded image; NA = nonadaptively restored image; GER = image restored using energy minimization method for constraint selection; KK = image restored using the Kang and Katsaggelos algorithm.

Kang and Katsaggelos [102]. LSMSE is defined as follows:

$$\text{LSMSE} = \frac{1}{NM} \sum_{x=0}^{N-1} \sum_{y=0}^{M-1} (\sigma_A(\mathbf{f}(x, y)) - \sigma_A(\mathbf{g}(x, y)))^2 \qquad (3.14)$$

where $\sigma_A(\mathbf{f}(x, y))$ is the standard deviation of the A by A local region surrounding pixel (x, y) in the first image and $\sigma_A(\mathbf{g}(x, y))$ is the standard deviation of the A by A local region surrounding pixel (x, y) in the image with which we wish to compare the first image. Unless otherwise stated, A is assumed to be set to 9 for the computation of LSMSE.

Figures 3.3 and 3.4 show the degraded and restored images of the flower image, while Figure 3.5 shows the degraded and restored images of the cat image in the case of a noise variance of 18.30. In Figures 3.3, 3.4, and 3.5, (a) is the degraded image, (b) is the image restored with a constant value of λ, (c) is the image restored using the greatest energy-reduction technique for selecting λ, and (d) uses the adaptive constraint technique of Kang and Katsaggelos [102].

From Figures 3.3, 3.4, and 3.5, the images (c) restored using the adaptive image-restoration method appear clearer and more visually pleasing when compared to the images produced by the nonadaptive algorithm (b).

FIGURE 3.3
Degraded and restored images for noise of variance 4.22.

FIGURE 3.4
Degraded and restored images for noise of variance 18.52.

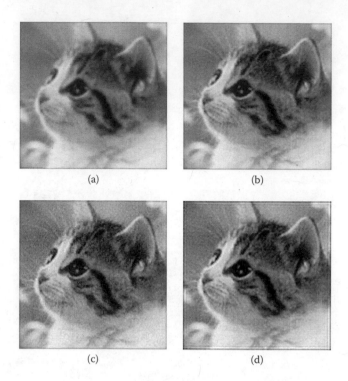

FIGURE 3.5
Degraded and restored images for noise of variance 18.30.

However, the results in Table 3.2 show that the algorithm is still less than perfect. Let us analyze this algorithm in more detail.

Analysis of the Gradient-Based Method

Based on the definition and other aforementioned assumptions, we present two theorems.

Theorem 3.1

If the GEM method is used to select a suitable λ value from a range of possible λ values, where $\lambda_a = 0$ and $\lambda_b = \infty$ in Definition 3.1, then λ_b will always be chosen.

Theorem 3.2

If the GEM method is used to select a suitable λ value, satisfying $\lambda_a \leq \lambda \leq \lambda_b$, where both λ_a and λ_b are finite, then either λ_a or λ_b will be chosen unless all available λ values would produce the same resultant decrease in energy.

To prove the above theorems, we must rearrange Equation (3.13). Expanding Equation (3.13) using Equation (3.1) we obtain

$$\Delta E_i = \left[\sum_{j=1}^{L} \sum_{p=1}^{L} h_{pi} h_{pj} \hat{f}_j - b_i \right] \Delta \hat{f}_i + \lambda_i \Delta \hat{f}_i \sum_{j=1}^{L} \sum_{p=1}^{L} d_{pi} d_{pj} \hat{f}_j$$

$$+ \frac{1}{2}(\Delta \hat{f}_i)^2 \sum_{p=1}^{L} h_{pi}^2 + \frac{1}{2}\lambda(\Delta \hat{f}_i)^2 \sum_{p=1}^{L} d_{pi}^2$$

$$= A\Delta \hat{f}_i + \lambda_i B \Delta \hat{f}_i + C(\Delta \hat{f}_i)^2 + D\lambda_i(\Delta \hat{f}_i)^2 \tag{3.15}$$

where $A = \sum_{j=1}^{L} \sum_{p=1}^{L} h_{pi} h_{pj} \hat{f}_j - b_i$, $B = \sum_{j=1}^{L} \sum_{p=1}^{L} d_{pi} d_{pj} \hat{f}_j$, $C = \frac{1}{2} \sum_{p=1}^{L} h_{pi}^2$, and $D = \frac{1}{2} \sum_{p=1}^{L} d_{pi}^2$.

It should be noted that $u_i = -A - \lambda B$ and

$$C + \lambda_i D = \frac{-w_{ii}}{2} \tag{3.16}$$

Consider the $\Delta \hat{f}_i$ that maximizes the energy reduction for pixel i. Let this term be defined as $\Delta \acute{f}_i$. To compute $\Delta \acute{f}_i$, we differentiate Equation (3.15) with respect to $\Delta \hat{f}_i$:

$$\frac{\delta \Delta E_i}{\delta \Delta \hat{f}_i} = A + \lambda_i B + 2C\Delta \acute{f}_i + 2D\Delta \lambda_i \Delta \acute{f}_i \tag{3.17}$$

and set Equation (3.17) to zero to obtain the optimal $\Delta \acute{f}_i$.

$$\Delta \acute{f}_i = \frac{-(A + \lambda_i B)}{2(C + \lambda_i D)} \tag{3.18}$$

Rearrangement of Equation (3.15) yields

$$\Delta E_i = (A + \lambda_i B)\Delta \acute{f}_i + (C + \lambda_i D)(\Delta \acute{f}_i)^2 \tag{3.19}$$

The substitution of Equation (3.18) into Equation (3.19) gives us

$$\Delta \acute{E}_i = \frac{-(A + \lambda_i B)^2}{2(C + \lambda_i D)} + \frac{(A + \lambda_i B)^2}{4(C + \lambda_i D)} = \frac{-(A + \lambda_i B)^2}{4(C + \lambda_i D)} \tag{3.20}$$

We can confirm from Equation (3.18) that $\Delta \acute{f}_i$ always has the same sign as u_i and that $\Delta \acute{E}_i$ is always negative for positive values of λ_i, as expected. The graph of $\Delta \acute{E}_i$ versus λ_i is sketched in Figure 3.6 for a particular example.

We can see from Equation (3.20) that for all acceptable values of ΔE_i, $\lim_{\lambda_i \to \infty} \Delta \acute{E}_i = -\infty$. This proves Theorem 3.1 and implies that a great energy reduction can always be obtained by choosing a very large value of λ_i. If we allow energy considerations alone to dictate the choice of λ_i, the results would be poor. An acceptable range of λ values must be set. Since C and D are always positive, and $\Delta \acute{E}_i$ is always negative, Equation (3.20) has a maximum at $\lambda_i = -(A/B)$.

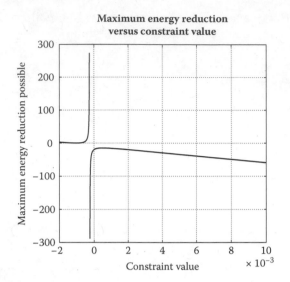

FIGURE 3.6
Graph of maximum energy reduction versus constraint value.

Since Equation (3.20) is a convex function when $\Delta \acute{E}_i < 0$, then we observe that if the set of acceptable λ values is finite and constrained between λ_a and λ_b ($\lambda_a < \lambda_b$), then Equation (3.20) indicates that under most circumstances, the best energy reduction will be obtained by choosing either λ_a or λ_b, and not any intermediate values of λ. Only under the conditions of $-A^2/(4C) \approx -B^2/(4D) \approx 0$ will numerical errors possibly enable the choice of intermediate values of λ. This proves Theorem 3.2.

From this analysis, a number of conclusions can be drawn. When A is low and B is high, the largest λ_i value gives the best energy minimization; however, when A is high and B is low, the smallest value of λ_i produces the best energy minimization.

The next question is which λ values will be chosen in low- and high-variance regions of the image to maximize the energy reduction. To clearly answer this question, it is necessary to examine how the factors $A, B, C,$ and D vary as a function of image statistics.

Let's look at each of the factors in turn.

Factor A

Factor A is given by

$$A = \sum_{j=1}^{L} \sum_{p=1}^{L} h_{pi} h_{pj} \hat{f}_j - b_i = K - b_i$$

where $K = \sum_{j=1}^{L} \sum_{p=1}^{L} h_{pi} h_{pj} \hat{f}_j = [\mathbf{H}^T \mathbf{H} \hat{\mathbf{f}}]_i$.

The factor K is the value of pixel $\hat{f}_i$ after a double application of the degrading PSF to the image estimate. Using the approximate formula for $\mathbf{H}$, then $\mathbf{H}$ is symmetric, $\mathbf{H}^T = \mathbf{H}$. Note that since $b_i = \sum_{p=1}^{L} h_{pi} g_p$, b_i is the value of pixel $\hat{f}_i$ after a single application of the degrading PSF to the degraded image, which is itself the result of applying the degrading PSF to the original image (ignoring noise added to the system). Therefore, in effect

$$b_i = \sum_{p=1}^{L} \sum_{j=1}^{L} h_{pi} h_{pj} f_j = [\mathbf{H}^T \mathbf{H} \mathbf{f}]_i \tag{3.21}$$

where $\mathbf{f}$ is the estimate of the original when noise is not considered. The algorithm would eventually return $\mathbf{f}$ as the restored image when $\lambda = 0$. Hence we obtain

$$\mathbf{A}_k = \mathbf{H}^T \mathbf{H} \hat{\mathbf{f}} - \mathbf{H}^T \mathbf{H} \mathbf{f} = \mathbf{H}^T \mathbf{H} (\hat{\mathbf{f}}^k - \mathbf{f}) \tag{3.22}$$

where $\mathbf{A}_k$ is the column vector whose element i is the value of factor A computed at neuron i during the kth iteration of the algorithm and $\hat{\mathbf{f}}^k$ is the estimate of the original image produced by the algorithm at the kth iteration.

Considering Equation (3.22), it is obvious that the entries in $\mathbf{A}_k$ may be positive or negative and will approach zero as the image estimate approaches the original image in the event of no noise. We can expect $\mathbf{A}_k$ to have its greatest values in the initial iteration of the algorithm.

If the initial image estimate is $\hat{\mathbf{f}} = \mathbf{g} = \mathbf{H} \mathbf{f}$, then

$$\mathbf{A}_1 = \mathbf{H}^T \mathbf{H} (\mathbf{H} \mathbf{f} - \mathbf{f}) = \mathbf{H}^T \mathbf{H} (\mathbf{H} - \mathbf{I}) \mathbf{f} \tag{3.23}$$

As long as the additive noise is not too severe, the double application of the low-pass degrading PSF given by $\mathbf{H}$ will remove most of the noise in the smooth regions of the image. Since $\mathbf{H} - \mathbf{I}$ is a high-pass filter, factor A would tend to be large in high-variance regions and small in low-variance regions of the image.

Factor B

For factor B we have $B = \sum_{j=1}^{L} \sum_{p=1}^{L} d_{pi} d_{pj} \hat{f}_j$, which can be rewritten as

$$B = [\mathbf{D}^T \mathbf{D} \hat{\mathbf{f}}]_i \tag{3.24}$$

Since $\mathbf{D}^T = \mathbf{D}$, factor B is hence the value of pixel $\hat{f}_i$ after a high-pass filter has been applied twice to the image estimate. On edges and high texture regions, we would expect the magnitude of factor B to be large. However, since the high-pass filter is applied twice, noise and very sharp edges would produce a larger magnitude of B than more gradual edges in the image. It is important to note that B may be positive or negative depending on whether the value of $\hat{f}_i$ is higher or lower than the mean of its neighbors; however, B will tend to zero in low-variance regions of the image.

Factors C and D

Factors C and D are given by $C = 1/2 \sum_{p=1}^{L} h_{pi}^2$ and $D = 1/2 \sum_{p=1}^{L} d_{pi}^2$. Both these factors are constant. C will always be quite small and in the case of a degrading PSF modeled by a 5 by 5 Gaussian blur of standard deviation 2.0 in both x and y directions, $C = 0.28$. Factor D depends upon the type of high-pass filter chosen and typical values are of the order of 50 to 500.

>   *Observation 1*: In high-texture regions of the image or near edges, both A and B can be expected to be large. However, the contribution from factor B tends to be less significant for two reasons:

1. The factor B in Equation (3.20) is multiplied by λ_i which is always $\ll 1$.
2. The factor B corresponding to a double application of a high-pass filter suggests that in the presence of gradual edges, B will not be very large. Since noise is added to the image after blurring, the presence of noise will contribute most to the value of B.

Therefore we can expect that $-A^2/(4C) < (-\lambda_i B^2)/(4D)$, and so the best energy reduction would be obtained by choosing the lowest value of λ available.

>   *Observation 2*: In low-texture regions of the image, A can be expected to be small. Factor B may also be small. However, the presence of noise in the initial estimate should make the effect of B on Equation (3.20) override that of A. Hence in low-texture regions of the image, choosing the largest available λ_i produces the best energy minimization.

This analysis has been verified in practice. Figure 3.7 shows the selection of λ values during the first iteration for the image in Figure 3.2a degraded by a 5 by 5 Gaussian PSF of standard deviation 2.0, with additive noise of variance 18.49. The darker regions in Figure 3.7 denote the selection of larger λ values, and the lighter regions denote the selection of smaller λ values.

From Figure 3.7 we can clearly see that to a great extent, only the lowest or the highest value of λ is chosen by the gradient descent method. Although mathematically elegant, using gradient descent to select the regularization parameter does not agree with biological vision. In addition, this method does not allow us to use a wide range of λ values to restore the image. Only two λ values are used. This effectively divides the image into low- and high-variance regions.

3.3.3   Local Statistics Analysis

The above analysis shows us that gradient descent is not an optimal approach for choosing λ_i in adaptive regularization. However, it leads us to the observation that in low-texture regions, a high value of λ_i results in a visually pleasing result, while in high-texture regions, a low λ_i value works best. These observations encourage us to use local statistics, since these are a good measure of

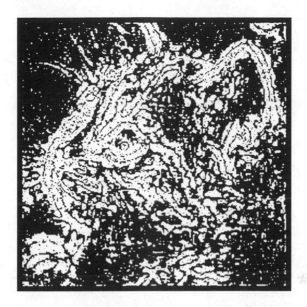

FIGURE 3.7
The λ values selected for each pixel during the initial iteration of the energy minimization-based algorithm for a typical image.

the image roughness to locally determine the λ_i value. Mathematically,

$$\lambda_i = Y(S_i) \tag{3.25}$$

where $Y(S_i)$ is a function of the local image statistics S_i at $\hat{f}_i$.

Since S_i has almost exactly the same value for all pixels, $\hat{f}_i$, in a statistically homogeneous area, the λ_i value is also almost exactly the same. Therefore, the structure of the processing model can be further modified such that, instead of assigning each pixel a different λ_i, we assign each statistically homogeneous area a λ_k.

Assume that there are K homogeneous areas. By first properly rearranging the pixels in such a way that the pixels in a homogeneous area are consecutively indexed in $\hat{f}$ to form a new vector $\hat{f}^\star$, Equation (3.7) can be rewritten as

$$E = \frac{1}{2}\{\|\mathbf{g}^\star - \mathbf{H}^\star\hat{f}^\star\|^2 + \hat{f}^{\star^T}(\mathbf{D}^{\star^T}\Lambda^\star\mathbf{D}^\star)\hat{f}^\star\} \tag{3.26}$$

where $\hat{f}^{\star^T} = [\mathbf{f}_1^T, \ldots, \mathbf{f}_K^T]$ with $\mathbf{f}_k$ being the vector consisting of the pixels in the kth homogeneous area,

$$\mathbf{H}^\star = \begin{bmatrix} \mathbf{H}_1 \\ \ldots \\ \mathbf{H}_K \end{bmatrix}, \quad \Lambda^\star = \begin{bmatrix} \Lambda_1 & 0 & 0 \\ 0 & \ldots & 0 \\ 0 & 0 & \Lambda_K \end{bmatrix}, \quad \mathbf{D}^\star = \begin{bmatrix} \mathbf{D}_1 \\ \ldots \\ \mathbf{D}_K \end{bmatrix}, \quad \mathbf{g}^\star = \begin{bmatrix} \mathbf{g}_1 \\ \ldots \\ \mathbf{g}_K \end{bmatrix} \tag{3.27}$$

with $\mathbf{H}_k$, Λ_k, $\mathbf{D}_k$ and $\mathbf{g}_k$ being the submatrices (vectors) of $\mathbf{H}^\star$, $\Lambda^\star$, $\mathbf{D}^\star$ and $\mathbf{g}^\star$, corresponding to $\mathbf{f}_k$, and $\Lambda_k = \lambda_k\mathbf{I}$, with $\mathbf{I}$ being the identity matrix. Define

$\mathbf{H}_k = [\mathbf{H}_{k l}, \ldots, \mathbf{H}_{k K}]$ and $\mathbf{D}_k = [\mathbf{D}_{k l}, \ldots, \mathbf{D}_{k K}]$, $k = 1, 2, \ldots, K$; then mathematical manipulation of Equation (3.26) leads to

$$E = -\frac{1}{2}\sum_{k=1}^{K}\{[\hat{\mathbf{f}}^{\star T}(\mathbf{H}_k^T\mathbf{H}_k + \lambda_k\mathbf{D}_k^T\mathbf{D}_k)\hat{\mathbf{f}}^{\star} + 2\mathbf{g}_k^T\mathbf{H}_k\hat{\mathbf{f}}^{\star}]\} + \|\mathbf{g}^{\star}\|^2$$

$$E = -\frac{1}{2}\sum_{k=1}^{K}\{\mathbf{f}_k^T(\mathbf{H}_{kk}^T\mathbf{H}_{kk} + \lambda_k\mathbf{D}_{kk}^T\mathbf{D}_{kk})\mathbf{f}_k$$

$$+ \sum_{l=1,l\neq k}^{K}\sum_{m=1,m\neq k}^{K}\mathbf{f}_l^T(\mathbf{H}_{kl}^T\mathbf{H}_{km} + \lambda_k\mathbf{D}_{kl}^T\mathbf{D}_{km})\mathbf{f}_m + 2\mathbf{g}_k^T\mathbf{H}_k\hat{\mathbf{f}}^{\star}\} + \|\mathbf{g}^{\star}\|^2 \quad (3.28)$$

Apparently, $\mathbf{H}_{kk}^T\mathbf{H}_{kk} + \lambda_k\mathbf{D}_{kk}^T\mathbf{D}_{kk}$ represents the intraconnections within area k, and $\mathbf{H}_{kl}^T\mathbf{H}_{km} + \lambda_k\mathbf{D}_{kl}^T\mathbf{D}_{km}$ represents the interarea contributions from areas l and m to area k. Equation (3.28) is the extension of a biologically motivated neural network: the network of networks (NoN) [103]. The significance of this mapping is as follows:

1. The human visual system pays little attention to individual pixels in an image. Instead it looks for areas of similar pixels in a statistically homogeneous sense, and it is more sensitive to edges [104].

2. The NoN is a computational model imitating a simplified representation of the human cortex, or part of it, the biological visual-processing machine. The neurons in the same cluster are similar to one another.

3. By representing a pixel with a neuron, a homogeneous image area is mapped to a cluster of neurons that are similar to one another. Local statistics is a good criterion to measure the level of similarity.

Therefore, using a NoN with a statistical criterion to adaptively determine the λ value and in turn the processing architecture may potentially simulate some aspects of human perception in recovering genuine information lost in recording.

Now the important issue is selecting the λ_i for each homogeneous area. In this section, we present a curve fitting approach [71,72,105,106]. In this approach, the largest and the smallest values of S_i, S_{imax}, and S_{imin} are identified and the corresponding λ_{imax} and λ_{imin} are determined.

Experiments performed by the authors found that if the variance in a region gradually increases by equal steps, then the change in variance level is much more noticeable when the overall variance levels are low rather than when the overall variance levels are high. Humans are less able to discern an increase in noise levels in high-variance regions than they are for low-variance regions. A log-linear relationship is therefore suggested

$$\lambda_i = a \, \log(S_i) + b \quad (3.29)$$

In Figure 3.8, six images with a constant level of pixel intensity are superimposed by noise of increasing variance. The increasing level of noise between

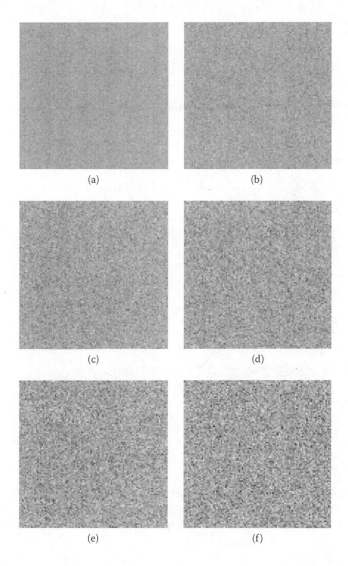

FIGURE 3.8
Images with varying levels of additive noise: (a) $\sigma^2 = 76$; (b) $\sigma^2 = 174$; (c) $\sigma^2 = 316$; (d) $\sigma^2 = 491$; (e) $\sigma^2 = 699$; (f) $\sigma^2 = 973$.

each image and its neighbor on the right is readily noticeable for the first three images when variance levels are low, compared to the last three when variance levels are high. The difference in apparent noise level between the last two images is the weakest of the entire set of images, despite the fact that the increase in variance between these two images is the greatest of the set.

The above observations lend weight to the use of constant λ values for each iteration based on those during the first iteration. Since the λ selected during the first iteration can often produce visually pleasing results (large λ in the

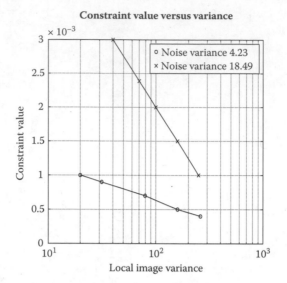

FIGURE 3.9
Constraint values chosen versus variance.

background and small λ on the edges), these can be held constant throughout the restoration procedure. The λ values may not act as favorably for later iterations in the algorithm. For example, the local variance levels in the areas surrounding edges in the image may increase during the restoration due to ripple effects from nearby edges such that they become large enough to cause a large value of λ to be applied to edges of the image, hence producing the opposite effect to that which we are attempting to achieve. Let us look now at some examples of this algorithm when applied to real imagery.

This example is an extension of the example given in Section 3.3.2. As with the example in Section 3.3.2, a PSF of size $P = 5$ was used with $\sigma_x = \sigma_y = 2.0$. Two levels of white noise, with variances approximately equal to 4 and 18, were added to each degraded image.

Table 3.3 is an extension of Table 3.2 and shows the SNR and LSMSE between the original and the degraded images, and images restored with a constant value of λ, the greatest energy-reduction technique for selecting λ, the adaptive constraint technique of Kang and Katsaggelos [102], and with an adaptive λ value based on local image variance and Equation (3.29). Figure 3.9 shows regularization parameters used in the experiment plotted against the level of local variance. The values of λ were associated with variance thresholds in a way consistent with Equation (3.29).

Figures 3.10 and 3.11 show the degraded and restored images of the flower image, while Figure 3.12 shows the degraded and restored images of the cat image in the case of a noise variance of 18.30. In Figures 3.10, 3.11, and 3.12, (a) is the degraded image, (b) is the image restored with a constant value of λ, (c) is the image restored using the greatest energy-reduction technique for selecting

TABLE 3.3

Statistics of Degraded and Restored Images for
Various Levels of Noise

Noise	SNR(dB)				
	DI	NA	GER	KK	VB
		Cat		*Image*	
4.24	12.61	13.30	12.55	11.38	13.19
18.30	12.25	12.81	11.02	10.83	12.58
		Flower		*Image*	
4.22	15.67	16.65	15.13	15.74	16.95
18.52	14.94	15.93	12.61	14.32	15.82

	LSMSE				
	DI	NA	GER	KK	VB
		Cat		*Image*	
4.24	42.98	14.58	14.98	30.57	13.51
18.30	38.65	17.45	30.43	30.46	15.79
		Flower		*Image*	
4.22	20.53	4.35	9.74	6.56	2.53
18.52	19.35	5.94	33.54	10.88	3.74

*Legend: DI = degraded image; NA = nonadaptively restored
image; GER = image restored using energy minimization
method for constraint selection; KK = image restored using
the Kang and Katsaggelos algorithm; VB = image restored
using variance based constraint selection.*

λ, (d) uses the adaptive constraint technique of Kang and Katsaggelos [102],
and (e) uses an adaptive λ value based on local image variance and Equation (3.29).

From Figures 3.10, 3.11, and 3.12, the images restored using the adaptive
image-restoration methods (c), (d), and (e), appear clearer and more visually pleasing despite a slight decrease in SNR when compared to the images
produced by the nonadaptive algorithm (b). Table 3.3 indicates that using
variance as a criterion to choose the value of the regularization parameter
produces images with an improved LSMSE as noise levels increase.

3.4 Correcting Spatially Variant Distortion Using Adaptive Constraints

In the previous section, the weights of the neural network were varied spatially in order to implement the adaptive regularization parameter. A similar
concept was considered in Section 3.2 to handle a simple form of spatially
variant distortion. In Section 3.2, we saw how multiple sets of weights could

FIGURE 3.10
Degraded and restored images for noise of variance 4.22.

be used to restore an image degraded by a cyclic variation of Gaussian PSFs of different standard deviations. In this section, we will examine the integration of the spatially variant restoration technique of Section 3.2 with the adaptive constraint technique of Section 3.3.

The method employed in Section 3.2 to handle spatially variant distortion was to precompute a number of sets of weights to handle the various PSFs degrading the image. If the spatial variation of the PSFs was known, the correct set of weights could be chosen to restore the image accurately. However, a fixed λ was used in Section 3.2. To implement a spatially variant restoration

FIGURE 3.11
Degraded and restored images for noise of variance 18.52.

technique with an adaptive constraint parameter, we produce as many sets of weights as needed to handle the space variant degradation and every possible choice of regularization parameter.

Let us consider as an example, the same type of distortion considered in Section 3.2. Restating the discussion in Section 3.2, the cyclic spatially variant distortion we consider is that obtained by the use of V PSFs $h_0(x, y), \ldots,$ $h_{V-1}(x, y)$. The pixels in any one row of the image are acted upon by the same PSF; however, the PSF applied to each row is varied cyclically through

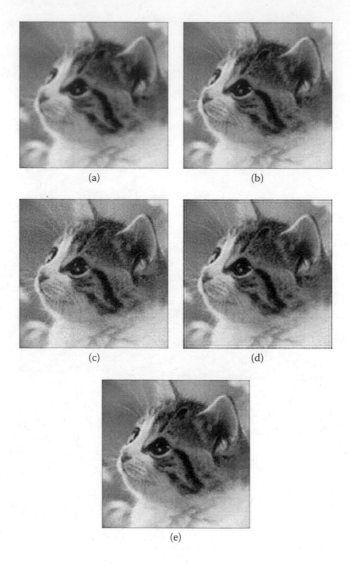

(a)   (b)

(c)   (d)

(e)

FIGURE 3.12
Degraded and restored images for noise of variance 18.30.

the sequence

$$S_H = \{h_0(x, y), h_1(x, y), \ldots, h_{V-1}(x, y), h_{V-2}(x, y), \ldots, h_1(x, y)\} \qquad (3.30)$$

The sequence S_H has a period of $2V - 2$, and hence $2V - 2$ unique sets of weights are required to restore the image when the same regularization parameter is chosen for every pixel. When additional sets of weights are created to implement an adaptive regularization parameter scheme, the following analysis applies.

Taking R as the number of choices of regularization parameter to be used in the restoration, then the sets of weights to be used to restore a row blurred by the ith element of sequence S_H form the set

$$\mathbf{WSM}_i = \{\mathbf{Wm}_0^i, \ldots, \mathbf{Wm}_{R-1}^i\} \tag{3.31}$$

where $\mathbf{Wm}_j^i$ is the set of weights required to restore a row degraded by PSF $[S_H]_i$ using the jth choice of regularization parameter being considered. The restoration problem becomes a problem of selecting the correct set of weights from the super set of

$$\mathbf{WMA} = \{\mathbf{WSM}_0, \ldots, \mathbf{WSM}_{2V-3}\} \tag{3.32}$$

Restoration of the image is accomplished by selecting the relevant set $\mathbf{WSM}_i$, based on the row coordinate of the current pixel being examined, then selecting the optimal $\mathbf{Wm}_j^i$ within that set based on the selection schemes described in Section 3.3 [71]. Let's look now at an example of implementing the spatially variant restoration scheme with an adaptive regularization parameter as described above. An image was created using a cyclic variation of 7 by 7 Gaussian PSFs. Using the analysis in Section 3.2, V was set to be 4. As before, Table 3.1 details the degrading PSFs used to blur the image as per Equation (3.30). Noise of variance 4.28 was added to the blurred image. The degraded image is shown in Figure 3.13a. First, the image was restored using the nonadaptive spatially invariant approach. A regularization parameter value of 0.0007 was used and the spatially variant distortion was approximated as a spatially invariant distortion by using a 7 by 7 Gaussian PSF of standard deviation 2.55. This image is shown in Figure 3.13b. The image was then restored using a nonadaptive spatially variant restoration method, with a regularization parameter value of 0.0007. This image is shown in Figure 3.13c. In the final approach, the degraded image was restored using a spatially variant, adaptive constraint method, with the variance-based λ value selection algorithm. The regularization value was selected in the same manner as the example in the previous section. This image is shown in Figure 3.13d. The statistics are summarized in Table 3.4.

We can see from Figure 3.13d that the space variant adaptive restoration approach has produced a visually more pleasing result with improved values of SNR and LSMSE compared to the other methods.

TABLE 3.4

Results of Spatially Variant Experiment

Image	SNR (dB)	LSMSE
Degraded	13.94	36.01
Space invariant nonadaptive	15.03	6.38
Space variant nonadaptive	15.87	6.04
Space variant adaptive	16.01	4.84

(a)   (b)

(c)   (d)

FIGURE 3.13
Results of experiments on an image degraded by spatially variant distortion.

3.5   Semiblind Restoration Using Adaptive Constraints

The previous sections have assumed that the nature of the degrading PSF is
known. In practical circumstances, however, this may not be the case. We may
have little or no knowledge of the degrading PSF in the case of a spatially
invariant distortion or may not know the exact nature of the space variance
in a case of spatially variant distortion. In addition, our estimate of the degrading PSF may be corrupted by noise. Restoring an image with incomplete
knowledge of the nature of the degradation is called *semiblind deconvolution*.
Figure 3.14 shows the effects of using an incorrect PSF estimate during the
restoration procedure.

Figure 3.14a is the original flower image. Figure 3.14b is the flower image degraded by a 5 by 5 Gaussian blur of standard deviation 2.0 with additive noise
of variance 4.22. Figure 3.14c shows the image restored using the parameters
of a 7 by 7 Gaussian blur of standard deviation 3.0, and Figure 3.14d shows
the image restored using the parameters of a 3 by 3 Gaussian blur of standard deviation 1.0. In Figure 3.14c, the degrading PSF estimate is too strong

FIGURE 3.14
Degraded image restored with various PSF estimates.

and so the image can be said to have been *over restored*. Although the edges and high-variance regions in Figure 3.14c appear sharp, ringing effects are apparent in the smooth (low-variance) regions of the image. In Figure 3.14d, on the other hand, the degrading PSF estimate is too weak and so the image can be said to have been *under restored*. In the smooth (low-variance) regions of Figure 3.14d, we see no artifacts; however, the edges and high-variance regions of Figure 3.14d appear blurry.

This would indicate that a too strong PSF estimate produces problems in low-variance regions whereas a too weak PSF estimate produces problems in high-variance regions. Figure 3.14d appears similar to the result obtained when a large value of λ is used during the restoration process. In fact, since λ controls the degree of smoothness in the solution, the effect of increasing λ is similar to performing the restoration with a weaker PSF estimate. In the adaptive λ image-restoration methods described in the previous sections, a small value of λ is used in high-variance regions and a larger value of λ is used in low-variance regions. This implies that the adaptive constraint restoration method described above may have an application in semiblind image restoration [72]. An example should illustrate this.

Assume that an image has suffered a spatially variant distortion and only an estimate of the average PSF is available. By using the adaptive constraint restoration methods developed in Section 3.3 we can consider that each different value of λ used in the methods approximates a different strength PSF estimate and hence we effectively have a range of available PSF estimates to choose from. In the high-variance regions of the image, the λ value is small, and this is similar to selecting the strongest PSF estimate. This is the preferable choice since over restoring the high-variance regions of the image is less visually disturbing than under restoring them. To reduce the effects of ripples in the low-variance regions of the image a weak PSF estimate is preferable. This is automatically achieved using the adaptive constraint methods described in Section 3.3 since these methods select the largest value of λ for low-variance regions. As mentioned above, larger values of λ produce results similar to using weaker PSF estimates. From this we can see that the adaptive λ restoration method we developed in Section 3.3 can be used to compensate for inaccuracies in the estimate of degrading PSF. Let's expand the example in Section 3.4 to examine this effect.

An image was created using a cyclic variation of 7 by 7 Gaussian PSFs. Using the analysis in Section 3.2, V was set to be 4. Table 3.1 details the degrading PSFs used to blur the image as per Equation (3.30).

Noise of variance 4.28 was added to the blurred image. The degraded image is shown in Figure 3.15a. This image was restored using four techniques. The first technique was a nonadaptive spatially invariant approach. As in the previous section, a regularization parameter value of 0.0007 was used and the spatially variant distortion was approximated as a spatially invariant distortion by using a 7 by 7 Gaussian PSF of standard deviation 2.55. This image is shown in Figure 3.15b. The second technique was an adaptive spatially invariant approach. The same approximation of the PSF, as in the previous experiment, was used; however, the regularization parameter was varied for each pixel using the same local variance levels and λ values as in the low-noise example in Section 3.3.3. This image is shown in Figure 3.15c. The image was then restored using a nonadaptive spatially variant restoration method, with a regularization parameter value of 0.0007. This image is shown in Figure 3.15d. In the final approach, the degraded image was restored using a spatially variant, adaptive constraint method, with the variance based λ value selection algorithm. The regularization value was selected in the same manner as the example in Section 3.3.3. This image is shown in Figure 3.15e. The statistics are summarized in Table 3.5. It is interesting to note that by using an adaptive approach we can compensate for a lack of knowledge regarding the degrading PSF. In the cases where the degraded images are restored by the space invariant approach, we can see that using the adaptive technique produces a much clearer image with a lower LSMSE although the cyclic blurring effect is marginally visible. In all cases, using the adaptive approach produces a clearer image with a lower LSMSE.

(a)

(b)

(c)

(d)

(e)

FIGURE 3.15
Results of experiments on an image degraded by spatially variant distortion.

TABLE 3.5

Results of Spatially Variant Experiment

Image	SNR (dB)	LSMSE
Degraded	13.94	36.01
Space invariant nonadaptive	15.03	6.38
Space invariant adaptive	14.92	5.45
Space variant nonadaptive	15.87	6.04
Space variant adaptive	16.01	4.84

3.6   Implementation Considerations

A major practical consideration is the number of choices of regularization value in the local statistics-based adaptive constraint restoration algorithm in Section 3.3. Taking R as the number of choices of regularization parameter to be used in the restoration, one would expect that a large value of R will give a better restoration quality; however, large values of R would slow the constraint precomputing stage of the algorithm and waste memory. It is desirable to have as low a value of R as possible. In practice setting $R = 3$ is usually sufficient, giving the algorithm the choice of doing nothing or selecting one of three constraint values for regions of high, medium, or low texture levels.

For the numerical results in this chapter, we associated each set of weights with a range of variance values computed in a certain neighborhood of size A by A of the current pixel. A variance threshold, κ, was set, below which the pixel being examined would not be updated. This is because for extremely low variance regions of an image blurring may not be noticeable; in this case restoration can only serve to enhance noise and waste time. This technique can yield improved results; however, the level of variance threshold and whether or not a variance threshold is used.at all depend on the image being restored and the degrading function. The variance constraint selection method has the advantage that it does not require the current pixel to be acted upon by each possible set of weights to compute the regularization parameter required. Another advantage to the variance method is the ability to fine-tune the variance thresholds, κ, or A, the area size, to suit a particular type of image being examined.

When we consider using variance to determine the regularization parameter, we expect that precomputation of the parameter, based on the degraded image statistics would produce similar results to computing the parameter in each iteration. High-variance regions in the degraded image should remain high variance in the restored image, and low-variance regions should likewise remain low variance; hence by this argument the chosen values of regularization parameter in the first iteration should remain approximately the same throughout the restoration procedure. This also has the added advantage that the algorithm is guaranteed to converge according to the analysis in Chapter 2.

In practice, precalculating the optimal set of weights for each pixel has further advantages:

- During restoration, the estimate may converge on the solution smoothly or with varying degrees of oscillation around the final value. During an oscillatory restoration, the image statistics may change in unpredictable ways, causing the regularization parameter chosen from iteration to iteration to vary also. This may not result in an optimal choice of regularization value during any one iteration or an optimal average value of the regularization parameter.

- Precomputation of the regularization parameter results in a faster restoration than that obtained by computation of the regularization parameter during each iteration. In fact, by precomputing the regularization values for each pixel, the adaptive constraint algorithm takes only slightly more time than a nonadaptive algorithm for large images, as will be shown in the next section.

3.7 More Numerical Examples

Let us look at two more examples of how the algorithms we have presented in this chapter perform on real imagery. The first example examines the processing efficiency. Following this, a practical example of the use of this method will be given. As with previous examples, the images will be compared by measuring their SNR and LSMSE [LSMSE is as defined in Equation (3.14)].

3.7.1 Efficiency

In this example, the time taken to restore an image was compared among the four different cases from the space variant example in Section 3.5. Each algorithm was run three times on a SUN Ultra 1 workstation; the average results for each algorithm are shown in Table 3.6.

It is worth noting that the execution times of the adaptive regularization parameter algorithms are similar to the execution times of the nonadaptive algorithms. The adaptive algorithms require the calculation of five times the number of weighting masks as the nonadaptive algorithms, and also require prerestoration calculations of local variance to precompute the regularization parameter for each pixel. The reason for the similar execution times is that the time lost through setting up the adaptive parameters of the network is offset by the time gained through the nonadjustment of pixels in low-variance regions of the image. The most important fact that we can observe from Table 3.6 is that the time required for a fully adaptive, spatially variant restoration is only double the time required for the much simpler nonadaptive spatially invariant restoration. This is much faster than any previously reported methods. In fact as the size of the image increases we would expect the time required for adaptive spatially variant restoration to approach the time required for

TABLE 3.6

Algorithm Run Times

Algorithm	Time (CPU Seconds)
Nonadaptive spatially invariant	259
Nonadaptive spatially variant	370
Adaptive spatially invariant	206
Adaptive spatially variant	667

nonadaptive, spatially invariant restoration. This is because the extra time required for the adaptive spatially variant restoration is primarily taken up by the initial extra weighting mask creation. The quality and speed of the adaptive spatially invariant approximation method offers a promising alternative to the traditional semiblind deconvolution methods, especially when the PSF is space variant.

3.7.2   An Application Example

The above neural network algorithm was applied to the problem of restoring images with an unknown level of blur. Images were supplied to us showing chromosomes in a solution imaged by a high-powered optical microscope. Limitations of the optical system had blurred the images of the chromosomes. To extract further information about the chromosomes in the images, the above neural network algorithm was used to enhance the images. Figure 3.16a shows one of the original degraded images. Figure 3.16b shows Figure 3.16a restored using the nonadaptive algorithm. Note that the restored image is sharper than the original image, but some noise has also been amplified as well as some ringing effects. Figure 3.16c shows Figure 3.16a restored using the adaptive algorithm. The level of sharpness is comparable to the results of the nonadaptive approach; however, the level of background noise and ringing effects have been greatly reduced. In this case the adaptive restoration algorithm has been successful at enhancing the detail present in the image.

3.8   Adaptive Constraint Extension of the Lagrange Model

Section 3.3 developed the adaptive constraint problem by considering the standard constrained least square cost function of Equation (3.6) as a simple penalty function. However, Equation (3.6) may also be considered in terms of the theory of Lagrange [94]. We will see in this section the consequences of this analysis. First we will expand the restoration problem in terms of the theory of Karush–Kuhn–Tucker (a generalization of the theory of Lagrange first presented in Chapter 2) and then show what solutions result. Finally, we will investigate what conditions on the problem need to be imposed for the analysis to be valid.

3.8.1   Problem Formulation

Consider the image degradation model

$$\mathbf{g} = \mathbf{Hf} + \mathbf{n} \tag{3.33}$$

where $\mathbf{g}$ and $\mathbf{f}$ are $L = MN$-dimensional lexicographically organized degraded and original images, respectively, $\mathbf{n}$ is an L-dimensional additive noise

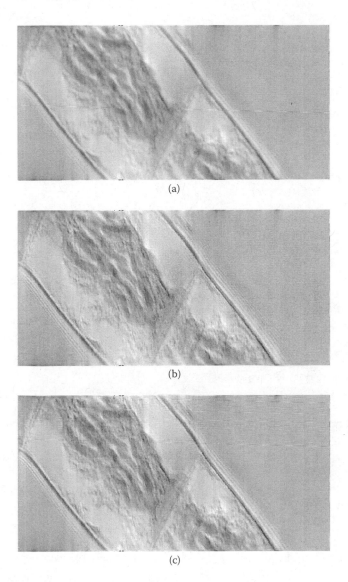

FIGURE 3.16
Degraded and restored chromosome images.

vector and **H** is a convolutional operator. Assume that the image **f** can be divided into K regions with common statistical properties, where each pixel belongs to one region and only one region. The regions are such that for any pixel in region i, any other pixel in region i may be reached by a path consisting only of pixels belonging to region i.

Let $\{i\}$ denote the set of lexicographic indices of pixels in region i. Let's define an operator that extracts only those pixels belonging to region i.

Define $\mathbf{K}_i \in \Re^{L \times L}$, $1 \le i \le K$ where

$$[\mathbf{K}_i]_{jl} = \begin{cases} 1, & \text{if } j = l \text{ and } j, l \in \{i\} \\ 0, & \text{if } j \notin \{i\} \text{ or } l \notin \{i\} \\ 0, & \text{if } j \ne l \end{cases} \qquad (3.34)$$

Hence $\mathbf{K}_i\mathbf{f}$ is a vector containing only pixels belonging to region i, that is,

$$[\mathbf{K}_i\mathbf{f}]_{jl} = \begin{cases} f_j, & \text{if } j \in \{i\} \\ 0, & \text{otherwise} \end{cases}$$

Note that $\mathbf{K}_i^T = \mathbf{K}_i\mathbf{K}_i = \mathbf{K}_i$, due to the simple structure of $\mathbf{K}_i$.

The problem we are considering is that given $\mathbf{g}$ and $\mathbf{H}$, we wish to recover $\mathbf{f}$. Assume that using the degraded image, $\mathbf{g}$, we can reasonably partition the expected result $\mathbf{f}$ into the K regions described above. The pixels in each region have similar statistical properties.

Restoring an image acted upon by the degradation model above is an ill-posed problem for the reasons discussed in Chapter 2. Hence, extra constraints will be required. One reasonable constraint is to require that the value of every pixel in the image be positive.

Consider a matrix $\mathbf{D} \in \Re^{L \times L}$ designed to act as a high-pass operator. $\mathbf{Df}$ is therefore the high-pass component of the estimated image. To reduce the effects of noise, we can constrain $\|\mathbf{Df}\|^2$ to be equal to a preset constant. If the constant is chosen too low, the image will be blurred, and if the constant is chosen too high, the image will contain unacceptable levels of noise. However, the best value of the constant to use for each region of the image may not be the same as every other region. A better approach is to constrain $\|\mathbf{Df}\|^2$ to be equal to different constants in different parts of the image.

This problem can then be stated as

$$\mathbf{f}^* = \arg \min_{\mathbf{f}} \|\mathbf{g} - \mathbf{Hf}\|^2 \text{ subject to } \mathbf{h}(\mathbf{f}^*) = 0 \text{ and } \mathbf{p}(\mathbf{f}^*) \le 0$$

where $\mathbf{h}(\mathbf{f}) \in \Re^K$, $h_i(\mathbf{f}) = \|\mathbf{K}_i\mathbf{Df}\|^2 - \sigma_i^2$, $1 \le i \le K$ and $\mathbf{p}(\mathbf{f}) \in \Re^L$, $\mathbf{p}(\mathbf{f}) = -\mathbf{f}$, $p_j(\mathbf{f}) = -f_j$, $1 \le j \le L$.

Note that $\mathbf{p}$ is the set of positivity constraints, while $\mathbf{h}$ comprises the constraints on the norm of the high-frequency components of each region.

Let us look now at how we can solve this equation. The following definition and theory (first presented in Chapter 2 and repeated here for reference) will be useful.

Definition 3.2
Let $\mathbf{f}^*$ satisfy $\mathbf{h}(\mathbf{f}^*) = 0$, $\mathbf{p}(\mathbf{f}^*) \le 0$ and let $J\{\mathbf{f}^*\}$ be the index set of active inequality constraints, that is, $J\{\mathbf{f}^*\} = \{j; p_j(\mathbf{f}^*) = 0\}$.

Then $\mathbf{f}^*$ is a regular point if $\nabla h_i(\mathbf{f}^*), \nabla p_j(\mathbf{f}^*)$ are linearly independent $\forall i, j$; $1 \leq i \leq K, j \in J\{\mathbf{f}^*\}$ [94].

Theorem of Karush–Kuhn–Tucker

Let $E, \mathbf{h}, \mathbf{p} \in C^1$. Let $\mathbf{f}^*$ be a regular point and a local minimizer for the problem of minimizing E subject to $\mathbf{h}(\mathbf{f}^*) = 0$ and $\mathbf{p}(\mathbf{f}^*) \leq 0$.

Then there exists $\lambda \in \Re^K$, $\mu \in \Re^L$ such that

1. $\mu \geq 0$
2. $\nabla E(\mathbf{f}^*) + \sum_{i=1}^{K} \lambda_i \nabla h_i(\mathbf{f}^*) + \sum_{j=1}^{L} \mu_j \nabla p_j(\mathbf{f}^*) = 0$
3. $\mu^T \mathbf{p}(\mathbf{f}^*) = 0$

3.8.2 Problem Solution

The theorem of Karush–Kuhn–Tucker can be used to solve this problem. Note the following equations:

$$h_i(\mathbf{f}^*) = \|\mathbf{K}_i \mathbf{D} \mathbf{f}^*\|^2 - \sigma_i^2 = \mathbf{f}^{*T} \mathbf{D}^T \mathbf{K}_i^T \mathbf{K}_i \mathbf{D} \mathbf{f}^* - \sigma_i^2$$

$$= \mathbf{f}^{*T} \mathbf{D}^T \mathbf{K}_i \mathbf{D} \mathbf{f}^* - \sigma_i^2 \tag{3.35}$$

$$\nabla h_i(\mathbf{f}^*) = 2 \mathbf{D}^T \mathbf{K}_i \mathbf{D} \mathbf{f}^* \tag{3.36}$$

$$p_j(\mathbf{f}^*) = -f_j^* \tag{3.37}$$

$$\nabla p_j(\mathbf{f}^*) = -\mathbf{e}_j \tag{3.38}$$

where

$$\mathbf{e}_j \in \Re^L; [\mathbf{e}_j]_i = \begin{cases} 1, & i = j \\ 0, & i \neq j \end{cases} \tag{3.39}$$

Note that $\nabla h_i(\mathbf{f}^*), \nabla p_j(\mathbf{f}^*) \in \Re^L$ for each $1 \leq i \leq K, j \in J\{\mathbf{f}^*\}$.

If we assume $\mathbf{f}^*$ is a regular point, the solution to the problem can be obtained as follows:

$$E(\mathbf{f}^*) = \|\mathbf{g} - \mathbf{H}\mathbf{f}^*\|^2 = (\mathbf{g} - \mathbf{H}\mathbf{f}^*)^T (\mathbf{g} - \mathbf{H}\mathbf{f}^*)$$

$$= \mathbf{g}^T \mathbf{g} - 2\mathbf{g}^T \mathbf{H}\mathbf{f}^* + \mathbf{f}^{*T} \mathbf{H}^T \mathbf{H}\mathbf{f}^* \tag{3.40}$$

Note that we have used the fact that $\mathbf{f}^{*T} \mathbf{H}\mathbf{g} = (\mathbf{g}^T \mathbf{H}\mathbf{f}^*)^T = \mathbf{g}^T \mathbf{H}\mathbf{f}^*$ since this quantity is a scalar.

$$\nabla E(\mathbf{f}^*) = 2\mathbf{H}^T \mathbf{H}\mathbf{f}^* - 2\mathbf{H}^T \mathbf{g} \tag{3.41}$$

By KKT condition 2 we have

$$\nabla E(\mathbf{f}^*) + \sum_{i=1}^{K} \lambda_i \nabla h_i(\mathbf{f}^*) + \sum_{j=1}^{L} \mu_j \nabla g_j(\mathbf{f}^*) = 0$$

$$2\mathbf{H}^T\mathbf{H}\mathbf{f}^* - 2\mathbf{H}^T\mathbf{g} + 2\sum_{i=1}^{K} \lambda_i \mathbf{D}^T \mathbf{K}_i \mathbf{D}\mathbf{f}^* - \sum_{j=1}^{L} \mu_j \mathbf{e}_j = 0$$

$$2\left(\mathbf{H}^T\mathbf{H} + \sum_{i=1}^{K} \lambda_i \mathbf{D}^T \mathbf{K}_i \mathbf{D}\right) \mathbf{f}^* = 2\mathbf{H}^T\mathbf{g} + \sum_{j=1}^{L} \mu_j \mathbf{e}_j$$

Therefore,

$$\mathbf{f}^* = \left[\mathbf{H}^T\mathbf{H} + \sum_{i=1}^{K} \lambda_i \mathbf{D}^T \mathbf{K}_i \mathbf{D}\right]^{-1} \left(\mathbf{H}^T\mathbf{g} + \frac{1}{2}\sum_{j=1}^{L} \mu_j \mathbf{e}_j\right) \qquad (3.42)$$

or alternately, $\mathbf{f}^* = [\mathbf{H}^T\mathbf{H} + \sum_{i=1}^{K} \lambda_i \mathbf{D}^T \mathbf{K}_i \mathbf{D}]^{-1}(\mathbf{H}^T\mathbf{g} + \frac{1}{2}\sum_{j\in J\{\mathbf{f}^*\}} \mu_j \mathbf{e}_j)$.

KKT condition 3 states that $\mu^T \mathbf{p}(\mathbf{f}^*) = -\sum_{j=1}^{L} \mu_j f_j^* = 0$. If $f_j^* \neq 0$ then $j \notin J\{\mathbf{f}^*\}$ and $\mu_j = 0$. If $f_j^* = 0$, then μ_j can be any value and satisfy KKT condition 3. In fact, by examining Equation (3.42) we discover that the nonzero values of μ are the solutions to

$$\frac{1}{2}\sum_{p\in J(\mathbf{f}^*)} \left[\left(\mathbf{H}^T\mathbf{H} + \sum_{i=1}^{K} \lambda_i \mathbf{D}^T \mathbf{K}_i \mathbf{D}\right)^{-1}\right]_{jp} \mu_p \qquad (3.43)$$

$$= -\left[\left(\mathbf{H}^T\mathbf{H} + \sum_{i=1}^{K} \lambda_i \mathbf{D}^T \mathbf{K}_i \mathbf{D}\right)^{-1} \mathbf{H}^T\mathbf{g}\right]_j$$

$$\forall j; \, j \in J(\mathbf{f}^*)$$

The complexity of this equation should not be surprising. A single pixel constrained to be positive by an active constraint will change the solution point of the problem and affects all the other pixels' values. In this way, it effects the degree to which other pixels need to be constrained in order to be positive. Hence the values of all the elements in μ are closely interrelated.

The equality constraint $\mathbf{h}(\mathbf{f}^*) = 0$ gives us the extra K equations required to solve for the values of λ. Substituting Equation (3.42) into Equation (3.35) we obtain λ as the solution to

$$\mathbf{X}^T\mathbf{D}^T\mathbf{K}_i \mathbf{D}\mathbf{X} = \sigma_i^2$$

where

$$\mathbf{X} = \left[\mathbf{H}^T\mathbf{H} + \sum_{i=1}^{K} \lambda_i \mathbf{D}^T \mathbf{K}_i \mathbf{D}\right]^{-1} \left(\mathbf{H}^T\mathbf{g} + \frac{1}{2}\sum_{j\in J\{\mathbf{f}^*\}} \mu_j \mathbf{e}_j\right) \quad \text{and} \quad 1 \leq i \leq K$$

$$(3.44)$$

In theory, Equations (3.43) and (3.44) form a set of $K + |J(\mathbf{f}^*)|$ simultaneous equations that may be solved to determine the elements of λ and μ (where $|J(\mathbf{f}^*)|$ is the number of active inequality constraints). Once the values of λ and μ are found, the solution to the problem can be found using Equation (3.42). It can be seen that the values of the elements in λ and μ are closely related. This is not surprising since constraints on the high-frequency components in each region can reasonably be expected to influence the degree to which pixels need to be constrained to positive. In turn, pixels constrained to be positive will influence the high-frequency components of the regions, and hence the elements in λ. In practice, the problem is not so complex. The value of λ_i for each region is set mainly by the constant σ_i assigned to the region. If the σ_i for each region is not known beforehand (as is often the case), then values of λ_i may be set in a trial-and-error fashion without using Equation (3.44). Once the optimal values of the λ_i are found, an iterative algorithm that operates on one pixel at a time and constrains each pixel to be positive during each iteration will implicitly calculate the μ_i.

3.8.3 Conditions for KKT Theory to Hold

The last subsection supplied the solution to this problem using the Karush–Kuhn–Tucker theory. This assumes that the preconditions of the theory are satisfied. To be specific, for the discussion in the last section to hold, $\mathbf{f}^*$ must be a regular point. In this section, we will examine the conditions that need to be imposed on the problem for this to be the case. But first, we require some definitions:

Definition 3.3
Define a distance measure $dis2(i, j)$, $1 \leq i, j \leq NM$, where $L = NM$, the two-dimensional size of the image. Take two pixels with coordinates $x(a, b)$ and $y(á, b́)$. The lexicographic coordinates of these pixels are given by

$$i = (a - 1)M + b \quad and \quad j = (á - 1)M + b́$$

Then we define $dis2(i, j) = \sqrt{(a - á)^2 + (b - b́)^2}$.

Definition 3.3 merely gives the two-dimensional distance between two pixels specified in lexicographic coordinates.

Next consider the high-pass operator, $\mathbf{D}$. Assuming this operator has an impulse response with a finite region of support, which is much less than $\min(N, M)$, then there exists a constant, Z, such that

$$[\mathbf{D}]_{ij} = 0, \quad if \quad dis2(i, j) > Z, \forall i, j; 1 \leq i, j \leq L$$

This inspires another definition.

Definition 3.4
Consider a pixel, f_j, belonging to region i. That is $j \in \{i\}$. We say that f_j is an interior point of region i in terms of $\mathbf{D}$ if and only if $\forall f_p; \; p \notin \{i\}$,

$dis2(j, p) > Z$, *where* Z *is the smallest number such that* $\forall i, j; 1 \leq i, j \leq L, dis2(i, j) > Z \Rightarrow [\mathbf{D}]_{ij} = 0$.

Definition 3.4 states that an interior point of a region in terms of a particular high-pass operator has no neighbors within the region of support of the operator that belong to a different region.

Using these two definitions, we have the following theory:

Theorem 3.3

$\mathbf{f}^*$ *is a regular point if each region* $1 \leq i \leq K$ *has at least one interior point,* r, *in terms of* $\mathbf{D}$ *such that*

$[\nabla h_i(\mathbf{f}^*)]_r \neq 0$ *and* $f_r^* \neq 0$

Note that $f_r^* \neq 0$ is the same as $r \notin J(\mathbf{f}^*)$.

Proof   Theorem 3.3 will be proven if we show that a solution point $\mathbf{f}^*$ satisfying the conditions of the theorem results in $\nabla h_i(\mathbf{f}^*), \nabla p_j(\mathbf{f}^*), 1 \leq i \leq K, j \in J(\mathbf{f}^*)$ being a set of linearly independent vectors.

The theorem can be proven in three parts:

1. Prove that the set $\nabla p_j(\mathbf{f}^*)$, $j \in J(\mathbf{f}^*)$ is a set of linearly independent vectors.

2. Prove that the set $\nabla h_i(\mathbf{f}^*), 1 \leq i \leq K$ is a set of linearly independent vectors.

3. Prove that each $\nabla p_j(\mathbf{f}^*)$ is linearly independent of the set $\nabla h_i(\mathbf{f}^*), 1 \leq i \leq K$.

Part (1)

Proof of part (1) is trivial since $\nabla p_j(\mathbf{f}^*) = -\mathbf{e}_j$ where $-\mathbf{e}_j$ is given by Equation (3.39).

Part (2)

Assume that r is an interior point of region i in terms of $\mathbf{D}$ and $[\nabla h_i(\mathbf{f}^*)]_r \neq 0$. Consider the rth component of $[\nabla h_j(\mathbf{f}^*)]_r$ where $j \neq i$.

$$[\nabla h_j(\mathbf{f}^*)]_r = 2\mathbf{D}^T \mathbf{K}_j \mathbf{D}\mathbf{f}^* = 2\mathbf{D}^T \mathbf{A}$$

where $\mathbf{A} = \mathbf{K}_j \mathbf{D}\mathbf{f}^*$.

$$[\nabla h_j(\mathbf{f}^*)]_r = \sum_{l=1}^{L} [\mathbf{D}]_{rl}\mathbf{A}_l \text{ since } \mathbf{D}^T = \mathbf{D}.$$

If $l \notin \{j\}$ then $[\mathbf{A}]_l = 0$ by the definition of $\mathbf{K}_j$ [Equation (3.34)]. Therefore $[\nabla h_j(\mathbf{f}^*)]_r = \sum_{l \in \{j\}} [\mathbf{D}]_{rl}\mathbf{A}_l$.

However, $dis2(r, l) > Z$ when $l \in \{j\}$ by the definition of an interior point in terms of $\mathbf{D}$; therefore $[\nabla h_j(\mathbf{f}^*)]_r = 0$. Since $[\nabla h_i(\mathbf{f}^*)]_r \neq 0$ and $[\nabla h_j(\mathbf{f}^*)]_r = 0$, $\forall j; j \neq i$, we conclude that $\nabla h_i(\mathbf{f}^*)$ is linearly independent of the other vectors in the set, since it will always have at least one nonzero component for which the other vectors in the set will have zero. If each region in the problem satisfies the conditions of the theorem, the set $\nabla h_i(\mathbf{f}^*)$, $1 \leq i \leq K$ will be a set of linearly independent vectors.

Part (3)

Consider a point $r \in \{i\}$ such that $[\nabla h_i(\mathbf{f}^*)]_r \neq 0$ and $r \notin J(\mathbf{f}^*)$.

Since r is not a part of the active inequality constraint set, then

$$\nabla h_i(\mathbf{f}^*) \neq \sum_{j \in J(\mathbf{f}^*)} \alpha_j \mathbf{e}_j$$

Since $\nabla p_j(\mathbf{f}^*) = -\mathbf{e}_j$, therefore $\nabla h_i(\mathbf{f}^*) \neq \sum_{j \in J(\mathbf{f}^*)} \alpha_j \nabla p_j(\mathbf{f}^*)$.

If every region i has a point that satisfies the conditions of Theorem 3.3, then the set $\nabla p_j(\mathbf{f}^*)$, $j \in J(\mathbf{f}^*)$ is linearly independent to the set $\nabla h_i(\mathbf{f}^*)$, $1 \leq i \leq K$.

Hence the proof is completed. We have shown that the set $\{\nabla h_i(\mathbf{f}^*), \nabla p_j(\mathbf{f}^*)\}$, $1 \leq i \leq K$, $j \in J(\mathbf{f}^*)$ is a set of linearly independent vectors. ∎

This section has shown that if an image can be divided into a number of regions each satisfying Theorem 3.3, the theorem of Karush–Kuhn–Tucker states that solutions to this problem will exist.

3.8.4 Discussion

Various schemes such as measuring local statistical properties may be used to select the regions and assign values of λ to these regions. To use the analysis in this section, we then need to make sure that the selected regions satisfy the conditions of Theorem 3.3. Note that the conditions imposed on the regions by Theorem 3.3 are *sufficient* but not *necessary*. This means that regions satisfying these conditions will be able to be analyzed by the approach in this section, but regions that do not satisfy these conditions may also be just as valid. Note that the algorithms derived by the penalty function approach in Section 3.3 are not compatible with the analysis in this section. The analysis in this section expands the single constraint problem into the problem of assigning a different constraint value to each of a number of predetermined regions and Theorem 3.3 is not valid if the regions are too small or too narrow. However, the analysis in Section 3.3 began by assigning each pixel a different value of λ and from this beginning moved into the concept of a region. The concept of regions in Section 3.3 was shown to arise naturally, but is not as integral a concept as it was in this section.

We will not delve any more into the Karush–Kuhn–Tucker analysis in this chapter, since although it is more mathematically justified than the penalty-function concept, we have shown in this chapter that algorithms based on the penalty-function concept produce acceptable results and have no conditions upon the size and shape of the regions created.

3.9 Summary

In this chapter, the use of spatially adaptive weights in constrained restoration methods was considered.

The first use of spatially adaptive weights we considered was adaptation of the neural weights to restore images degraded by spatially variant distortions. We showed how the neural network approach could be used to solve this problem without a great deal of extra computation.

The second use of spatially adaptive weights we examined was in order to implement a spatially adaptive constraint parameter. Since the human visual system favors the presence of edges and boundaries, rather than more subtle differences in intensity in homogeneous areas [104], noise artifacts may be less disturbing in high-contrast regions than in low-contrast regions. It is then advantageous to use a stronger constraint in smooth areas of an image than in high-contrast regions. While traditional restoration methods find it difficult to implement an adaptive restoration spatially across an image, neural network-based image-restoration methods are particularly amenable to spatial variance of the restoration parameters.

We expanded upon the basic restoration function by looking at it as having a restoration term followed by a penalty term. From this analysis, two methods of selecting the best constraint parameter for each pixel were investigated. The best method was based on using local image statistics to select the optimal value of regularization parameter. This method imitates the human visual system and produces superior results when compared to nonadaptive methods. In addition it was found that no disadvantage occurred when the values of regularization parameter for each pixel were chosen by the restoration algorithm before starting the restoration, rather than during each iteration of the restoration procedure. In fact, precomputing of the regularization parameter further increased the restoration speed.

In the next section of this chapter, the work was expanded upon to adaptively restore images degraded by a spatially variant PSF. It was shown how adaptive regularization techniques can compensate for insufficient knowledge of the degradation in the case of spatially variant distortions. Moreover, an adaptive spatially variant restoration is shown to be able to be completed in the same order of magnitude of time as a much simpler nonadaptive spatially invariant restoration.

In the course of this chapter, we have examined some implementation considerations and tested the algorithms developed on real imagery.

We finished our analysis of the adaptive constraint concept by looking at how the basic restoration function may be expanded upon using the theory of Lagrange. We looked at what conditions the theorem of Lagrange applied to the adaptive constraint problem and how these conditions related back to the penalty function-based approaches.

4

Perceptually Motivated Image Restoration

4.1 Introduction

In Chapter 3, we mentioned the problems caused by considering the image to be a stationary process. Any restoration process based on this concept can only ever produce suboptimal results. However, there is another consideration. When a restoration algorithm is designed with the aim of creating an image that will be more pleasing to the human eye, we must incorporate some kind of model of the human visual system. The basis of most restoration algorithms is some form of image error measure which is being minimized.

The most common method to compare the similarity of two images is to compute their mean square error (MSE). However, the MSE relates to the power of the error signal and has little relationship to human visual perception. An important drawback to the MSE and any cost function that attempts to use the MSE to restore a degraded image is that the MSE treats the image as a stationary process. All pixels are given equal priority regardless of their relevance to human perception. This suggests that information is ignored. When restoring images for the purpose of better clarity as perceived by humans the problem becomes acute. Considerations regarding human perception have been examined in the past [104,107–116]. When humans observe the differences between two images, they do not give much consideration to the differences in individual pixel-level values. Instead humans are concerned with matching edges, regions, and textures between the two images. This is contrary to the concepts involved in the MSE. From this it can be seen that any cost function which treats an image as a stationary process can only produce a suboptimal result.

In Chapter 1, we looked at a new error measure based on comparing local standard deviations that examines the image in a regional sense rather than a pixel-by-pixel sense. The formula for this error measure was given in Equation (3.14). This error measure uses some simple concepts behind human perception without being too complex. In this chapter, we further develop these concepts by examining two restoration algorithms that incorporate versions of this image error measure. The cost functions that these algorithms are based on are nonlinear and cannot be efficiently implemented by conventional methods. In this chapter, we introduce extended neural network algorithms to iteratively perform the restoration. We show that the new cost functions and processing algorithms perform very well when applied to both color and grayscale images. One important property of the methods we will examine in this chapter, compared with the neural network implementation of the constrained least square filter, is that they are very fault-tolerant in the sense that when some of the neural connections are damaged, these algorithms can still produce very satisfactory results. Comparison with some of the conventional methods will be provided to justify the new methods.

This chapter is organized as follows. Section 4.2 describes the motivation for incorporating the error measure described in Chapter 1 into a cost function. Section 4.3 presents the restoration cost function incorporating a version of the proposed image error measure from Chapter 1. Section 4.4 builds on the previous section to present an algorithm based on a more robust variant of the novel image error measure. Section 4.5 describes implementation considerations. Section 4.6 presents some numerical data to compare the algorithms presented in this chapter with those of previous chapters. Section 4.7 examines the new error measures in terms of Lagrangian theory, while Section 4.8 summarizes this chapter.

4.2   Motivation

In the introduction to this chapter, we considered the problems inherent in using the mean square error (MSE) and signal to noise ratio (SNR) to compare two images. It was seen that the MSE and SNR have little relationship to the way that humans perceive the differences between two images. Although incorporating concepts involved in human perception may seem a difficult task, a new image error measure was presented in Chapter 1 that, despite its simplicity, incorporates some concepts involved in human appraisal of images.

In Chapter 3, the basic neural network restoration algorithm described in Chapter 2 was expanded to restore images adaptively using simple human visual concepts. These adaptive algorithms obtained superior results when compared to the nonadaptive algorithm, and were shown to produce a more robust restoration when errors occurred due to insufficient knowledge of the degrading function. Despite the improved performance of the adaptive algorithms, it is still not simple to choose the correct values of the constraint parameter, λ. In the case of the adaptive algorithms described in Chapter 3, the problem is compounded by the fact that many values of λ must be selected, rather than just one value as in the nonadaptive case. In addition, the selected λ values must be related to local variance levels.

We desire to create an algorithm that can adaptively restore an image, using simple concepts involved in human perception, with only a few free parameters to be set. Such an algorithm would be more robust and easier to use than the variance selection algorithm in Chapter 3. In the previous adaptive algorithm, minimizing the MSE was still at the base of the restoration strategy. However, Chapter 1 provides us with a simple alternative to the MSE. It seems logical to create a new cost function that minimizes an LSMSE-related term. In this way, the adaptive nature of the algorithm would be incorporated in the local standard deviation mean equare error (LSMSE) term rather than imposed by the external selection of λ values.

4.3   LVMSE-Based Cost Function

The discussion in the previous section prompts us to structure a new cost function that can properly incorporate LSMSE into restoration. Since the formula for the neural network cost function has a term that attempts to minimize the MSE, let us look at restoring images using a cost function with a term that endeavors to minimize an LSMSE-related error measure. Let's start by adding an additional term to Equation (2.3). The new term evaluates the local variance mean square error (LVMSE) [72,117]:

$$\text{LVMSE} = \frac{1}{NM} \sum_{x=0}^{N-1} \sum_{y=0}^{M-1} \left(\sigma_A^2(\mathbf{f}(x, y)) - \sigma_A^2(\mathbf{g}(x, y)) \right)^2 \qquad (4.1)$$

where $\sigma_A^2(\mathbf{f}(x, y))$ is the variance of the local region surrounding pixel (x, y) in the first image and $\sigma_A^2(\mathbf{g}(x, y))$ is the variance of the local region surrounding pixel (x, y) in the image with which we wish to compare the first image.

Hence the new cost function we suggest is

$$E = \frac{1}{2}\|\mathbf{g} - \mathbf{H}\hat{\mathbf{f}}\|^2 + \frac{\lambda}{2}\|\mathbf{D}\hat{\mathbf{f}}\|^2 + \frac{\theta}{NM} \sum_{x=0}^{N-1} \sum_{y=0}^{M-1} \left(\sigma_A^2(\hat{\mathbf{f}}(x, y)) - \sigma_A^2(\mathbf{g}(x, y))^\star \right)^2 \quad (4.2)$$

In this case, $\sigma_A^2(\hat{\mathbf{f}}(x, y))$ is the variance of the local region surrounding pixel (x, y) in the image estimate and $\sigma_A^2(\mathbf{g}(x, y))^\star$ is the variance of the local region surrounding pixel (x, y) in the degraded image scaled to predict the variance in the original image. The comparison of local variances rather than local standard deviations was chosen for the cost function since it is easier and more efficient to calculate. The first two terms in Equation (4.2) ensure a globally balanced restoration, whereas the added LVMSE term enhances local features. In Equation (4.2), $\sigma_A^2(\mathbf{g}(x, y))^\star$ is determined based on the following logic.

Since the degraded image has been blurred, image variances in $\mathbf{g}$ will be lower than the corresponding variances in the original image. In this case, the variances $\sigma_A^2(\mathbf{g}(x, y))^\star$ would be scaled larger than $\sigma_A^2(\mathbf{g}(x, y))$ to reflect the decrease in variance due to the blurring function. In general, if we consider an image degraded by a process that is modeled by Equation (2.2), then we find that a useful approximation is

$$\sigma_A^2(\mathbf{g}(x, y))^* = K(x, y)(\sigma_A^2(\mathbf{g}(x, y)) - J(x, y)) \qquad (4.3)$$

where $J(x, y)$ is a function of the noise added to the degraded image at point (x, y) and $K(x, y)$ is a function of the degrading point spread function at point (x, y). Although it may appear difficult to accurately determine the optimal values of $K(x, y)$, in fact, as we will demonstrate later, the algorithm is extremely tolerant of variations in this factor and only a rough estimate is required. For example, if the image degradation is a moderate blurring function, with a region of support of around 5 or 7, then $K(x, y)$ would be set to 2 for all pixels in the image. This indicates that the local variances in the original image are on average approximately twice that of the degraded image. A high degree of accuracy is not required. In highly textured regions of the image where the preservation of image details is most important, the LVMSE term requires that the variance of the region be large, and the first two terms of Equation (4.2) ensure the sharpness and accuracy of the image features.

4.3.1 Extended Algorithm for the LVMSE-Modified Cost Function

The LVMSE-modified cost function does not fit easily into the neural network energy function as given by Equation (2.26); however, an efficient algorithm can be designed to minimize this cost function. One of the first considerations when attempting to implement the LVMSE cost function is prompted by a fundamental difference in the cost function that occurs due to the addition of the new term. In the case of a cost function based on minimizing mean square error alone, any changes in an individual pixel's value affect the entire image MSE in a simple way. The square error between any pixel in the image estimate and the corresponding pixel in the original image does not affect the square error of its neighbors. This simplifies the implementation of the cost function. In the case of the LVMSE-modified cost function, it is different. When a pixel's value is altered by the algorithm, the total change in the LVMSE is not a simple function of the current pixel's change in value alone. Changing the current pixel's value changes its own local variance, and the local variances of all of its neighbors within an A by A proximity of the current pixel. Hence to truly calculate the total change in LVMSE for the entire image, the algorithm must calculate how changing the current pixel's value effects the local variances of all its neighbors and how these changes effect the overall LVMSE. This approach is computationally prohibitive. To resolve this problem, we must go back to the fundamental justification for adding the LVMSE term in the first place. The justification for adding this term was the fact that we wished to create a cost function that matched the local statistics of pixels in the original image to that of the image estimate. In this case it is sufficient that the algorithm considers only minimizing the difference in the local variance of the estimated image pixel to the corresponding original image pixel and not minimizing the total LVMSE of the image. The great benefit arising from this approximation will become apparent when explained below.

The first step in the development of the algorithm is a change in notation. For an N by M image let $\mathbf{f}$ represent the lexicographically organized image vector of length NM as per the model given by Equation (2.2) and the algorithm for the unmodified neural network cost function [Equation (2.26)]. The translation between the two indices x and y of $f(x, y)$ and the single index i of f_i is given by

$$i = x + yN \tag{4.4}$$

Define x^i and y^i as the two-dimensional x and y values associated with pixel i by Equation (4.4).

Define the two-dimensional distance between pixels i and j as

$$\text{dis2}(i, j) = |x^i - x^j| + |y^i - y^j| \tag{4.5}$$

Note that $\text{dis2}(i, j)$ has a lot in common with $\text{dis2}(i, j)$ in Definition 3.2. However, $\text{dis2}(i, j)$ describes the *city block* distance between two pixels, while $\text{dis2}(i, j)$ describes the *Euclidean* distance between the same two pixels.

Let Φ^i represent the NM by NM matrix which has the following property:

$$\text{Let } \mathbf{f}^i = \Phi^i \mathbf{f} \tag{4.6}$$

then

$$[\mathbf{f}^i]_j = \begin{cases} 0, & \text{dis2}(i, j) > \frac{A-1}{2} \\ f_j, & \text{dis2}(i, j) \le \frac{A-1}{2} \end{cases} \tag{4.7}$$

Φ^i has the effect of setting to zero all pixels not within the A by A neighborhood centered on the pixel with coordinates x^i, y^i. As a shortened notation, we will denote $[\mathbf{f}^i]_j$ as f_j^i. Using this notation, the average pixel value in the A by A region surrounding pixel i is given by

$$M_A(i) = \frac{1}{A^2} \sum_{j=1}^{MN} f_j^i$$

Let $\beta_i = \sum_{j=1}^{NM} (f_j^i)^2$ and $\gamma_i = \sum_{j=1}^{NM} f_j^i$. Then the estimated variance of the A by A region surrounding pixel i is given by

$$V^i = \frac{\beta_i}{A^2} - \frac{\gamma_i^2}{A^4} \tag{4.8}$$

Note that, strictly speaking, V^i is an estimate of the variance of this region given the available pixel values. The true variance of this region is the expectation of the second moment. However, Equation (4.8) is a suitable approximation given the available data. In the rest of this analysis, Equation (4.8) will be called the *local variance* and the term *estimated local variance* will be used to refer to $\sigma_A^2(\mathbf{g}(x, y))^*$.

The LVMSE between the image estimate, $\hat{f}$, and the original image, f, may then be written as

$$\text{LVMSE}(\hat{f}, f) = \frac{1}{NM} \sum_{i=1}^{NM} (V^i(\hat{f}) - V^i(f))^2 \qquad (4.9)$$

Let $V^i(f)$ be approximated by Vf^i. Vf^i is the estimate of the local variance of pixel i in the original image based on the degraded image and knowledge of the degrading point spread function as per Equation (4.3). Vf^i is calculated before the algorithm commences and remains a constant throughout the restoration procedure.

The algorithm we examine in this section first computes the negative direction of the gradient that gives an indication of whether increasing or decreasing the current neuron's value will result in a net decrease in energy. Once the negative gradient is found the neuron's value is changed in unit steps and the resultant energy decrease after each step is computed. This ends when no further energy minimization is possible. In Chapter 2, we showed that the negative gradient of the unmodified cost function, Equation (2.3), is in fact the input to the neuron. Hence the negative gradient of the modified cost function will therefore be the input to the neuron minus the derivative of Equation (4.9).

The gradient of Equation (4.9) is given by

$$\frac{\delta}{\delta \hat{f}_i} \text{LVMSE} = \frac{2}{NM} (V^i(\hat{f}) - Vf^i) \frac{\delta}{\delta \hat{f}_i} (V^i(\hat{f}) - Vf^i) \qquad (4.10)$$

Note that this formula is an approximation of the gradient that ignores the contributions of the local variances of the pixels adjacent to i to the overall LVMSE of the image.

$$\frac{\delta V^i(\hat{f})}{\delta \hat{f}_i} = \frac{2\hat{f}_i}{A^2} - \frac{2\gamma_i}{A^4} \qquad (4.11)$$

Note that $\hat{f}_i^i = \hat{f}_i$. Substituting Equation (4.11) into Equation (4.10) we obtain

$$\frac{\delta}{\delta \hat{f}_i} \text{LVMSE} = \frac{2}{NM} \left(\frac{\beta_i}{A^2} - \frac{\gamma_i^2}{A^4} - Vf^i \right) \left(\frac{2\hat{f}_i}{A^2} - \frac{2\gamma_i}{A^4} \right)$$

Therefore,

$$\frac{\delta}{\delta \hat{f}_i} \text{LVMSE} = \frac{4}{NMA^2} \left(\frac{\hat{f}_i \beta_i}{A^2} - \frac{\hat{f}_i \gamma_i^2}{A^4} - \hat{f}_i Vf^i - \frac{\beta_i \gamma_i}{A^4} + \frac{\gamma^3}{A^6} + \frac{\gamma_i Vf^i}{A^2} \right)$$

$$(4.12)$$

Multiplying Equation (4.12) by θ and subtracting it from the input to the neuron gives us the negative gradient of the cost function. Given a change in the value of pixel i, the resultant change in energy is the previous change in energy given by Equation (2.40) plus θ times the change in LVMSE. The change in LVMSE is given by

$$\Delta \text{LVMSE} = \frac{1}{NM}\left((V_{new}^i - Vf^i)^2 - (V_{old}^i - Vf^i)^2\right) \qquad (4.13)$$

where

$$\gamma_i^{new} = \gamma_i^{old} + \Delta \hat{f}_i$$

$$\beta_i^{new} = \beta_i^{old} + 2\hat{f}_i \Delta \hat{f}_i + (\Delta \hat{f}_i)^2$$

$$V_{new}^i = \frac{\beta_i^{new}}{A^2} - \frac{(\gamma_i^{new})^2}{A^4}$$

$$= V_{old}^i + \frac{\left(\Delta \hat{f}_i\right)^2}{A^2} + \frac{2\hat{f}_i \Delta \hat{f}_i}{A^2} - \frac{2\gamma_i^{old}\Delta \hat{f}_i}{A^4} - \frac{\left(\Delta \hat{f}_i\right)^2}{A^4} \qquad (4.14)$$

where β_i^{old}, γ_i^{old} and V_{old}^i are the values of these parameters before the change in the state of neuron i occurred. The LVMSE algorithm is therefore

Algorithm 4.1

repeat

{

  for $i = 1, \ldots, L$ do

  {

    $u_i = b_i + \sum_{j=1}^{L} w_{ij} \hat{f}_j$

    $\beta_i^{old} = \sum_{j=1}^{L} \left(\hat{f}_j\right)^2$

    $\gamma_i^{old} = \sum_{j=1}^{L} \hat{f}_j$

    $V_{old}^i = \frac{\beta_i^{old}}{A^2} - \frac{(\gamma_i^{old})^2}{A^4}$

    $-\frac{\delta E}{\delta \hat{f}_i} = u_i$

$$-\frac{4\theta}{NMA^2}\left(\frac{\hat{f}_i \beta_i^{old}}{A^2} - \frac{\hat{f}_i (\gamma_i^{old})^2}{A^4} - \hat{f}_i Vf^i - \frac{\beta_i^{old}\gamma_i^{old}}{A^4} + \frac{(\gamma_i^{old})^3}{A^6} + \frac{\gamma_i^{old} Vf^i}{A^2}\right)$$

$$\Delta \hat{f}_i = G\left(-\frac{\delta E}{\delta \hat{f}_i}\right) \text{ where } G(u) = \begin{cases} 1, & u > 0 \\ 0, & u = 0 \\ -1, & u < 0 \end{cases}$$

$$V_{new}^i = V_{old}^i + \frac{(\Delta \hat{f}_i)^2}{A^2} + \frac{2\hat{f}_i \Delta \hat{f}_i}{A^2} - \frac{2\gamma_i^{old}\Delta \hat{f}_i}{A^4} - \frac{(\Delta \hat{f}_i)^2}{A^4}$$

$$\Delta E = -\frac{1}{2}w_{ii}\left(\Delta \hat{f}_i\right)^2 - u_i \Delta \hat{f}_i + \frac{\theta}{NM}\left([V_{new}^i - Vf^i]^2 - [V_{old}^i - Vf^i]^2\right)$$

    repeat

    {

      $\hat{f}_i(t'+1) = K(\hat{f}_i(t') + \Delta \hat{f}_i)$

      where $K(u) = \begin{cases} 0, & u < 0 \\ u, & 0 \le u \le S \\ S, & u > S \end{cases}$

      $u_i = u_i + w_{ii}\Delta \hat{f}_i$

$$V_{old}^i = V_{new}^i$$

$$\gamma_i^{old} = \gamma_i^{old} + \Delta \hat{f}_i$$

$$V_{new}^i = V_{old}^i + \frac{\left(\Delta \hat{f}_i\right)^2}{A^2} + \frac{2\hat{f}_i \Delta \hat{f}_i}{A^2} - \frac{2\gamma_i^{old} \Delta \hat{f}_i}{A^4} - \frac{\left(\Delta \hat{f}_i\right)^2}{A^4}$$

$$\Delta E = -\frac{1}{2} w_{ii} (\Delta \hat{f}_i)^2 - u_i \Delta \hat{f}_i + \frac{\theta}{NM} ([V_{new}^i - Vf^i]^2 - [V_{old}^i - Vf^i]^2)$$

$$t' = t' + 1$$

}

until $\Delta E \geq 0$

$t = t + 1$

}

}

until $\left(\hat{f}_i(t) = \hat{f}_i(t-1) \forall i = 1, \ldots, L \right)$

Note that Algorithm 4.1 still utilizes some features of Algorithm 2.3, specifically the use of bias inputs and interconnection strength matrices.

4.3.2   Analysis

It is important to verify that Algorithm 4.1 acts upon a pixel in the intended manner. To verify this, we must examine the LVMSE term in Algorithm 4.1 more closely.

According to Equation (4.12), the gradient of the LVMSE term of the cost function, when all pixels except $\hat{f}_i$ are held constant, is given by

$$\frac{\delta}{\delta \hat{f}_i} \text{LVMSE} = \frac{2}{NM} \left(\frac{\beta_i}{A^2} - \frac{\gamma_i^2}{A^4} - Vf^i \right) \left(\frac{2\hat{f}_i}{A^2} - \frac{2\gamma_i}{A^4} \right)$$

Note that $\beta_i = \sum_{j=1}^{L} (f_j^i)^2$ and $\gamma_i = \sum_{j=1}^{L} f_j^i$ can be rewritten as $\beta_i = \acute{\beta}_i + \hat{f}_i^2$ and $\gamma_i = \acute{\gamma}_i + \hat{f}_i$ where $\acute{\beta}_i = \sum_{j=1, j\neq i}^{L} (f_j^i)^2$ and $\acute{\gamma}_i = \sum_{j=1, j\neq i}^{L} f_j^i$. In this way, we can extract the elements of Equation (4.12) that depend on $\hat{f}_i$. Hence we obtain

$$\frac{\delta}{\delta \hat{f}_i} \text{LVMSE} = \frac{2}{NM} \left(\frac{\hat{f}_i^2}{A^2} + \frac{\acute{\beta}_i}{A^2} - \frac{(\hat{f}_i + \acute{\gamma}_i)^2}{A^4} - Vf^i \right) \left(\frac{2\hat{f}_i}{A^2} - \frac{2\hat{f}_i}{A^4} - \frac{2\acute{\gamma}_i}{A^4} \right)$$

$$= \frac{2}{NM} \left(\frac{\hat{f}_i^2}{A^2} - \frac{\hat{f}_i^2}{A^4} - \frac{2\hat{f}_i \acute{\gamma}_i}{A^4} + \frac{\acute{\beta}_i}{A^2} - \frac{\acute{\gamma}_i^2}{A^4} - Vf^i \right)$$

$$\times \left(\frac{2\hat{f}_i}{A^2} - \frac{2\hat{f}_i}{A^4} - \frac{2\acute{\gamma}_i}{A^4} \right) \tag{4.15}$$

Observe that $(\acute{\beta}_i / A^2)/(\acute{\gamma}_i^2 A^4)$ is an approximation to the local variance at pixel i neglecting the contribution of the value of pixel i itself. Let $\acute{V} = (\acute{\beta}_i / A^2)/(\acute{\gamma}_i^2 A^4)$. As A increases in value, $\acute{V}$ approaches the value of the local variance at pixel i. Similarly, we can define an approximation to the local mean of pixel i as

$$\acute{M} = \frac{1}{A^2} \sum_{j=1, j \neq i}^{NM} \hat{f}^i_j = \frac{\acute{\gamma}_i}{A^2} \qquad (4.16)$$

If A is greater than 3, then pixel i contributes less than 15% of the value of the local mean and local variance of its neighborhood. If A is 7 then the contribution is only 2%. Hence approximating $\acute{V}$ as the local variance is valid.

This leaves us with

$$\frac{\delta}{\delta \hat{f}_i} \text{LVMSE} = \frac{4}{NMA^2} \left(\hat{f}_i^2 \left[\frac{1}{A^2} - \frac{1}{A^4} \right] - \frac{2\hat{f}_i \acute{M}}{A^2} + J \right) \left(\hat{f}_i \left[1 - \frac{1}{A^2} \right] - \acute{M} \right) \qquad (4.17)$$

where $J = \acute{V} - V f^i$.

The points for which Equation (4.17) is equal to zero are the stationary points of the LVMSE term in Equation (4.2). Equation (4.17) is zero when $\hat{f}_i = \acute{M}/(1 - \frac{1}{A^2})$. This corresponds to the case where $\hat{f}_i$ is approximately equal to its local mean. Equation (4.17) also has zeros when $\hat{f}_i$ satisfies

$$\hat{f}_i = \frac{\frac{2\acute{M}}{A^2} \pm \sqrt{\frac{4\acute{M}^2}{A^4} - 4J \left(\frac{1}{A^2} - \frac{1}{A^4} \right)}}{2 \left(\frac{1}{A^2} - \frac{1}{A^4} \right)}$$

If A is greater than or equal to 5, then the error resulting from approximating $1/A^2 - 1/A^4$ as $1/A^2$ is less than 4%. Therefore, if we assume that A is large enough that $1/A^2 \gg 1/A^4$ then Equation (4.17) has zeros at

$$\hat{f}_i \approx \acute{M} \pm A \sqrt{\frac{\acute{M}^2}{A^2} - J} \qquad (4.18)$$

Note that Equation (4.18) indicates that if $J > 0$, Equation (4.18) may not have a real-valued solution. There are three cases to examine.

Case 4.1 ($J < 0$)
When $J < 0$, then local variance of the region surrounding pixel i is less than the estimated local variance. Equations (4.18) and (4.17) indicate that the LVMSE term in the cost function will have three stationary points. The function will then appear as Figure 4.1. The stationary point given by $\hat{f}_i = \acute{M}/(1 - (1/A^2))$ becomes a maximum and the function has two minima given by Equation (4.18). At least one of these minima will be in the domain $\hat{f}_i > 0$, and so as long as J is not excessively negative it is possible to minimize the LVMSE term in the cost function in this case.

This is as one would expect since if $J < 0$, then the variance of the local region surrounding pixel i needs to be increased to meet the target variance estimate. It is always possible to increase the local variance of pixel i by moving that pixel's value further from the mean. This is why the stationary point $\hat{f}_i = \acute{M}/(1 - 1/A^2)$, which is the point where $\hat{f}_i$ is approximately equal to the mean of the local region, becomes a maximum.

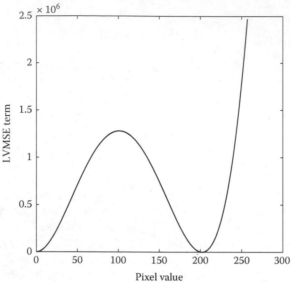

FIGURE 4.1
Graph of function for Case 1.

Case 4.2 $(0 < J \leq \acute{M}^2/A^2)$

When $0 < J \leq \acute{M}^2/A^2$, then the local variance of the region surrounding pixel i is greater than the estimated local variance, but the difference is not great. Equations (4.18) and (4.17) indicate that the LVMSE term in the cost function will again have three stationary points. The function will then appear as Figure 4.2.

The stationary point given by $\hat{f}_i = \acute{M}/(1 - (1/A^2))$ becomes a maximum and the function has two minima given by Equation (4.18). At least one of these minima will be in the domain $\hat{f}_i > 0$, and it is again possible to minimize the LVMSE term in the cost function in this case.

When $J > 0$, the local variance of pixel i needs to be decreased to match the target variance estimate. If $J \leq \acute{M}^2/A^2$, it is possible to match the local variance of pixel i with the target variance estimate by moving that pixel's value toward the mean pixel value in the local region. The stationary point $\hat{f}_i = \acute{M}/(1 - (1/A^2))$, which is the point where $\hat{f}_i$ is approximately equal to the mean of the local region, is again a maximum, unless $J = \acute{M}^2/A^2$. When $J = \acute{M}^2/A^2$, all three stationary points correspond to the same value and the function has one minimum at $\hat{f}_i = \acute{M}/(1 - (1/A^2)) \approx \acute{M}$.

Case 4.3 $(J > \acute{M}^2/A^2)$

When $J > \acute{M}^2/A^2$, then the local variance of the region surrounding pixel i is greater than the estimated local variance, but the difference is too large for equality to be

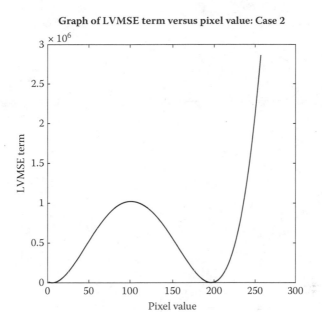

FIGURE 4.2
Graph of function for Case 2.

reached by changing a single pixel's value. Equation (4.18) will have no solutions and the LVMSE term in the cost function will have one stationary point. The function will then appear as Figure 4.3.

The stationary point given by $\hat{f}_i = \acute{M}/(1 - (1/A^2))$ becomes a minimum, where $\acute{M} > 0$ since all pixel values are constrained to be positive. So the minimum will be in the domain $\hat{f}_i > 0$, and it is possible to minimize the LVMSE term in the cost function in this case.

When $J > \acute{M}^2/A^2$, the local variance of pixel i needs to be decreased to match the target variance estimate. The minimum local variance is obtained by changing the value of the current pixel to the mean value of the local region. In this case, the minimum local variance obtainable by altering one pixel's value is still greater than the estimate. The stationary point $\hat{f}_i = \acute{M}/(1 - (1/A^2))$ is the point where $\hat{f}_i$ is approximately equal to the mean of the local region. Thus, the pixel will move toward the mean value of the local region, hence reducing the local variance as much as possible.

By examining each of the three cases above, we can see that the LVMSE term in the cost function is well behaved despite its nonlinear nature. In all cases a minimum exists and the local variance of the current pixel's region will always move closer to the target variance estimate.

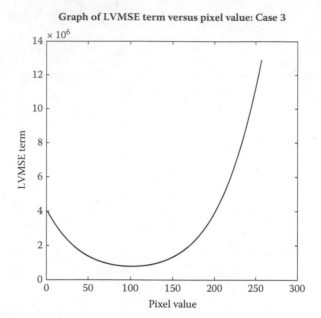

FIGURE 4.3
Graph of function for Case 3.

4.4   Log LVMSE-Based Cost Function

The previous section showed that the LVMSE cost term in Equation (4.2) is well behaved, and results in an algorithm that does indeed work as intended to match the local variance of the region surrounding the current pixel to that of a target variance estimate of the original image. The LVMSE term in Equation (4.2) has its greatest effect when the difference between the actual local variance and the target variance estimate is large. When the difference is small, the LVMSE term in Equation (4.2) has little effect. The strength of the LVMSE term is proportional to the square of the absolute difference between the two variances and does not depend on the level of the variances. The dependence on the absolute difference between the variances is in fact a disadvantage. When the local variance of a region is large and the target variance for that region is also large, then noise will not be readily noticed. In this case, the first term in the cost function that ensures sharpness of edges should be allowed to dominate the restoration. However, in this case, the difference between the target variance and actual variance may be large, causing the LVMSE term to dominate instead. On the other hand, when the local variance of a region is small and the target variance for that region is also small, we would want the LVMSE term to dominate, since this term would keep the local variance low and suppress noise. However, in this case since the target variance and the actual variance are small, the LVMSE term can also be too small and have an insufficient effect.

This prompts us to move away from absolute differences in local variances, and, instead, compare the ratio of the local variance and its estimate. Taking this idea one step further, we notice that taking the log of this ratio provides additional emphasis of small differences in variance at low-variance levels and deemphasis of large differences in variance at high-variance levels.

Hence a new cost function is introduced [72]:

$$E = \frac{1}{2}\|\mathbf{g} - \mathbf{H}\hat{\mathbf{f}}\|^2 + \frac{\lambda}{2}\|\mathbf{D}\hat{\mathbf{f}}\|^2 + \frac{\theta}{NM} \sum_{x=0}^{N-1} \sum_{y=0}^{M-1} \left[\ln\left(\frac{\sigma_A^2(\hat{\mathbf{f}}(x, y))}{\sigma_A^2(\mathbf{g}(x, y))^{\star}} \right) \right]^2 \quad (4.19)$$

We will denote this new term in Equation (4.19) the log local variance ratio (log LVR) term.

4.4.1 Extended Algorithm for the Log LVR-Modified Cost Function

The algorithm to minimize Equation (4.19) has the same basic strategy as the algorithm created in Section 4.3. First, the negative direction of the gradient is computed. Once the negative gradient is found, the neuron value is changed in unit steps and the resultant energy decrease after each step is computed. This ends when no further energy reduction is possible. As in the last section, multiplying the partial derivative of the log LVR term in Equation (4.19) by θ and subtracting it from the input to the neuron gives us the negative gradient of the cost function. As defined in Section 4.3, the local variance of the A by A region centered on pixel i is given by

$$V^i = \frac{\beta_i}{A^2} - \frac{\gamma_i^2}{A^4}$$

where $\beta_i = \sum_{j=1}^{L}(f_j^i)^2$ and $\gamma_i = \sum_{j=1}^{L} f_j^i$.

The gradient of the log LVR term in Equation (4.19) is given by

$$\frac{\delta}{\delta \hat{f}_i} \text{Log LVR} = \frac{1}{NM} 2 \ln\left(\frac{V^i}{Vf^i} \right) \frac{Vf^i}{V^i} \frac{1}{Vf^i} \left(\frac{2\hat{f}_i}{A^2} - \frac{2\gamma_i}{A^4} \right)$$

$$= \frac{4}{NM} \ln\left(\frac{\frac{\beta_i}{A^2} - \frac{\gamma_i^2}{A^4}}{Vf^i} \right) \frac{\left(\hat{f}_i - \frac{\gamma_i}{A^2} \right)}{\left(\beta_i - \frac{\gamma_i^2}{A^2} \right)} \quad (4.20)$$

Similarly, given a change in the value of pixel i, the resultant change in energy is the previous change in energy given by Equation (2.40) plus θ times the change in log LVR. The change in log LVR is given by

$$\Delta \text{Log LVR} = \frac{1}{NM} \left(\ln\left[\frac{V_{new}^i}{Vf^i} \right]^2 - \ln\left[\frac{V_{old}^i}{Vf^i} \right]^2 \right) \quad (4.21)$$

where

$$\gamma_i^{new} = \gamma_i^{old} + \Delta \hat{f}_i$$

$$\beta_i^{new} = \beta_i^{old} + 2\hat{f}_i \Delta \hat{f}_i + (\Delta \hat{f}_i)^2$$

$$V_{new}^i = \frac{\beta_i^{new}}{A^2} - \frac{(\gamma_i^{new})^2}{A^4}$$

$$= V_{old}^i + \frac{(\Delta \hat{f}_i)^2}{A^2} + \frac{2\hat{f}_i \Delta \hat{f}_i}{A^2} - \frac{2\gamma_i^{old} \Delta \hat{f}_i}{A^4} - \frac{(\Delta \hat{f}_i)^2}{A^4}$$

The new algorithm is, therefore,

Algorithm 4.2
repeat
{
  for $i = 1, \ldots, L$ do
  {

$$u_i = b_i + \sum_{j=1}^{L} w_{ij} \hat{f}_j$$

$$\beta_i^{old} = \sum_{j=1}^{L} \left(\hat{f}_j^i\right)^2$$

$$\gamma_i^{old} = \sum_{j=1}^{L} \hat{f}_j^i$$

$$V_{old}^i = \frac{\beta_i^{old}}{A^2} - \frac{(\gamma_i^{old})^2}{A^4}$$

$$-\frac{\delta E}{\delta \hat{f}_i} = u_i - \frac{4\theta}{NM} \ln\left(\frac{\frac{\beta_i}{A^2} - \frac{\gamma_i^2}{A^4}}{Vf^i}\right) \frac{\left(\hat{f}_i - \frac{\gamma_i}{A^2}\right)}{\left(\beta_i - \frac{\gamma_i^2}{A^2}\right)}$$

$$\Delta \hat{f}_i = G\left(-\frac{\delta E}{\delta \hat{f}_i}\right) \text{ where } G(u) = \begin{cases} 1, & u > 0 \\ 0, & u = 0 \\ -1, & u < 0 \end{cases}$$

$$V_{new}^i = V_{old}^i + \frac{(\Delta \hat{f}_i)^2}{A^2} + \frac{2\hat{f}_i \Delta \hat{f}_i}{A^2} - \frac{2\gamma_i^{old} \Delta \hat{f}_i}{A^4} - \frac{(\Delta \hat{f}_i)^2}{A^4}$$

$$\Delta E = -\frac{1}{2} w_{ii} \left(\Delta \hat{f}_i\right)^2 - u_i \Delta \hat{f}_i + \frac{\theta}{NM}\left(\ln\left[\frac{V_{new}^i}{Vf^i}\right]^2 - \ln\left[\frac{V_{old}^i}{Vf^i}\right]^2\right)$$

  repeat
  {

$$\hat{f}_i(t'+1) = K(\hat{f}_i(t') + \Delta \hat{f}_i)$$

$$\text{where } K(u) = \begin{cases} 0, & u < 0 \\ u, & 0 \le u \le S \\ S, & u > S \end{cases}$$

$$u_i = u_i + w_{ii} \Delta \hat{f}_i$$

$$V_{old}^i = V_{new}^i$$

$$\gamma_i^{old} = \gamma_i^{old} + \Delta \hat{f}_i$$

$$V_{new}^i = V_{old}^i + \frac{(\Delta \hat{f}_i)^2}{A^2} + \frac{2\hat{f}_i \Delta \hat{f}_i}{A^2} - \frac{2\gamma_i^{old} \Delta \hat{f}_i}{A^4} - \frac{(\Delta \hat{f}_i)^2}{A^4}$$

$$\Delta E = -\frac{1}{2}w_{ii}\left(\Delta\hat{f}_i\right)^2 - u_i\Delta\hat{f}_i + \frac{\theta}{NM}\left(\ln\left[\frac{V_{new}^i}{Vf^i}\right]^2 - \ln\left[\frac{V_{old}^i}{Vf^i}\right]^2\right)$$

$$t' = t' + 1$$

$$\}$$

until $\Delta E \geq 0$

$$t = t + 1$$

$$\}$$

$$\}$$

until $\left(\hat{f}_i(t) = \hat{f}_i(t-1)\forall i = 1,\ldots,L\right)$

Note that Algorithm 4.2 is almost identical to Algorithm 4.1 and still utilizes some features of Algorithm 2.3, specifically the use of bias inputs and interconnection strength matrices.

4.4.2 Analysis

As in the previous section, we will verify the correct operation of Algorithm 4.2. To verify this, we must examine the log LVR term in Equation (4.19) more closely.

The gradient of the log LVR term of the cost function when all pixels except $\hat{f}_i$ are held constant is given by

$$\frac{\delta}{\delta\hat{f}_i}\,\text{Log LVR} = \frac{4}{NM}\ln\left(\frac{\frac{\beta_i}{A^2} - \frac{\gamma_i^2}{A^4}}{Vf^i}\right)\left(\frac{\frac{f_i}{A^2} - \frac{\gamma_i}{A^4}}{\frac{\beta_i}{A^2} - \frac{\gamma_i^2}{A^4}}\right) \tag{4.22}$$

Note that $\beta_i = \sum_{j=1}^{L}(f_j^i)^2$ and $\gamma_i = \sum_{j=1}^{L}f_j^i$ can be rewritten as $\beta_i = \hat{\beta}_i + \hat{f}_i^2$ and $\gamma_i = \hat{\gamma}_i + \hat{f}_i$, where $\hat{\beta}_i = \sum_{j=1,j\neq i}^{L}(f_j^i)^2$ and $\hat{\gamma}_i = \sum_{j=1,j\neq i}^{L}f_j^i$. In this way, we can extract the elements of Equation (4.22) that depend on $\hat{f}_i$. Hence we obtain

$$\frac{\delta}{\delta\hat{f}_i}\,\text{Log LVR}$$

$$= \frac{4}{NM}\frac{\left(\ln\left[\hat{f}_i^2(\frac{1}{A^2}-\frac{1}{A^4}) - \frac{2\hat{f}_i\hat{\gamma}_i}{A^4} + \frac{\hat{\beta}_i}{A^2} - \frac{\hat{\gamma}_i^2}{A^4}\right] - \ln(Vf^i)\right)\left(\frac{\hat{f}_i}{A^2} - \frac{\hat{\gamma}_i}{A^4} - \frac{\hat{f}_i}{A^4}\right)}{\hat{f}_i^2(\frac{1}{A^2}-\frac{1}{A^4}) - \frac{2\hat{f}_i\hat{\gamma}_i}{A^4} + \frac{\hat{\beta}_i}{A^2} - \frac{\hat{\gamma}_i^2}{A^4}}$$

If we assume that A is large enough so that $(1/A^2) \gg (1/A^4)$, then

$$\frac{\delta}{\delta\hat{f}_i}\,\text{Log LVR}$$

$$= \frac{4}{NMA^2}\frac{\left(\ln\left[\frac{\hat{f}_i^2}{A^2} - \frac{2\hat{f}_i\hat{\gamma}_i}{A^4} + \frac{\hat{\beta}_i}{A^2} - \frac{\hat{\gamma}_i^2}{A^4}\right] - \ln(Vf^i)\right)\left(\hat{f}_i[1-\frac{1}{A^2}] - \frac{\hat{\gamma}_i}{A^2}\right)}{\frac{\hat{f}_i^2}{A^2} - \frac{2\hat{f}_i\hat{\gamma}_i}{A^4} + \frac{\hat{\beta}_i}{A^2} - \frac{\hat{\gamma}_i^2}{A^4}}$$

Observe that $(\hat{\beta}_i/A^2) - (\hat{\gamma}_i^2/A^4)$ is an approximation to the local variance at pixel i neglecting the contribution of the value of pixel i itself. Let

$\acute{V} = (\acute{\beta}_i/A^2) - (\acute{\gamma}_i^2/A^4)$. As A increases in value, $\acute{V}$ approaches the value of the local variance at pixel i. Similarly, we can define an approximation to the local mean of pixel i as

$$\acute{M} = \frac{1}{A^2} \sum_{j=1, j \neq i}^{NM} \hat{f}_j^i = \frac{\acute{\gamma}_i}{A^2}$$

This leaves us with

$$\frac{\delta}{\delta \hat{f}_i} \text{Log LVR}$$

$$= \frac{4}{NMA^2} \frac{\left(\ln\left[\frac{\hat{f}_i^2}{A^2} - \frac{2\hat{f}_i\acute{M}}{A^2} + \acute{V} \right] - \ln(Vf^i) \right) \left(\hat{f}_i \left[1 - \frac{1}{A^2} \right] - \acute{M} \right)}{\frac{\hat{f}_i^2}{A^2} - \frac{2\hat{f}_i\acute{M}}{A^2} + \acute{V}} \quad (4.23)$$

There are some points for which Equation (4.23) is undefined. These points are given by

$$\hat{f}_i \approx \acute{M} \pm A\sqrt{\frac{\acute{M}^2}{A^2} - \acute{V}}$$

At these points the function will have an infinitely negative gradient. In the event that $\acute{V} > \acute{M}^2/A^2$ then the function will be defined for all values of $\hat{f}_i$. Fortunately this will almost always happen. This can be seen by examining the condition for undefined points to exist

$$\acute{V} \leq \frac{\acute{M}^2}{A^2}$$

Using the formulas for $\acute{V}$ and $\acute{M}$ we get

$$\frac{\acute{\beta}_i}{A^2} - \frac{\acute{\gamma}_i^2}{A^4} \leq \frac{1}{A^2} \left(\frac{\acute{\gamma}_i}{A^2} \right)^2$$

$$\frac{\acute{\beta}_i}{A^2} - \frac{\acute{\gamma}_i^2}{A^4} \leq \frac{1}{A^6} \acute{\gamma}_i^2$$

Hence we require that

$$\frac{\acute{\beta}_i}{A^2} \leq \acute{\gamma}_i^2 \left(\frac{1}{A^4} + \frac{1}{A^6} \right) \quad (4.24)$$

However, the variance of a set of numbers is always greater than or equal to zero. $\acute{V}$ is the variance of the local region of pixel i obtained when the value of pixel i is set to zero. This gives us

$$\acute{V} = \frac{\acute{\beta}_i}{A^2} - \frac{\acute{\gamma}_i^2}{A^4} \geq 0$$

which means that

$$\frac{\hat{\beta}_i}{A^2} \geq \frac{\hat{\gamma}_i^2}{A^4} \qquad (4.25)$$

Equation (4.25) of course means that condition Equation (4.24) will only be satisfied when the local variance is very close to zero. As long as steps are taken to ensure that this does not occur, the function will be well defined for all values of $\hat{f}_i$.

The points for which equation (4.23) is equal to zero are the stationary points of the log LVR term in Equation (4.19). Equation (4.23) is zero when $\hat{f}_i = \acute{M}/(1 - 1/A^2)$. This corresponds to the case where $\hat{f}_i$ is approximately equal to its local mean. Equation (4.23) also has zeros when $\hat{f}_i$ satisfies

$$\ln\left(\frac{\hat{f}_i^2}{A^2} - \frac{2\hat{f}_i \acute{M}}{A^2} + \acute{V}\right) - \ln(Vf^i) = 0$$

This is equivalent to

$$\frac{\hat{f}_i^2}{A^2} - \frac{2\hat{f}_i \acute{M}}{A^2} + J = 0$$

where $J = \acute{V} - Vf^i$.

The stationary points are thus given by

$$\hat{f}_i \approx \acute{M} \pm A\sqrt{\frac{\acute{M}^2}{A^2} - J} \qquad (4.26)$$

Note that Equation (4.26) indicates that if $J > 0$, Equation (4.26) may not have a real-valued solution. Equation (4.26) is identical to Equation (4.18) in the previous section and so the case-by-case analysis in Section 4.3 is identical for a log LVR-modified cost function as it was for the LVMSE-modified cost function.

As with the LVMSE term, the log LVR term in the cost function is well behaved despite its nonlinear nature. In all cases, a minimum exists and by minimizing the log LVR term in the cost function, the local variance of the current pixel's region will always move closer to the target variance estimate.

4.5   Implementation Considerations

A problem to be overcome is that the third terms in the LVMSE-based cost functions are not quadratic in nature. When the local variance in the image estimate is much lower than the projected local variances of the original image, the LVMSE term in Algorithm 4.1 becomes large and may force the pixel values to an extreme of the range of acceptable values in order to create a high-variance region. The LVMSE term should never completely dominate over the first term in Equation (4.2), since the LVMSE term only attempts to match regions, not pixels, and fine structure within the region will be lost. To remedy this situation, the pixel values are not allowed to change by more than a set amount per iteration. This method appears to work well in practice and the pixel values converge to a solution after a finite number of iterations. This method, however, is not required to the same degree in Algorithm 4.2. Algorithm 4.2 was designed to avoid this effect; however, this method may still be employed to improve results.

The addition of the LVMSE term into the cost function allows a powerful optimization to be made to Algorithm 4.1. In regions where the degraded image is very smooth and the variance estimate of the original image is very small, improvement in image-processing speed can be achieved by not restoring these pixels. This will not affect the quality of processing since attempting to deconvolve regions where the blurring effect is not noticeable by humans can only serve to amplify noise. It is logical not to attempt to restore such regions when using Algorithm 4.1 since the LVMSE-based term in the cost function for this algorithm has little effect at low-variance regions. Algorithm 4.2 on the other hand was designed to smooth these regions and so it is not necessary to avoid attempting to restore these regions.

4.6   Numerical Examples

In this section, a number of examples are given to show the performance of the methods we examined in this chapter. Comparisons will be made with some well-known methods in the literature.

4.6.1   Color Image Restoration

For the first experiment, color images were used consisting of three color planes, red, green, and blue in the sRGB format. The image was degraded by a 5-by-5 Gaussian PSF of standard deviation 2.0 applied to each of the color planes. In addition, additive noise of variance 369.31 was also added to each color plane. Figure 4.4a shows the original image and Figure 4.4b shows the degraded image. The degraded image has an SNR of 19.81 dB and an LSMSE of 313.05. LSMSE is as defined in Equation (3.14). The SNR was calculated by adding together the SNR of each color plane:

$$\text{SNR} = 20 \log \left(\frac{\sigma_o^r}{\sigma_n^r} + \frac{\sigma_o^g}{\sigma_n^g} + \frac{\sigma_o^b}{\sigma_n^b} \right) \qquad (4.27)$$

Similarly the LSMSE for the entire image was calculated by summing the LSMSEs of each color plane. A 9-by-9 neighborhood was used for calculating

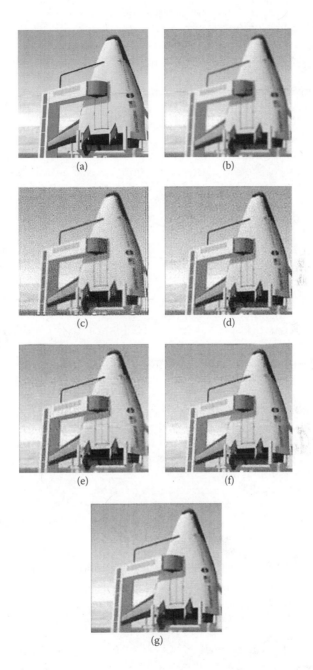

FIGURE 4.4
Color images restored using various algorithms. (*Adapted from [118], with permission of publisher SPIE—The International Society for Optical Engineering.*)

the local variance and local standard deviation. In this example, we assumed that each color plane in our test image does not have a high level of correlation and so the filters can be applied to each color plane separately. A Wiener filter restored image is shown in Figure 4.4c and has an SNR of 16.65 dB and an LSMSE of 859.80 [4,21]. The image was also restored using Algorithm 2.3, without the LSMSE term. A constraint factor of $\lambda = 0.001$ was chosen. The CLS restored image is shown in Figure 4.4d and has an SNR of 17.26 dB and an LSMSE of 634.04. The image was also restored using the variance selection adaptive constraint algorithm from Chapter 3. This image is shown in Figure 4.4e and has an SNR of 19.19 dB and an LSMSE of 195.68. The same degraded image was also restored using the LSMSE-modified cost function, Algorithm 4.1. In the LSMSE-modified cost function, the value of λ was set to 0.0005. The factor θ was set to be 0.00001 and the image local variance estimate was computed as

$$\sigma_A^2(\mathbf{g}(x, y))^\star = 2\left(\sigma_A^2(\mathbf{g}(x, y)) - 200\right)$$

This image is shown in Figure 4.4f and has an SNR of 19.89 dB and an LSMSE of 180.81. Finally, the degraded image was restored using the log LVR-modified cost function, Algorithm 4.2. In the log LVR-modified cost function, the value of λ was set to 0.0005. The factor θ was set to be 50 and the image local variance estimate was computed as for Algorithm 4.1. This image is shown in Figure 4.4g and has an SNR of 21.65 dB and an LSMSE of 88.43.

By visual observation it can be seen that Figures 4.4f and 4.4g, produced by the LSMSE and log LVR-based cost functions, display better noise suppression in background regions and are at the same time sharper than Figures 4.4c and 4.4d, produced by the Wiener and the CLS approaches. Figures 4.4f and 4.4g also display a better SNR and LSMSE than Figures 4.4c, 4.4d, and 4.4e. Although the LSMSE-restored images are visually closer to the original image than the degraded image, their SNRs are only slightly higher than the degraded image. This is not surprising in view of the arguments above that SNR does not correspond well with human visual perception. However, LSMSE does match with human observation and assigns a much lower value to Figures 4.4f and 4.4g. Comparing the two different forms of the LSMSE-based cost functions, we find that Algorithm 4.2 (Figure 4.4g) is superior, with a similar level of sharpness when compared to Figure 4.4f, yet better noise suppression in background regions.

We see that the variance selection adaptive constraint method produces a similar result to Algorithm 4.1. This is primarily because both algorithms use the concept of a variance threshold. As mentioned in Section 4.5, if the local variance is below the threshold, the pixel is not adjusted. Both algorithms use identical thresholds and so have similar LSMSEs. Algorithm 4.2, however, was designed not to require the variance threshold and instead provides additional smoothing to background regions and hence a much lower LSMSE.

4.6.2 Grayscale Image Restoration

For the second example, a grayscale image was degraded by a 5-by-5 Gaussian PSF of standard deviation 2.0. Additive noise of variance 87.62 was also added. Figure 4.5a shows the original image and Figure 4.5b shows the degraded image. The degraded image has an SNR of 12.58 dB and an LSMSE of 28.13. The degraded image was first restored using a Wiener filter approach. The Wiener restored image is shown in Figure 4.5c and has an SNR of 11.66 dB and an LSMSE of 38.69. The image was also restored using the CLS algorithm (Algorithm 2.3). Figure 4.5d shows the image restored using the CLS algorithm with a constant factor of $\lambda = 0.001$. Figure 4.5d has an SNR of 8.76 dB and an LSMSE of 128.09. Figure 4.5e shows the image restored using the CLS algorithm with a constant factor of $\lambda = 0.002$. Figure 4.5e has an SNR of 11.93 dB and an LSMSE of 36.91. Figure 4.5f shows the image restored using the adaptive constraint algorithm presented in Chapter 3 using a range of constraint values from 0.02 to 0.0015 associated with levels of local variance. Figure 4.5f has an SNR of 11.97 dB and an LSMSE of 22.28. The degraded image was also restored using the LVMSE-modified cost function implemented using Algorithm 4.1. Figure 4.5g shows this image which has an SNR of 12.15 dB and an LSMSE of 22.71. Finally, the degraded image was restored using the log LVR-modified cost function implemented using Algorithm 4.2. Figure 4.5h shows this image that has an SNR of 12.07 dB and an LSMSE of 20.59. By observation, it can be seen that Figure 4.5h is visually closest to the original image. LSMSE confirms visual inspection and indicates that Figure 4.5h is the most well restored. Note that once again the adaptive algorithm from Chapter 3 performs similarly to the LSMSE-based algorithms. The advantage of the LSMSE algorithms is that they have fewer free variables to set up.

4.6.3 LSMSE of Different Algorithms

For the third example, the original flower image was blurred using a 5-by-5 Gaussian blur of standard deviation 2.0. A number of images were created, each suffering a different value of noise. The images were restored using Algorithms 2.3, 4.1, 4.2, a Wiener filter, and the adaptive constraint algorithm from Chapter 3. For each image, the same value of λ was used in Algorithm 2.3, Algorithm 4.1, and Algorithm 4.2. This meant that the restored images from Algorithm 2.3, Algorithm 4.1, and Algorithm 4.2 had the same degree of sharpness, but differed in the level of noise suppression. In this way the effects of the LSMSE-based terms in Equations (4.2) and (4.19) could be examined in isolation. Figure 4.6 shows the results of this experiment. It can be clearly seen that in terms of the LSMSE, Algorithms 4.1 and 4.2 outperform the other algorithms, especially the standard CLS approach for the same level of sharpness.

FIGURE 4.5
Grayscale images restored using various algorithms. (*Adapted from [118], with permission of publisher SPIE—The International Society for Optical Engineering.*)

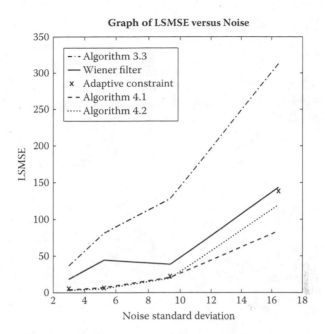

FIGURE 4.6
Graph of LSMSE for various algorithms and levels of noise.

4.6.4 Robustness Evaluation

For the fourth example, the original flower image was blurred using a 5-by-5 Gaussian blur of standard deviation 2.0. Additive noise of variance 87.62 was also added. The degraded image was restored using Algorithm 2.3, Algorithm 4.1, and Algorithm 4.2. In each algorithm, λ was set to 0.001 to maintain the same level of sharpness. Figure 4.7a shows the results of Algorithm 2.3. This image is identical to Figure 4.5d and has an SNR of 8.76 dB and an LSMSE of 128.09. Figure 4.7b shows the results of Algorithm 4.1. Figure 4.7b has an SNR of 12.45 dB and an LSMSE of 20.76. Figure 4.7c shows the results of Algorithm 4.2. Figure 4.7c has an SNR of 12.25 dB and an LSMSE of 19.76. Next we severed one of the neural interconnections to a neighboring neuron for every neuron in the network. The same connection was severed for each neuron in the network. This would be expected to degrade the performance of the network. Using the same parameters, the restorations were performed again. Figure 4.7d shows the results of restoring the image using Algorithm 2.3 with a faulty network. The SNR is minus 4.21 dB and the LSMSE is 5889.42. Figure 4.7e shows the results of restoring the image using Algorithm 4.1 with a faulty network. The SNR is 12.15 dB and the LSMSE is 23.23. Figure 4.7f shows the results of restoring the image using Algorithm 4.2 with a faulty network. The SNR is 10.06 dB and the LSMSE is 40.13. From these results we can see that Algorithm 2.3 is not very tolerant of errors in weights. The image produced by the faulty network is very degraded and has poor values

FIGURE 4.7

Images restored using correct and faulty networks. (*Adapted from* [118], *with permission of publisher SPIE—The International Society for Optical Engineering.*)

of SNR and LSMSE. On the other hand, Algorithm 4.1 and Algorithm 4.2 have almost no visual differences between images restored using the correct network and images restored using the faulty network. The images restored using the faulty network have only slightly worse values of SNR and LSMSE compared to the image restored using the correct network. The reason that Algorithms 4.1 and 4.2 are more fault-tolerant than Algorithm 2.3 is due to the LSMSE-related terms in these algorithms. The damaged weights in Algorithm 2.3 produced streaks in the image. These streaks would cause the pixels in their vicinity to have very high local variances. Since Algorithm 2.3 does not consider the local regional statistics of the image, the streaks are not suppressed. However, Algorithm 4.1 and Algorithm 4.2 attempt to match local variances in the restored image with an estimate of the original image. The streaks are therefore suppressed by Algorithms 4.1 and 4.2. It is clear that Algorithms 4.1 and 4.2 are very robust and are not greatly affected by errors in the network. Algorithm 4.1 is more robust than Algorithm 4.2 on the edges because of the fact that the log-ratio relationship between the local variance and the target variance used by Algorithm 4.2 was developed to deemphasize the LSMSE effect on edges. Algorithm 4.1 has its greatest effect on edges, whereas Algorithm 4.2 was specifically designed to have the least effect on edges and the greatest effect on smooth regions. However, both algorithms are still quite tolerant of network errors. This is because the LVMSE-based terms in these algorithms can never be affected by severed neural connections.

4.6.5 Subjective Survey

For the fifth experiment a subjective survey of the proposed method was conducted [118]. Six human subjects without experience in the image processing field participated in the evaluation. The two restoration methods compared were constrained least squares (CLS) with the neural network implementation (Algorithm 2.3), and the LVMSE restoration method (Algorithm 4.1). A Gaussian-shaped PSF was used in the experiment; however, the region of support of the PSF was kept at 5 for all of the experiments, making the PSF more of a uniform PSF at standard deviations of 4.0 and 6.0. Two images were used in the experiment, and each was blurred by five different sets of distortion parameters. The first image is shown in Figure 3.2a. The results of the subjective survey for this image are given in Table 4.1. The second image is shown in Figure 3.2b. The results of the subjective survey for this image are given in Table 4.2. In Tables 4.1 and 4.2, the bolded numbers in columns 3 and 4

TABLE 4.1

Image Cat, Size of the PSF = 5×5

PSF Standard Deviation / Noise Variance	Restoration Method	MSE Measurement	LSMSE Measurement
PSF SD = 2	LVMSE	**17.14**	**25.80**
Noise variance = 100	CLS	17.75	37.50
PSF SD = 2	LVMSE	**18.53**	**36.14**
Noise variance = 150	CLS	20.21	64.00
PSF SD = 2	LVMSE	**21.11**	**52.34**
Noise variance = 200	CLS	22.23	94.45
PSF SD = 4	LVMSE	**18.04**	**28.32**
Noise variance = 100	CLS	24.87	145.66
PSF SD = 6	LVMSE	**18.23**	**29.33**
Noise variance = 100	CLS	26.50	186.38

TABLE 4.2

Image Flower, Size of the PSF = 5 × 5

PSF Standard Deviation / Noise Variance	Restoration Method	MSE Measurement	LSMSE Measurement
PSF SD = 2	LVMSE	13.85	25.79
Noise variance = 100	CLS	14.44	42.83
PSF SD = 2	LVMSE	15.21	40.00
Noise variance = 150	CLS	17.18	77.19
PSF SD = 2	LVMSE	17.16	60.78
Noise variance = 200	CLS	22.33	116.67
PSF SD = 4	LVMSE	14.96	30.34
Noise variance = 100	CLS	22.36	180.04
PSF SD = 6	LVMSE	15.28	31.23
Noise variance = 100	CLS	24.15	220.89

represent the unanimous selection of the most visually pleasing image by the subjective survey participants. From these results a number of observations become apparent. Firstly, all of the human subjects picked restorations by Algorithm 4.1 as superior to the restorations by Algorithm 2.3 under each level of noise and every standard deviation value of the PSF. For all of the experiments, the LVMSE restoration always gave smaller MSE/LSMSE measurements. An important observation from the results is that as the PSF standard deviation gets larger (and hence the degrading blur becomes more severe), the difference in MSE/LSMSE between the two methods gets very significant. This indicates that LVMSE restoration is an apparently better choice for severe blur.

4.7   Local Variance Extension of the Lagrange Model

As was shown in Chapter 2, the restoration problem may be considered in terms of the theory of Lagrange [94]. We will see in this section the consequences of this analysis for LVMSE-based restoration. In a similar way to the analysis in Chapter 3, we will expand the restoration problem in terms of the theory of Karush–Kuhn–Tucker and then show what solutions result in. A less detailed version of the following analysis was presented in [119].

4.7.1   Problem Formulation

Consider the image degradation model:

$$\mathbf{g} = \mathbf{Hf} + \mathbf{n} \tag{4.28}$$

where **g** and **f** are $L = MN$-dimensional lexicographically organized degraded and original images, respectively, **n** is an L-dimensional additive noise vector and **H** is a convolutional operator.

The problem we are considering is that given **g** and **H**, we wish to recover **f**. Assume that using the degraded image, **g**, we can reasonably estimate the local variance of each pixel in the original image, **f**. The local variance of a pixel is defined to be the sample variance of the collection of pixels centered at the pixel being considered. We will call the pixels over which the local variance of a pixel is calculated to be that pixel's *local neighborhood*.

Let $\{i\}$ denote the set of lexicographic indices of all pixels who are local neighbors of pixel i. Let's define an operator that extracts only the local neighbors of i, including i.

Define $\mathbf{K}_i \in \Re^{L \times L}, 1 \le i \le L$ where

$$[\mathbf{K}_i]_{jl} = \begin{cases} 1, & \text{if } j = l \text{ and } j, l \in \{i\} \\ 0, & \text{if } j \notin \{i\} \text{ or } l \notin \{i\} \\ 0, & \text{if } j \ne l \end{cases} \tag{4.29}$$

Hence $\mathbf{K}_i \mathbf{f}$ is a vector containing only pixels that are local neighbors of i, that is,

$$[\mathbf{K}_i \mathbf{f}]_j = \begin{cases} f_j, & \text{if } j \in \{i\} \\ 0, & \text{otherwise} \end{cases}$$

The restoration problem is ill-posed as was discussed in Chapter 2. For this reason, extra constraints will be required. Consider a matrix $\mathbf{D} \in \Re^{L \times L}$ designed to act as a high-pass operator. $\mathbf{Df}$ is therefore the high-pass component of the estimated image. To reduce the effects of noise, we can constrain $\|\mathbf{Df}\|^2$ to be equal to a preset constant. If the constant is chosen too low, the image will be blurred, and if the constant is chosen too high, the image will contain unacceptable levels of noise. An additional constraint is to require that the local variance of each pixel in the estimate matches our expectation of what they should be in the original image.

This problem can then be stated as

$$\mathbf{f}^* = \arg \min_f \|\mathbf{g} - \mathbf{Hf}\|^2 \text{ subject to } h(\mathbf{f}^*) = 0 \text{ and } \mathbf{p}(\mathbf{f}^*) \le 0$$

where $h(\mathbf{f}) \in \Re$, $h(\mathbf{f}) = \|\mathbf{Df}\|^2 - \sigma^2$ and $\mathbf{p}(\mathbf{f}) \in \Re^L$, $p_i(\mathbf{f}) = (V^i - \alpha^i)^2 - \epsilon, 1 \le i \le L$, where V^i is the local variance of pixel i and α^i is the estimate of the local variance of pixel i in the original image. Calculation of this estimate is considered in Section 4.6 but in terms of this problem, the α^i are constants. The constant ϵ is a threshold level. We desire the square error between the estimate's local variance and the estimated local variance of the original image to be less than ϵ.

Note that **p** is the set of LVMSE constraints, while h comprises the constraint on the norm of the high-frequency components of each region.

Let us look now at how we can solve this equation. The following definition and theory presented previously in Chapter 2 will be useful.

Definition 4.1

Let $\mathbf{f}^*$ satisfy $h(\mathbf{f}^*) = 0$, $\mathbf{p}(\mathbf{f}^*) \le 0$ and let $J\{\mathbf{f}^*\}$ be the index set of active inequality constraints, that is, $J\{\mathbf{f}^*\} = \{j;\ p_j(\mathbf{f}^*) = 0\}$.

    Then $\mathbf{f}^*$ is a regular point if $\nabla h(\mathbf{f}^*)$, $\nabla p_j(\mathbf{f}^*)$ are linearly independent $\forall j \in J\{\mathbf{f}^*\}$. [94]

Theorem of Karush–Kuhn–Tucker

    Let $E, h, \mathbf{p} \in C^1$. Let $\mathbf{f}^*$ be a regular point and a local minimizer for the problem of minimizing E subject to $h(\mathbf{f}^*) = 0$ and $\mathbf{p}(\mathbf{f}^*) \le 0$.

    Then there exists $\lambda \in \Re$, $\mu \in \Re^L$ such that

1. $\mu \ge 0$
2. $\nabla E(\mathbf{f}^*) + \lambda \nabla h(\mathbf{f}^*) + \sum_{j=1}^{L} \mu_j \nabla p_j(\mathbf{f}^*) = 0$
3. $\mu^T \mathbf{p}(\mathbf{f}^*) = 0$

4.7.2   Computing Local Variance

For the sake of simplicity, we will restrict the local neighborhood of i to be a square A by A region centered on pixel i. In this case, the local variance of pixel i can be computed as

$$V^i = \frac{\beta_i}{A^2} - \frac{\gamma_i^2}{A^4} \tag{4.30}$$

where $\beta_i = \sum_{j=1}^{L}(f_j^i)^2$ and $\gamma_i = \sum_{j=1}^{L} f_j^i$ and $\mathbf{f}^i = \mathbf{K}_i \mathbf{f}$.

    Using matrix notation, β_i can be rewritten

$$\beta_i = (\mathbf{f}^i)^T \mathbf{f}^i = \mathbf{f}^T \mathbf{K}_i^T \mathbf{K}_i \mathbf{f} = \mathbf{f}^T \mathbf{K}_i \mathbf{f} \tag{4.31}$$

since $\mathbf{K}_i = (\mathbf{K}_i)^T = \mathbf{K}_i \mathbf{K}_i$.

    Similarly,

$$\gamma_i = (\mathbf{K}_i \mathbf{f})^T \mathbf{1} = \mathbf{f}^T \mathbf{K}_i \mathbf{1} \tag{4.32}$$

where $\mathbf{1}$ is the L by 1 vector containing only ones. Merging the above results together, we obtain

$$V^i = \frac{\mathbf{f}^T \mathbf{K}_i \mathbf{f}}{A^2} - \frac{\mathbf{f}^T \mathbf{K}_i \mathbf{1} \mathbf{1}^T \mathbf{K}_i \mathbf{f}}{A^4} \tag{4.33}$$

4.7.3 Problem Solution

The theorem of Karush–Kuhn–Tucker can be used to solve this problem. Note the following equations:

$$h(\mathbf{f}^\star) = \|\mathbf{D}\mathbf{f}^\star\|^2 - \sigma^2 = \mathbf{f}^{\star T}\mathbf{D}^T\mathbf{D}\mathbf{f}^\star - \sigma^2 \tag{4.34}$$

$$\nabla h(\mathbf{f}^\star) = 2\mathbf{D}^T\mathbf{D}\mathbf{f}^\star \tag{4.35}$$

$$p_i(\mathbf{f}^\star) = (V^i - \alpha^i)^2 - \epsilon \tag{4.36}$$

$$\nabla p_i(\mathbf{f}^\star) = 2(V^i - \alpha^i)\nabla V^i \tag{4.37}$$

$$\nabla p_i(\mathbf{f}^\star) = 4\left(\frac{(\mathbf{f}^\star)^T\mathbf{K}_i\mathbf{f}^\star}{A^2} - \frac{(\mathbf{f}^\star)^T\mathbf{K}_i\mathbf{11}^T\mathbf{K}_i\mathbf{f}^\star}{A^4} - \alpha^i\right)\left(\frac{\mathbf{K}_i\mathbf{f}^\star}{A^2} - \frac{\mathbf{K}_i\mathbf{11}^T\mathbf{K}_i\mathbf{f}^\star}{A^4}\right) \tag{4.38}$$

$$\nabla p_i(\mathbf{f}^\star) = 2y^i(\mathbf{f}^\star)\mathbf{Y}^i\mathbf{f}^\star \tag{4.39}$$

where

$$y^i(\mathbf{f}^\star) = 2\left(\frac{(\mathbf{f}^\star)^T\mathbf{K}_i\mathbf{f}^\star}{A^2} - \frac{(\mathbf{f}^\star)^T\mathbf{K}_i\mathbf{11}^T\mathbf{K}_i\mathbf{f}^\star}{A^4} - \alpha^i\right) \in \Re \tag{4.40}$$

is a scalar proportional to the difference in local variance and local variance estimate at pixel i, and

$$\mathbf{Y}^i = \left(\frac{\mathbf{K}_i}{A^2} - \frac{\mathbf{K}_i\mathbf{11}^T\mathbf{K}_i}{A^4}\right) \in \Re^{L \times L} \tag{4.41}$$

is an L by L filter that is a constant for each i and in fact, except at boundaries, has the same form for all i in the image. Note that $\nabla h(\mathbf{f}^\star)$, $\nabla p_j(\mathbf{f}^\star) \in \Re^L$ for each $j \in J\{\mathbf{f}^\star\}$.

If we assume $\mathbf{f}^\star$ is a regular point, the solution to the problem can be obtained as follows:

$$E(\mathbf{f}^\star) = \|\mathbf{g} - \mathbf{H}\mathbf{f}^\star\|^2 = (\mathbf{g} - \mathbf{H}\mathbf{f}^\star)^T(\mathbf{g} - \mathbf{H}\mathbf{f}^\star)$$
$$= \mathbf{g}^T\mathbf{g} - 2\mathbf{g}^T\mathbf{H}\mathbf{f}^\star + \mathbf{f}^{\star T}\mathbf{H}^T\mathbf{H}\mathbf{f}^\star \tag{4.42}$$

Note that we have used the fact that $\mathbf{f}^{\star T}\mathbf{H}^T\mathbf{g} = (\mathbf{g}^T\mathbf{H}\mathbf{f}^\star)^T = \mathbf{g}^T\mathbf{H}\mathbf{f}^\star$ since this quantity is a scalar.

$$\nabla E(\mathbf{f}^\star) = 2\mathbf{H}^T\mathbf{H}\mathbf{f}^\star - 2\mathbf{H}^T\mathbf{g} \tag{4.43}$$

By KKT condition 2 we have

$$\nabla E(\mathbf{f}^*) + \lambda \nabla h(\mathbf{f}^*) + \sum_{i=1}^{L} \mu_i \nabla p_i(\mathbf{f}^*) = 0$$

$$2\mathbf{H}^T\mathbf{H}\mathbf{f}^* - 2\mathbf{H}^T\mathbf{g} + 2\lambda \mathbf{D}^T\mathbf{D}\mathbf{f}^* + 2\sum_{i=1}^{L} \mu_i y^i(\mathbf{f}^*)\mathbf{Y}^i \mathbf{f}^* = 0$$

$$2\left(\mathbf{H}^T\mathbf{H} + \lambda \mathbf{D}^T\mathbf{D} + \sum_{i=1}^{L} \mu_i y^i(\mathbf{f}^*)\mathbf{Y}^i\right) \mathbf{f}^* = 2\mathbf{H}^T\mathbf{g}$$

Therefore,

$$\mathbf{f}^* = \left[\mathbf{H}^T\mathbf{H} + \lambda \mathbf{D}^T\mathbf{D} + \sum_{i=1}^{L} \mu_i y^i(\mathbf{f}^*)\mathbf{Y}^i\right]^{-1} (\mathbf{H}^T\mathbf{g}) \qquad (4.44)$$

Note the form of Equation (4.44). It implies a Hopfield neural network with bias inputs and weights given by

$$\mathbf{b} = \mathbf{H}^T\mathbf{g} \qquad (4.45)$$

$$\mathbf{W} = -\mathbf{H}^T\mathbf{H} - \lambda \mathbf{D}^T\mathbf{D} - \sum_{i=1}^{L} \mu_i y^i(\mathbf{f}^*)\mathbf{Y}^i \qquad (4.46)$$

The bias inputs as given by Equation (4.45) are identical to the old Hopfield network formulation. The weight matrix [Equation (4.46)] is the old matrix with the addition of a new set of weights. The new set of weights consists of a variance modifying filter that comes into effect when the difference between the local variance in the estimated image and the local variance expected in the original exceeds a threshold. The degree to which these new weights modify the old set of weights is dependent on the difference in variance.

4.7.4   Conditions for KKT Theory to Hold

The last subsection supplied the solution to this problem using the Karush–Kuhn–Tucker theory. This assumes that the preconditions of the theory are satisfied. To be specific, for the discussion in the last section to hold, $\mathbf{f}^*$ must be a regular point. In this section, we will examine the conditions that need to be imposed on the problem for this to be the case. But first, we require some definitions:

Theorem 4.1
$\mathbf{f}^*$ is a regular point if it satisfies the following three conditions:

1. *$\mathbf{f}^* \geq 0$.*
2. *Every pixel, j, for which $p_j(\mathbf{f}^*) = 0$ has at least one nonzero local neighbor (in terms of a square A by A region centered on j) that is also not a local*

neighbor of another pixel k for which $p_k(\mathbf{f}^*) = 0$ and is not equal to the sample mean of all local neighbors of j.

3. There exists at least one pixel, r, for which $[\nabla h(\mathbf{f}^*)]_r \neq 0$ that is also not a local neighbor of another pixel k for which $p_k(\mathbf{f}^*) = 0$.

Proof   Theorem 4.1 will be proven if we show that a problem satisfying the conditions of the theorem results in $\nabla h(\mathbf{f}^*), \nabla p_j(\mathbf{f}^*), j \in J(\mathbf{f}^*)$ being a set of linearly independent vectors.

The theorem can be proven in three parts:

1. Prove that the set $\nabla p_j(\mathbf{f}^*), j \in J(\mathbf{f}^*)$ is a set of linearly independent vectors.

2. Prove that each $\nabla p_j(\mathbf{f}^*)$ is linearly independent of $\nabla h(\mathbf{f}^*)$. ■

Part (1)

Consider a solution, $\mathbf{f}^*$, that satisfies the conditions of Theorem 4.1. In addition, consider a pixel, i, that satisfies $p_i(\mathbf{f}^*) = 0$. Now consider a second pixel, r that is a nonzero local neighbor of i (i.e., $r \in \{i\}$). Based on Equation (4.39) we have

$$\nabla [p_i(\mathbf{f}^*)]_r = 2y^i(\mathbf{f}^*)[\mathbf{Y}^i \mathbf{f}^*]_r \tag{4.47}$$

$$= \pm 4\sqrt{\epsilon}[\mathbf{Y}^i \mathbf{f}^*]_r \tag{4.48}$$

$$= \pm 4\sqrt{\epsilon} \sum_{p=1}^{L} [\mathbf{Y}^i]_{rp}[\mathbf{f}^*]_p \tag{4.49}$$

Careful examination of $\mathbf{Y}^i$ shows us that

$$[\mathbf{Y}^i]_{rp} = \begin{cases} \neq 0, & \text{if } r \in \{i\} \text{ and } p \in \{i\} \\ 0, & \text{if } r \notin \{i\} \text{ or } p \notin \{i\} \end{cases} \tag{4.50}$$

Using Equation (4.50) we can see that $\nabla[p_j(\mathbf{f}^*)]_r = 0$ if $r \notin \{j\}$. Since the matrix $\mathbf{Y}^i$ contains both positive and negative elements, then $\nabla[p_i(\mathbf{f}^*)]_r$ is not guaranteed to be nonzero. However, if the pixel r further satisfies the conditions of Theorem 4.1 and is not equal to the sample mean of the local neighborhood of i, then $[\nabla p_i(\mathbf{f}^*)]_r \neq 0$.

Since $[\nabla p_i(\mathbf{f}^*)]_r \neq 0$ and $[\nabla p_j(\mathbf{f}^*)]_r = 0, \forall j; j \neq i, j \in J\{\mathbf{f}^*\}$, we conclude that $\nabla p_i(\mathbf{f}^*)$ is linearly independent of the other vectors in the set, since it will always have at least one nonzero component for which the other vectors in the set will have zero. If $\mathbf{f}^*$ satisfies the conditions of the theorem, the set $\nabla p_i(\mathbf{f}^*), i \in J\{\mathbf{f}^*\}$ will be a set of linearly independent vectors.

Part (2)

Consider a solution, $\mathbf{f}^*$, that satisfies the conditions of Theorem 4.1. According to condition 3 of Theorem 4.1, there exists at least one pixel, r, for which

$[\nabla h(\mathbf{f}^*)]_r \neq 0$ that is also not a local neighbor of another pixel k for which $p_k(\mathbf{f}^*) = 0$.

As mentioned in Part (1) of the proof, $\nabla[p_k(\mathbf{f}^*)]_r = 0$ if $r \notin \{k\}$. Remember that the set $\{k\}$ is the set of local neighbors of pixel k. Hence we can deduce that $\nabla[p_k(\mathbf{f}^*)]_r = 0$, $\forall k$ such that $p_k(\mathbf{f}^*) = 0$. Since $[\nabla h(\mathbf{f}^*)]_r \neq 0$, then $\nabla h(\mathbf{f}^*)$ cannot be described by any linear combination of the set $\nabla p_i(\mathbf{f}^*), i \in J\{\mathbf{f}^*\}$. Hence we have completed Part (2) of the proof.

If the solution, $\mathbf{f}^*$ satisfies the third condition of Theorem 4.1, then the set $\nabla p_j(\mathbf{f}^*)$, $j \in J(\mathbf{f}^*)$ is linearly independent to $\nabla h(\mathbf{f}^*)$.

Hence the proof is completed. We have shown that for solutions satisfying Theorem 4.1, the set $\{\nabla h(\mathbf{f}^*), \nabla p_j(\mathbf{f}^*)\}$, $j \in J(\mathbf{f}^*)$ is a set of linearly independent vectors. Note that Theorem 4.1 provides sufficient but not necessary conditions for $\mathbf{f}^*$ to be a regular point.

4.7.5 Implementation Considerations for the Lagrangian Approach

In the previous section, it was shown that expanding the restoration problem in terms of KKT theory resulted in bias inputs and a weight matrix that are very similar to the basic constrained least squares penalty function in Equation (2.3). In fact, the bias weights and the weight matrix are the same but with an added term that acts as a variance-modifying filter. We have shown above that LVMSE restoration significantly improves the restoration of an image. The improvement is in terms of the human perception of the image rather than the SNR value. The KKT approach is more compatible with the neural network approaches introduced in Chapter 2 in the sense that unlike the LVMSE approaches presented above, the KKT approach integrates the constraints on local variance directly into the weight matrix for the neural network [Equation (4.46)], rather than as a separate term from the weights as in Algorithms 4.1 and 4.2. However, the previously presented approaches and the KKT approach would be expected to give different solutions to this problem. To examine the differences in implementing the KKT approach better, let us examine the new term in the weight matrix [Equation (4.46)]. The purpose of this new term is to match the variance of our image with that of our estimate of the original. The new term has an effect when the difference in variance between the image and the estimate of the original exceeds a threshold, ϵ [Equation (4.35)]. The threshold value in Equation (4.35) determines whether the new term is included in the weight matrix. The degree to which the new term modifies the old set of weights depends on the difference in variance [Equation (4.40)]. To simplify the computations we chose a single value for all of the μ_j, this value acts as a step size for matching the variances of the two images. Therefore, attention to choosing the value of μ is needed. If the value of μ is chosen to be too high, too much emphasis on variance equalization will produce noisy results (by dominating the image fidelity terms in the cost function). On the other hand, if μ is chosen to be too low a value, then the variance modifying filter will have little effect on the solution. Suitable values for μ may be found by trial and error.

FIGURE 4.8
Images restored with Algorithm 4.1 and the LVMSE-KKT approach.

4.7.6 Numerical Experiment

In Reference [119], an experiment conducted to compare the results of the KKT algorithm against the previously presented LVMSE-based algorithms. A grayscale image was degraded by a 5×5 Gaussian PSF of standard deviation 2.0. Additive noise of variance 87.62 was also added. Figures 4.8a and 4.8b show the original and degraded images, respectively. The image was restored using a constraint value of $\lambda = 0.003$ and $\theta = 0.0005$ (for Algorithm 4.1), the image local variance estimate was computed as

$$\sigma_A^2(\mathbf{g}(x, y))^\star = 2\big(\sigma_A^2(\mathbf{g}(x, y)) - 200\big)$$

The same value of λ was used for both restoration algorithms in order to compare the results of the new KKT-based restoration algorithm with that of Algorithm 4.1. The value μ in the KKT algorithm was chosen to be 0.001. For each of the restoration results, the LSMSE was computed as per Equation (3.14). The restored image using Algorithm 4.1 had an LSMSE of 21.05 (Figure 4.8c). The restored image using the KKT-based algorithm had an LSMSE of 20.33 (Figure 4.8d). When we compare Figures 4.8c and 4.8d, the results are very similar. Visually the only differences are that the KKT-based restoration method had slightly sharper edges. This suggests that the two different ways of viewing this problem are for the most part equivalent.

4.8 Summary

In Chapter 1, a novel error measure was introduced which compared two images by consideration of their regional statistical differences rather than their pixel-level differences. It was found that this error measure more closely corresponds to human visual perception of image quality. Based on the new error measure, two cost functions were presented in this chapter. The first cost function was based closely on the LVMSE error measure introduced in Chapter 1. This cost function was analyzed and shown to be well behaved. The analysis of the first modified cost function suggested that improvements could be made by incorporating a logarithmic version of the LVMSE into the standard cost function. A second cost function was hence introduced and shown to be well behaved. Algorithms to optimize these cost functions were designed based on adaptation of the neural network approach to optimize constrained least square error.

The introduced algorithms were shown to suppress noise strongly in low variance regions while still preserving edges and highly textured regions of an image. The algorithms were shown to perform well when applied to both grayscale and color images. The results of the proposed algorithm were overwhelmingly favored by subjects of a subjective survey conducted by the authors. It was also shown that the proposed iterative algorithms are very robust.

Finally, the LVMSE restoration problem was considered in terms of the theorem of Karush–Kuhn–Tucker. The results of this consideration motivated a new algorithm and it was shown that the LVMSE-KKT-based algorithm was quite similar to the basic constraint least squares (CLS) algorithm with the exception of an added variance modifying term in the weight matrix. The LVMSE-KKT-based restoration algorithm was shown to have comparable results to that of the LVMSE-based restoration.

Acknowledgments

The authors would like to thank the participants in the subjective survey that gave valuable insights to the mechanism of the LVMSE algorithms.

5

Model-Based Adaptive Image Restoration

5.1 Model-Based Neural Network

In this chapter, we introduce the technique of model-based neural networks, which is to be applied to our problem of adaptive regularization in image restoration. Instead of adopting the general neural network architecture introduced in Chapter 1 for our adaptive image-processing applications, we propose the use of a *modular model-based* neural network for our purpose. We used the term "model-based" in the sense of Caelli et al. [73], where expert knowledge in the problem domain is explicitly incorporated into a neural network by restricting the domain of the network weights to a suitable subspace, in which the solution of the problem resides. In this way, a single weight vector can be uniquely specified by a small number of parameters and the "curse of dimensionality" problem is partially alleviated.

To appreciate this formulation more readily, we review some fundamental concepts of artificial neuron computation, where each such neuron is the elementary unit of computation in a neural network [34,35]. In general, the sth neuron in the network implements a mapping $f_s : \mathbf{R}^N \longrightarrow \mathbf{R}$ which is given by

$$
\begin{aligned}
y_s &= f_s(\mathbf{x}) \\
&= g(\mathbf{p}_s^T \mathbf{x}) \\
&= g\left(\sum_{n=1}^{N} p_{qs} x_n \right)
\end{aligned}
\tag{5.1}
$$

where

$$
\mathbf{x} = [x_1, \dots, x_N]^T \in \mathbf{R}^N \quad \text{and} \quad \mathbf{p}_s = [p_{s1}, \dots, p_{sN}]^T \in \mathbf{R}^N
$$

are the input vector and the weight vector for the neuron, respectively. g is usually a nonlinear sigmoid function that limits the output dynamic range of the neuron. We will extend this concept and define model-based neuron in the next section.

5.1.1   Weight-Parameterized Model-Based Neuron

The main assumption in this weight-parameterized model-based formulation [73] is that for a specific domain of knowledge, the corresponding weight domain is restricted to a low-dimensional submanifold of $\mathbf{R}^N$.

Denoting this weight domain by $\mathcal{W}_p$, the formulation thus assumes the existence of a mapping $\mathcal{M} : \mathbf{R}^M \longrightarrow \mathcal{W}_p \subset \mathbf{R}^N$ such that

$$\mathbf{p} = \mathcal{M}(\mathbf{z}) \tag{5.2}$$

where

$$\mathbf{z} = [z_1, \dots, z_M]^T \in \mathbf{R}^M$$
$$\mathbf{p} = [p_1, \dots, p_N]^T \in \mathbf{R}^N$$
$$= [\mathcal{M}_1(\mathbf{z}), \dots, \mathcal{M}_N(\mathbf{z})]^T$$

with $M < N$. The mappings $\mathcal{M}_n : \mathbf{R}^M \longrightarrow \mathbf{R}, n = 1, \dots, N$ are the component functions of $\mathcal{M}$. The structure of a typical weight-parameterized model-based neuron is shown in Figure 5.1.

Assuming that each component function is differentiable with respect to $\mathbf{z}$, the steepest descent update rule for the components of $\mathbf{z}$ is as follows:

$$
\begin{aligned}
z_m(t+1) &= z_m(t) - \eta \frac{\partial \mathcal{E}_t}{\partial z_m} \\
&= z_m(t) - \eta \sum_{n=1}^{N} \frac{\partial \mathcal{E}_t}{\partial p_n} \frac{\partial p_n}{\partial z_m}
\end{aligned}
\tag{5.3}
$$

where $\mathcal{E}_t$ is an instantaneous error measure between the network output and the desired output.

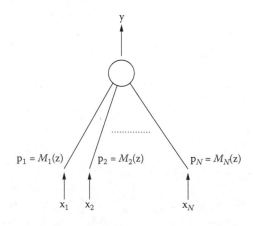

FIGURE 5.1
The weight-parameterized model-based neuron.

As a result, if we possess prior knowledge of our problem domain in the form of the mapping $\mathcal{M}$ and if $M \ll N$, the optimization can proceed within a subspace of greatly reduced dimensionality and the problem of "curse of dimensionality" is partially alleviated. In the next section, we will present an alternative formulation of the adaptive regularization problem in terms of this model-based neuron.

5.2   Hierarchical Neural Network Architecture

While the previous model-based formulation describes the operations of individual neurons, the term "modular neural network" [47,120–125] refers to the overall architecture, where specific computations are localized within certain structures known as *subnetworks*. This class of networks serves as natural representations of training data arising from several distinct classes, where each such class is represented by a single subnetwork. Specific implementations of modular neural networks include the various hierarchical network architecture described in [48,126–129].

We are especially interested in the particular hierarchical implementation of the modular network structure by Kung and Taur [48]. Their proposed architecture associates a *subnetwork output* $\phi(\mathbf{x}, \mathbf{w}_r)$ with the rth subnetwork. In addition, we define lower-level *neurons* within each subnetwork with their corresponding *local neuron output* $\psi_r(\mathbf{x}, \mathbf{w}_{s_r})$, $s_r = 1, \ldots, S_r$. The subnetwork output is defined as the linear combination of the local neuron outputs as follows:

$$\phi(\mathbf{x}, \mathbf{w}_r) = \sum_{s_r=1}^{S_r} c_{s_r} \psi_r(\mathbf{x}, \mathbf{w}_{s_r}) \tag{5.4}$$

where c_{s_r} are the combination coefficients. The hierarchical architecture and the structure of its subnetwork is shown in Figure 5.2.

5.3   Model-Based Neural Network with Hierarchical Architecture

Given the previous description of the model-based neuron, which specifies the computational operations at the neuronal level, and the overall hierarchical structure, which specifies the macroscopic network architecture, it is natural to combine the two in a single framework. More specifically, we can incorporate model-based neuron computation at the lowest hierarchy, that is, at the level of a single neuron within a subnetwork, of the hierarchical neural network.

Formally, if we assume $\mathbf{w}_{s_r} \in \mathbf{R}^N$, the local neuron output $\psi_r(\mathbf{x}, \mathbf{w}_{s_r})$ can be specified as follows:

$$\psi_r(\mathbf{x}, \mathbf{w}_{s_r}) \equiv \psi_r(\mathbf{x}, \mathcal{M}(\mathbf{z}_{s_r})) \tag{5.5}$$

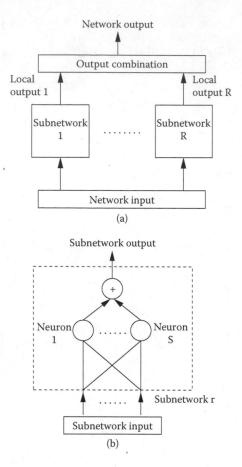

FIGURE 5.2
Hidden-node hierarchical network architecture: (a) global network architecture; (b) subnetwork architecture.

where $\mathbf{z}_{s_r} \in \mathbf{R}^M$, with $M < N$, is the model-based weight vector, and $\mathcal{M} : \mathbf{R}^M \longrightarrow \mathbf{R}^N$ is the mapping relating the model-based weight vector to the neuronal weight vector. In this case, the lower-dimensional model-based vector $\mathbf{z}_{s_r}$ is embedded in the higher-dimensional weight vector $\mathbf{w}_{s_r}$, and the weight optimization is carried out in the lower-dimensional weight space $\mathbf{R}^M$.

5.4   HMBNN for Adaptive Image Processing

Our current formulation of the adaptive regularization problem requires the partition of an image into disjoint regions, and the assignment of the optimal regularization parameter value λ to the respective regions. It is natural, therefore, to assign a single subnetwork to each such region, and regard the regularization parameter λ as a model-based weight to be optimized using a special set of training examples. More specifically, for the rth subnetwork, it will be shown in the next section that the image restoration process is characterized by the evaluation of the local neuron output as a linear operation as follows:

$$\psi_r(\mathbf{x}, \mathbf{p}_{s_r}) = \mathbf{p}_{s_r}^T \mathbf{x} \tag{5.6}$$

where $\mathbf{x} \in \mathbf{R}^N$ denotes a vector of image gray-level values in a local neighborhood and $\mathbf{p}_{s_r}$ is the *image-restoration convolution mask* derived from the point spread function (PSF) of the degradation mechanism. The dimension N of the vectors $\mathbf{x}$ and $\mathbf{p}_{s_r}$ depends on the size of the PSF. It will be shown that the convolution mask coefficient vector $\mathbf{p}_{s_r}$ can be expressed as a function of the regularization parameter λ. In other words, there exists a mapping $\mathcal{M} : \mathbf{R} \longrightarrow \mathbf{R}^N$ such that

$$\mathbf{p}_{s_r} = \mathcal{M}(\lambda) \tag{5.7}$$

which is equivalent to the embedding of the scalar parameter λ in the high-dimensional space $\mathbf{R}^N$, and thus corresponds to a weight-parameterized model-based neuron.

5.5 Hopfield Neural Network Model for Image Restoration

We recall that, in regularized image restoration, the cost function consists of a data-conformance evaluation term and a model-conformance evaluation term [24,26]. The model-conformance term is usually specified as a continuity constraint on neighboring gray-level values in image-processing applications. The contribution of these two terms is adjusted by the so-called regularization parameter or the Lagrange multiplier that allows the two terms to combine additively:

$$E = \frac{1}{2}\|\mathbf{y} - \mathbf{H}\hat{\mathbf{f}}\|^2 + \frac{1}{2}\lambda\|\mathbf{D}\hat{\mathbf{f}}\|^2 \tag{5.8}$$

where the vector $\mathbf{y}$, with y_i, $i = 1, \ldots, N_I$ as components and N_I as the number of image pixels, denotes the blurred image with its pixel values lexicographically ordered. The vectors $\hat{\mathbf{f}}$ with components $\hat{f}_i$ is the corresponding restored image and the matrix $\mathbf{H}$, with components h_{ij}, $i, j = 1, \ldots, N_I$, is the blur function. The matrix $\mathbf{D}$ with components d_{ij} is a linear operator on $\hat{\mathbf{f}}$ and λ is the regularization parameter. For image-restoration purpose, $\mathbf{D}$ is usually a differential operator, and the minimization of the above cost function effectively limits the local variations of the restored image.

In the current work, the optimization of the primary image-restoration cost function E was performed within the framework of neural network optimization, where a modified neural network architecture based on the Hopfield neural network model in Chapter 2 was employed. This *primary* Hopfield image-restoration neural network, which is mainly responsible for optimizing the value of E, is to be carefully distinguished from the HMBNN described previously, which serves as a *secondary* neural network to optimize the regularization parameter λ in E.

It was shown in Chapter 2 that the neuron output can be updated using the following equation:

$$\Delta \hat{f}_i = - \frac{u_i(t)}{w_{ii}} \tag{5.9}$$

and the final output of the neuron is given by

$$\hat{f}_i(t+1) = \hat{f}_i(t) + \Delta \hat{f}_i(t) \tag{5.10}$$

We can see that Equation (5.9) expresses the change in neuron output in terms of w_{ii}, which is in turn a function of the regularization parameter λ. This dependency is essential for the reformulation of the current adaptive regularization problem into a learning problem for a set of model-based neurons.

5.6 Adaptive Regularization: An Alternative Formulation

In this section, we propose an alternative formulation of adaptive regularization that centers on the concept of *regularization parameters as model-based neuronal weights*. By adopting this alternative viewpoint, a neat correspondence is found to exist between the mechanism of adaptive regularization and the computation of an artificial model-based neuron.

Referring to Equation (5.9), we can express the instantaneous restoration stepsize $\Delta \hat{f}_i(t)$ in Equation (5.9) in the form of a neural computational process as follows:

$$\begin{aligned}
\Delta \hat{f}_i &= - \frac{u_i(t)}{w_{ii}} \\
&= - \frac{\sum_{j=1}^{N_l} w_{ij} \hat{f}_j + b_i}{w_{ii}} \\
&= \sum_{j=1}^{N_l} p_{ij} \hat{f}_j + q_i b_i \\
&= \mathbf{p}_i^T \hat{\mathbf{f}}_i
\end{aligned} \tag{5.11}$$

where

$$\mathbf{p}_i = [p_{i1}, \ldots, p_{iN_I}, q_i]^T \in \mathbf{R}^{N_I+1}$$
$$\hat{\mathbf{f}}_i = [\hat{f}_1, \ldots, \hat{f}_{N_I}, b_i]^T \in \mathbf{R}^{N_I+1}$$

and the weights of this hypothetical neuron are defined as

$$p_{ij} = -\frac{w_{ij}}{w_{ii}} \tag{5.12}$$

$$q_i = -\frac{1}{w_{ii}} \tag{5.13}$$

The weights w_{ij} of the primary Hopfield neuron (as opposed to those of this hypothetical neuron) are in turn given by the following equation (see Chapter 2):

$$w_{ij} = -\sum_{p=1}^{N_I} h_{pi} h_{pj} - \lambda \sum_{p=1}^{N_I} d_{pi} d_{pj}$$
$$= g_{ij} + \lambda l_{ij} \tag{5.14}$$

where we define

$$g_{ij} = -\sum_{p=1}^{N_I} h_{pi} h_{pj} \tag{5.15}$$

$$l_{ij} = \sum_{p=1}^{N_I} d_{pi} d_{pj} \tag{5.16}$$

From these equations, the weights p_{ij} of the hypothetical neuron can be expressed as a function of the regularization parameter λ as follows

$$p_{ij} = -\frac{g_{ij} + \lambda l_{ij}}{g_{ii} + \lambda l_{ii}} \tag{5.17}$$

$$q_i = -\frac{1}{g_{ii} + \lambda l_{ii}} \tag{5.18}$$

To reinterpret these equations as the computation of a model-based neuron, we recast them into its two-dimensional form from the previous lexicographical ordered form. Defining the *lexicographical mapping* $\mathcal{L}$ as follows:

$$i = \mathcal{L}(i_1, i_2) = i_1 N_x + i_2 \tag{5.19}$$

where (i_1, i_2) is the corresponding 2D position of the ith lexicographical vector entry in the image lattice, and N_x is the width of the image lattice. With this mapping, we can reinterpret Equation (5.11) in a 2D setting

$$\Delta \tilde{f}_{i_1, i_2} = \sum_{k=-L_c}^{L_c} \sum_{l=-L_c}^{L_c} \tilde{p}_{i_1, i_2, k, l} \tilde{f}_{i_1+k, i_2+l} + \tilde{q}_{i_1, i_2} \tilde{b}_{i_1, i_2} \tag{5.20}$$

where the tilded form of the variables indicate that the current quantity is indexed by its 2D position in the lattice, rather than its 1D position in the lexicographically ordered vector. The summation in the equation is taken over the support of a *2D neuronal weight mask*, the size of which depends on the extent of the original point spread function [46] and which is $(2L_c + 1)^2$ in this case. For $i = \mathcal{L}(i_1, i_2)$, we define

$$\widetilde{f}_{i_1,i_2} = \hat{f}_i \tag{5.21}$$

$$\widetilde{p}_{i_1,i_2,k,l} = p_{i,\mathcal{L}(i_1+k,i_2+l)} \tag{5.22}$$

and the variables $\triangle\widetilde{f}_{i_1,i_2}$, $\widetilde{q}_{i_1,i_2}$ and $\widetilde{b}_{i_1,i_2}$ are similarly defined as $\widetilde{f}_{i_1,i_2}$. For spatially invariant degradation, the variables $\widetilde{p}_{i_1,i_2,k,l}$ and $\widetilde{q}_{i_1,i_2}$ are independent of the position (i_1, i_2) in the image lattice, and the above equation can be re-written as

$$\triangle\widetilde{f}_{i_1,i_2} = \sum_{k=-L_c}^{L_c}\sum_{l=-L_c}^{L_c} \widetilde{p}_{k,l}\widetilde{f}_{i_1+k,i_2+l} + \widetilde{q}\widetilde{b}_{i_1,i_2}$$
$$= \widetilde{\mathbf{p}}^T\,\widetilde{\mathbf{f}}_{i_1,i_2} \tag{5.23}$$

where

$$\widetilde{\mathbf{p}} = [\widetilde{p}_{-L_c,-L_c}, \dots, \widetilde{p}_{0,0}, \dots, \widetilde{p}_{L_c,L_c}, \widetilde{q}]^T \in \mathbf{R}^{N_c+1}$$

$$\widetilde{\mathbf{f}}_{i_1,i_2} = [\widetilde{f}_{i_1-L_c,i_2-L_c}, \dots, \widetilde{f}_{i_1,i_2}, \dots, \widetilde{f}_{i_1+L_c,i_2+L_c}, \widetilde{b}_{i_1,i_2}]^T \in \mathbf{R}^{N_c+1}$$

and $N_c = (2L_c + 1)^2$.

5.6.1 Correspondence with the General HMBNN Architecture

For spatially invariant degradation with small support, we have the condition $N_c \ll N_I$, where N_I is the number of pixels in the image, and we can view Equation (5.23) as a local convolution operation over a selected neighborhood of $\widetilde{f}_{i_1,i_2}$. On the other hand, due to the invariant nature of $\widetilde{p}_{k,l}$ and $\widetilde{q}$ for all i_1 and i_2, this operation can alternatively be viewed as the computational process of a model-based neuron with input vector $\widetilde{\mathbf{f}}_{i_1,i_2}$ and weight vector $\widetilde{\mathbf{p}}$. In other words, if we assume that the image is subdivided into regions $\mathcal{R}_r, r = 1, \dots, R$, and we assign a subnetwork with a single model-based neuron to each region, that is, $S_r = 1$ for each r according to the notation adopted previously, then the local model-based neuron output corresponding to $\mathcal{R}_r$ can be represented as

$$\psi_r(\widetilde{\mathbf{f}}_{i_1,i_2}, \widetilde{\mathbf{p}}_{k,l}(\lambda_r)) = \widetilde{\mathbf{p}}_{k,l}^T(\lambda_r)\widetilde{\mathbf{f}}_{i_1,i_2} \tag{5.24}$$

where λ_r, the regional regularization parameter, can be considered the *scalar* model-based weight associated with each region $\mathcal{R}_r$. As a result it can be classified into the class of hidden-node hierarchical networks. The architecture of this HMBNN for adaptive regularization is shown in Figure 5.3.

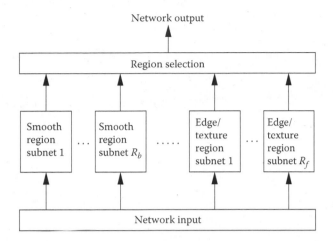

FIGURE 5.3
The model-based neural network with hierarchical architecture for adaptive regularization.

Adopting the weight-parameterized model-based neuron structure due to the natural way in which $\lambda_r \in \mathbf{R}$ is embedded in the high-dimensional weight vector $\tilde{\mathbf{p}}_{k,l}(\lambda_r) \in \mathbf{R}^{N_c+1}$, and referring to Equation (5.17), we can obtain the component mappings $\mathcal{M}_n$ of the model-based neuron for the rth subnetwork as follows:

$$\tilde{p}_{k,l}(\lambda_r) = \mathcal{M}_n(\lambda_r) = -\frac{\tilde{g}_{k,l} + \lambda_r \tilde{l}_{k,l}}{\tilde{g}_{0,0} + \lambda_r \tilde{l}_{0,0}} \quad n = 1, \ldots, N_c \tag{5.25}$$

$$\tilde{q}(\lambda_r) = \mathcal{M}_{N_c+1}(\lambda_r) = -\frac{1}{\tilde{g}_{0,0} + \lambda_r \tilde{l}_{0,0}} \tag{5.26}$$

where the mapping index n corresponds to some specific arrangement of the indices (k, l) in the 2D weight mask $\tilde{p}_{k,l}$, and the values of $\tilde{g}_{k,l}$ and $\tilde{l}_{k,l}$ can be obtained from their lexicographical counterparts g_{ij} and l_{ij} using Equation (5.22). The resulting concatenation $\mathcal{M}$ of the components is thus a mapping from $\mathbf{R}$ to $\mathbf{R}^{N_c+1}$. Corresponding to the notation of Section 5.1, we have $M = 1$ and $N = N_c + 1$ for the current adaptive regularization problem. In addition, each mapping is in the form of a *first-order rational polynomial* of λ_r, the differentiability of which ensures that the weights $\tilde{p}_{k,l}$ are trainable. The structure of the model-based neuron adapted for restoration is shown in Figure 5.4.

To complete the formulation, we define a training set for the current model-based neuron. Assuming that for a certain region $\mathcal{R}_r$ in the image lattice, there exists a desired restoration behavior $\triangle \tilde{f}^d_{i_1,i_2}$ for each pixel $(i_1, i_2) \in \mathcal{R}_r$. We can then define the training set $\mathcal{V}_r$ as follows:

$$\mathcal{V}_r = \{(\tilde{\mathbf{f}}_{i_1,i_2}, \triangle \tilde{f}^d_{i_1,i_2}) : (i_1, i_2) \in \mathcal{R}_r\} \tag{5.27}$$

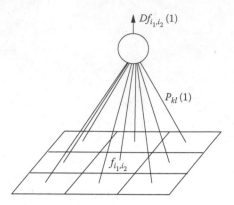

FIGURE 5.4
The model-based neuron for adaptive regularization.

where the input vector $\tilde{\mathbf{f}}_{i_1,i_2}$ contains the gray-level values of pixels in the N_c-neighborhood of (i_1, i_2). Defining the instantaneous cost function $\mathcal{E}_t$ of the neuron as follows:

$$\mathcal{E}_t = \frac{1}{2}\left(\Delta \tilde{f}^d_{i_1,i_2} - \Delta \tilde{f}_{i_1,i_2}\right)^2 \tag{5.28}$$

we can then apply steepest descent to modify the value of the regularization parameter λ_r.

$$\lambda_r(t+1) = \lambda_r(t) - \eta \frac{\partial \mathcal{E}_t}{\partial \lambda_r} \tag{5.29}$$

where

$$\frac{\partial \mathcal{E}_t}{\partial \lambda_r} = -\left(\Delta \tilde{f}^d_{i_1,i_2} - \Delta \tilde{f}_{i_1,i_2}\right) \frac{\partial(\Delta \tilde{f}_{i_1,i_2})}{\partial \lambda_r} \tag{5.30}$$

To evaluate the partial derivative in Equation (5.30), we first define the following two quantities:

$$\tilde{\alpha}_{i_1,i_2} = \sum_{k=-L_c}^{L_c} \sum_{l=-L_c}^{L_c} \tilde{g}_{k,l} \tilde{f}_{i_1+k,i_2+l} \tag{5.31}$$

$$\tilde{\beta}_{i_1,i_2} = \sum_{k=-L_c}^{L_c} \sum_{l=-L_c}^{L_c} \tilde{l}_{k,l} \tilde{f}_{i_1+k,i_2+l} \tag{5.32}$$

From Equation (5.23), the change in neuron output, $\triangle \tilde{f}_{i_1,i_2}$, is then expressed in terms of these quantities as follows:

$$
\begin{aligned}
\triangle \tilde{f}_{i_1,i_2} &= \sum_{k=-L_c}^{L_c} \sum_{l=-L_c}^{L_c} \tilde{p}_{k,l} \tilde{f}_{i_1+k,i_2+l} + \tilde{q}\tilde{b}_{i_1,i_2} \\
&= -\frac{\sum_{k=-L_c}^{L_c} \sum_{l=-L_c}^{L_c} (\tilde{g}_{k,l} + \lambda_r \tilde{l}_{k,l}) \tilde{f}_{i_1+k,i_2+l} + \tilde{b}_{i_1,i_2}}{\tilde{g}_{0,0} + \lambda_r \tilde{l}_{0,0}} \\
&= -\frac{\tilde{\alpha}_{i_1,i_2} + \lambda_r \tilde{\beta}_{i_1,i_2} + \tilde{b}_{i_1,i_2}}{\tilde{g}_{0,0} + \lambda_r \tilde{l}_{0,0}}
\end{aligned}
\tag{5.33}
$$

From this equation, we can evaluate the derivative in Equation (5.30):

$$
\begin{aligned}
\frac{\partial(\triangle \tilde{f}_{i_1,i_2})}{\partial \lambda_r} &= \frac{\partial}{\partial \lambda_r} \left(-\frac{\tilde{\alpha}_{i_1,i_2} + \lambda_r \tilde{\beta}_{i_1,i_2} + \tilde{b}_{i_1,i_2}}{\tilde{g}_{0,0} + \lambda_r \tilde{l}_{0,0}} \right) \\
&= \frac{(\tilde{\alpha}_{i_1,i_2} + \tilde{b}_{i_1,i_2})\tilde{l}_{0,0} - \tilde{\beta}_{i_1,i_2}\tilde{g}_{0,0}}{(\tilde{g}_{0,0} + \lambda_r \tilde{l}_{0,0})^2}
\end{aligned}
\tag{5.34}
$$

We can see from Equation (5.34) that the evaluation of the derivatives depend on variables that have already been precomputed for the purpose of the primary restoration. For example, the weights $\tilde{g}_{0,0}$, $\tilde{l}_{0,0}$ and the bias $\tilde{b}_{i_1,i_2}$ are precomputed at the beginning of the restoration process, and the quantities $\tilde{\alpha}_{i_1,i_2}$ and $\tilde{\beta}_{i_1,i_2}$ are already evaluated for the purpose of determining the instantaneous pixel value change $\triangle \tilde{f}_{i_1,i_2}$ in Equation (5.23). As a result, the adaptive regularization does not involve excessive computational overhead.

The size of the training set V_r depends on the extent of the region $\mathcal{R}_r$ in Equation (5.27). It is possible to define $\mathcal{R}_r$ to include the entire image lattice, which would amount to a fixed regularization scheme where we will search for the optimum global regularization parameter value using a single model-based subnetwork, but for the problem of adaptive regularization, the region $\mathcal{R}_r$ is usually restricted to a subset of the image lattice, and several such regions are defined for the entire lattice to form an image partition. We associate each such region with a subnetwork, the totality of which forms a model-based neural network with hierarchical architecture. In general, we would expect the emergence of regional λ_r values that results in the improved visual quality for the associated region through the training process. This in turn depends critically on the definition of our desired output $\triangle \tilde{f}_{i_1,i_2}^d$ in the regional training set and our image partition. These two issues will be addressed in Sections 5.7 and 5.8.

5.7   Regional Training Set Definition

To complete the definition of the regional training set $\mathcal{V}_r$, we should supply the desired output $\triangle \tilde{f}^d_{i_1,i_2}$ for each input vector $\tilde{\mathbf{f}}_{i_1,i_2}$ in the set. The exact value of $\triangle \tilde{f}^d_{i_1,i_2}$ which would lead to an optimal visual quality for the particular region concerned is normally unknown due to the usually unsupervised nature of image-restoration problems. Nevertheless, an appropriate approximation of this value can usually be obtained by employing a *neighborhood-based prediction scheme* to estimate $\triangle \tilde{f}^d_{i_1,i_2}$ for the current pixel. This is in the same spirit as the neighborhood-based estimation technique widely used in nonlinear filtering applications where a nonlinear function defined on a specified neighborhood of the current pixel is used to recover the correct gray-level value from its noise-corrupted value [86]. The nonlinear filters are usually designed with the purpose of noise suppression in mind. The resulting operations thus have a tendency to oversmooth the edge and textured regions of the image. Remedies to this oversmoothing problem include various edge-adaptive filtering schemes where the prediction is performed along the current edge orientation to avoid filtering across the edges [19,86]. In this work, we would similarly adopt two different prediction schemes for the smooth and textured regions and use the resulting estimated values as the desired outputs for the respective regional training sets.

For combined edge/textured regions, we shall adopt a prediction scheme that emphasizes the dynamic range of the associated region, or equivalently, a scheme that biases toward large values of $\triangle \tilde{f}^d_{i_1,i_2}$, but at the same time suppresses excessive noise occurring in those regions. For the edge/texture prediction scheme, the following *prediction neighborhood set* for each pixel was adopted.

$$\mathcal{N}_p = \{(k,l) : k,l = -L_p, \ldots, 0, \ldots, L_p\} \tag{5.35}$$

In addition, we define the following mean gray-level value $\overline{\tilde{f}}_{i_1,i_2}$ with respect to this neighborhood set as follows:

$$\overline{\tilde{f}}_{i_1,i_2} = \frac{1}{N_p} \sum_{k=-L_p}^{L_p} \sum_{l=-L_p}^{L_p} \tilde{f}_{i_1+k,i_2+l} \tag{5.36}$$

where $\tilde{f}_{i_1,i_2}$ denotes the gray-level value at (i_1, i_2) and $N_p = (2L_p + 1)^2$.

To avoid the problem of oversmoothing in the combined edge/textured regions while suppressing excessive noise at the same time, we have adopted the concept of *weighted order statistic (WOS) filter* [130,131] in deriving a suitable desired network output $\triangle \tilde{f}^d_{i_1,i_2}$ for the training set. The set of order statistics corresponding to the prediction neighborhood set can be defined as follows: For the n_pth order statistic $\tilde{f}^{(n_p)}_{i_1,i_2}$,

$$n_p = P(k,l) \quad \exists (k,l) \in \mathcal{N}_p \tag{5.37}$$

where $P : \mathcal{N}_p \longrightarrow \{1, \ldots, N_p\}$ is a one-to-one mapping such that the following condition is satisfied:

$$\tilde{f}_{i_1,i_2}^{(1)} \leq \cdots \leq \tilde{f}_{i_1,i_2}^{(n_p)} \leq \cdots \leq \tilde{f}_{i_1,i_2}^{(N_p)} \tag{5.38}$$

The output of a weighted order statistic filter is defined as the linear combination of the order statistics

$$\tilde{f}_{i_1,i_2}^d = \sum_{n_p=1}^{N_p} \omega^{(n_p)} \tilde{f}_{i_1,i_2}^{(n_p)} \tag{5.39}$$

For odd N_p, the simplest example of a WOS filter is the median filter [86] where

$$\omega^{(M_p)} = 1$$
$$\omega^{(n_p)} = 0 \qquad n_p \neq M_p \tag{5.40}$$

and

$$M_p = \frac{N_p + 1}{2} \tag{5.41}$$

Therefore, the WOS filter can be considered as a generalization of the median filter where the information from all the order statistics are combined to provide an improved estimate of a variable. In general, for the purpose of noise filtering, the filter weights $\omega^{(1)}$ and $\omega^{(N_p)}$ are chosen such that $\omega^{(1)} \approx 0$ and $\omega^{(N_p)} \approx 0$, since the corresponding order statistics $\tilde{f}_{i_1,i_2}^{(1)}$ and $\tilde{f}_{i_1,i_2}^{(N_p)}$ usually represent outliers.

Adopting the value $L_p = 1$ and $N_p = (2 \cdot 1 + 1)^2 = 9$ as the size of the prediction neighborhood set, we define the *predicted gray-level value* $\tilde{f}_{i_1,i_2}^d$ and the corresponding desired network output $\triangle \tilde{f}_{i_1,i_2}^d$, according to the operation of a WOS filter as follows:

$$\tilde{f}_{i_1,i_2}^d = \begin{cases} \tilde{f}_{i_1,i_2}^{(3)} & \tilde{f}_{i_1,i_2} < \overline{\tilde{f}}_{i_1,i_2} \\ \tilde{f}_{i_1,i_2}^{(7)} & \tilde{f}_{i_1,i_2} \geq \overline{\tilde{f}}_{i_1,i_2} \end{cases} \tag{5.42}$$

$$\triangle \tilde{f}_{i_1,i_2}^d = \tilde{f}_{i_1,i_2}^d - \tilde{f}_{i_1,i_2} \tag{5.43}$$

or equivalently

$$\tilde{f}_{i_1,i_2}^d = \sum_{n_p=1}^{9} \omega^{(n_p)} \tilde{f}_{i_1,i_2}^{(n_p)} \tag{5.44}$$

where

$$\omega^{(3)} = 1, \omega^{(n_p)} = 0, n_p \neq 3 \quad \text{for } \tilde{f}_{i_1,i_2} < \overline{\tilde{f}}_{i_1,i_2}$$
$$\omega^{(7)} = 1, \omega^{(n_p)} = 0, n_p \neq 7 \quad \text{for } \tilde{f}_{i_1,i_2} \geq \overline{\tilde{f}}_{i_1,i_2}$$

The motivation for choosing the respective order statistics for the two different cases is that, for the case $\widetilde{f}_{i_1,i_2} \geq \overline{\widetilde{f}}_{i_1,i_2}$, we assume that the true gray-level value lies in the interval $[\overline{\widetilde{f}}_{i_1,i_2}, \widetilde{f}^{(9)}_{i_1,i_2}]$. For a blurred or partially restored image, this corresponds approximately to the interval $[\widetilde{f}^{(5)}_{i_1,i_2}, \widetilde{f}^{(9)}_{i_1,i_2}]$ with its endpoints at the median and maximum gray-level values.

Within this interval, we cannot choose $\widetilde{f}^{d}_{i_1,i_2} = \widetilde{f}^{(5)}_{i_1,i_2} \approx \overline{\widetilde{f}}_{i_1,i_2}$, as this will result in excessive smoothing for the combined edge/textured regions. Neither can we choose $\widetilde{f}^{d}_{i_1,i_2} = \widetilde{f}^{(9)}_{i_1,i_2}$ with the corresponding order statistic being usually considered an outlier which does not accurately reflect the correct gray-level value of the current pixel. A possible candidate could be $\widetilde{f}^{d}_{i_1,i_2} = 0.5(\widetilde{f}^{(5)}_{i_1,i_2} + \widetilde{f}^{(9)}_{i_1,i_2})$, but the presence of the outlier $\widetilde{f}^{(9)}_{i_1,i_2}$ in the linear combination will still result in nonrepresentative predictions, especially for high levels of noise. To ensure the comparative noise immunity of the resulting estimate while avoiding excessive smoothing, the choice $\widetilde{f}^{d}_{i_1,i_2} = \widetilde{f}^{(7)}_{i_1,i_2}$ represents a compromise that offers the additional advantage that the gray-level value is among one of those in the prediction neighborhood set. The adoption of this value thus implicitly imposes a continuity constraint between the current and the neighboring pixels. The choice of $\widetilde{f}^{d}_{i_1,i_2} = \widetilde{f}^{(3)}_{i_1,i_2}$ for the case $\widetilde{f}_{i_1,i_2} < \overline{\widetilde{f}}_{i_1,i_2}$ is similarly justified.

On the other hand, we should adopt a prediction scheme that biases toward small values of $\triangle \widetilde{f}^{d}_{i_1,i_2}$ for the smooth regions to suppress the more visible noise and ringing there. In view of this, the following prediction scheme is adopted for the smooth regions

$$\widetilde{f}^{d}_{i_1,i_2} = \overline{\widetilde{f}}_{i_1,i_2} \tag{5.45}$$

$$\triangle \widetilde{f}^{d}_{i_1,i_2} = \widetilde{f}^{d}_{i_1,i_2} - \widetilde{f}_{i_1,i_2} \tag{5.46}$$

This prediction scheme essentially employs the local mean, which serves as a useful indicator of the correct gray-level values in smooth regions, as an estimate for the current gray-level value. Alternatively, it can be viewed as the operation of a filter mask with all its coefficients being $1/N_p$.

The essential difference between the current approach and traditional adaptive nonlinear filtering techniques [86] is that, whereas the traditional filtering techniques *replace* the current gray-level value with the above predicted value, the current scheme uses this predicted value as a training *guidance* by incorporating it as the desired output in the regional training set. We then apply steepest descent to change the corresponding regional regularization parameter according to the information in these training patterns. In this way, both the information of the degradation mechanism (in the form of the model-based neuronal weights $\widetilde{p}_{k,l}$) and the regional image model (in the form of the regional training set $\mathcal{V}_r$) are exploited to achieve a more accurate restoration.

5.8 Determination of the Image Partition

Conforming to the description of an image as a combination of edge/textured and smooth components, we denote the edge/textured components by $\mathcal{F}_{r_f}, r_f = 1, \ldots, R_f$ and the smooth components by $\mathcal{B}_{r_b}, r_b = 1, \ldots, R_b$. We further define the following combinations of these components:

$$\mathcal{F} = \bigcup_{r_f} \mathcal{F}_{r_f} \tag{5.47}$$

$$\mathcal{B} = \bigcup_{r_b} \mathcal{B}_{r_b} \tag{5.48}$$

$$\mathcal{P}_F = \{\mathcal{F}_{r_f}, r_f = 1, \ldots, R_f\} \tag{5.49}$$

$$\mathcal{P}_B = \{\mathcal{B}_{r_b}, r_b = 1, \ldots, R_b\} \tag{5.50}$$

$$\mathcal{P}_R = \mathcal{P}_F \cup \mathcal{P}_B \tag{5.51}$$

Our partitioning strategy is to first classify each pixel as belonging to the region $\mathcal{F}$ or $\mathcal{B}$ and then derive the partitions $\mathcal{P}_F$ and $\mathcal{P}_B$ using connectivity criteria. We perform the preliminary classification by adopting the following local activity measure δ_{i_1,i_2} for each pixel

$$\delta_{i_1,i_2} = \ln(\sigma_{i_1,i_2}) \tag{5.52}$$

where

$$\sigma_{i_1,i_2} = \left(\frac{1}{N_p} \sum_{k=-L_p}^{L_p} \sum_{l=-L_p}^{L_p} (\tilde{f}_{i_1+k,i_2+l} - \overline{\tilde{f}}_{i_1,i_2})^2 \right)^{\frac{1}{2}} \tag{5.53}$$

and $\overline{\tilde{f}}_{i_1,i_2}$ is defined in equation (5.36). The logarithm mapping is adopted to approximate the nonlinear operation in the human vision system which transforms intensity values to perceived contrasts [2]. We then assign the current pixel to either $\mathcal{F}$ or $\mathcal{B}$ according to the value of δ_{i_1,i_2} relative to a threshold T

$$\mathcal{F} = \{(i_1, i_2) : \delta_{i_1,i_2} > T\} \tag{5.54}$$

$$\mathcal{B} = \{(i_1, i_2) : \delta_{i_1,i_2} \leq T\} \tag{5.55}$$

The threshold T is usually selected according to some optimality criteria based on the particular image content. In this work, we have chosen $T = T^*$, where T^* is the threshold which minimizes the total within-class variance ϱ^2 of δ_{i_1,i_2}, that is

$$T^* = \arg\min_T \varrho^2(T) \tag{5.56}$$

where

$$\varrho^2 = \frac{1}{|\mathcal{B}|} \sum_{(i_1,i_2)\in\mathcal{B}} (\sigma_{i_1,i_2} - \sigma_{\mathcal{B}})^2 + \frac{1}{|\mathcal{F}|} \sum_{(i_1,i_2)\in\mathcal{F}} (\sigma_{i_1,i_2} - \sigma_{\mathcal{F}})^2 \tag{5.57}$$

and

$$\sigma_{\mathcal{B}} = \frac{1}{|\mathcal{B}|} \sum_{(i_1, i_2) \in \mathcal{B}} \sigma_{i_1, i_2} \tag{5.58}$$

$$\sigma_{\mathcal{F}} = \frac{1}{|\mathcal{F}|} \sum_{(i_1, i_2) \in \mathcal{F}} \sigma_{i_1, i_2} \tag{5.59}$$

The partitions $\mathcal{P}_F$ and $\mathcal{P}_B$ are in turn extracted from the sets $\mathcal{F}$ and $\mathcal{B}$ by considering each element of $\mathcal{P}_F$ or $\mathcal{P}_B$ to be a maximally connected component of $\mathcal{F}$ or $\mathcal{B}$. To be more precise, if we adopt the usual eight-path connectivity relation $\mathcal{C}_8$ as our connectivity criterion in the case of $\mathcal{F}$ [4], then the partition $\mathcal{P}_F$ will be given by the *quotient set* $\mathcal{F}/\mathcal{C}_8$ of $\mathcal{F}$ under the equivalence relation $\mathcal{C}_8$. The partition $\mathcal{P}_B$ is similarly defined.

In view of the simplicity of this preliminary segmentation scheme and the fact that the relative proportion of the two region types would vary during the course of image restoration, we should adopt some resegmentation procedures throughout the restoration process to account for this variation. In this work, we have used a modified version of the *nearest neighbor classification procedure* [132] to perform the reassignment, and we have restricted the reassignment to pixels on the region boundary. The following *neighboring region set* $\mathcal{N}_{i_{b1}, i_{b2}}^{\mathcal{R}} \subset \mathcal{P}_R$ is defined for each such boundary pixel

$$\mathcal{N}_{i_{b1}, i_{b2}}^{\mathcal{R}} = \{\mathcal{R}_q, q = 1, \dots, Q\} \tag{5.60}$$

where each region $\mathcal{R}_q$ in the set is adjacent to the boundary pixel (i_{b1}, i_{b2}), or more precisely, there exists at least one pixel in each $\mathcal{R}_q$ which is in the eight-neighborhood of (i_{b1}, i_{b2}). Corresponding to each $\mathcal{R}_q$ we can define the following regional activity $\overline{\sigma}_q$

$$\overline{\sigma}_q = \frac{1}{|\mathcal{R}_q|} \sum_{(i_1, i_2) \in \mathcal{R}_q} \sigma_{i_1, i_2} \tag{5.61}$$

where σ_{i_1, i_2} is defined in Equation (5.53). With the definition of these variables, we can proceed with the reclassification of the boundary pixels by adopting the following nearest neighbor decision rule for (i_{b1}, i_{b2}):

$$(i_{b1}, i_{b2}) \in \mathcal{R}_{q^*} \quad \text{if } |\sigma_{i_{b1}, i_{b2}} - \overline{\sigma}_{q^*}| < |\sigma_{i_{b1}, i_{b2}} - \overline{\sigma}_q|, \quad q = 1, \dots, Q \tag{5.62}$$

The application of this decision rule is manifested as a continual change in the boundary of each region in such a way as to increase the homogeneity of the activities of the regions. Finally, with the full determination of the various texture and smooth components in the image by the above scheme, we can apply the steepest descent rule of Equation (5.29) to each regional regularization parameter λ_r and use the prediction schemes for the respective regions to achieve adaptive regularization.

5.9 Edge-Texture Characterization Measure

In this section, we introduce the edge-texture characterization (ETC) measure, which is a scalar quantity summarizing the degree of resemblance of a particular pixel-value configuration to either textures or edges. In other words, pixel-value arrangements corresponding to textures and edges will in general exhibit different values for this measure. This is unlike the case where the local variance or the edge magnitude [2] is adopted for the image activity measure. Due to the possibility that both edges and textures may exhibit similar levels of image activities in terms of gray-level variations around their neighborhoods, it is usually not possible to distinguish between these two feature types using the conventional image activity measures.

On the other hand, the current ETC measure is derived based on the correlational properties of individual pixel-value configurations. In general, we may expect that configurations corresponding to edges and textures will possess significantly different correlational properties, with the individual pixels in a texture configuration being far less correlated with each other than those in an edge configuration. This is described in a quantitative way using the current measure. More importantly, we have analytically established intervals of ETC measure values corresponding to those pixel configurations that visually more resemble textures than edges, and vice versa.

This measure is especially useful for distinguishing between edges and textures in image restoration such that we can specify different levels of regularization to each of them. Due to the different noise-masking capabilities of these two feature types, it is usually not desirable to apply similar values of regularization parameters to both of them. With the incorporation of the newly formulated ETC measure into our HMBNN approach, we are able to separately estimate two different parameter values that are optimal to edges and textures, respectively, and apply the correct parameter to the current pixel in accordance with its associated ETC measure value.

The starting point of our formulation is as follows: we consider the gray-level values of image pixels in a local region as independent and identically distributed (i.i.d.) random variables with variance σ^2. If we apply a local $K \times K$ averaging operation to each pixel, the variance σ'^2 of the smoothed random variables is given by

$$\sigma'^2 = \frac{1}{K^4} \mathbf{u}^T \mathbf{R} \mathbf{u} = \frac{\sigma^2}{K^2} \tag{5.63}$$

where $\mathbf{R} = diag[\sigma^2, \ldots, \sigma^2]$ is the $K^2 \times K^2$ covariance matrix of the K^2 random variables in the $K \times K$ averaging window, and $\mathbf{u}$ is a $K^2 \times 1$ vector with all entries equal to one. The diagonal structure of the covariance matrix is due to the independence of the random variables. The i.i.d. assumption above is in general not applicable to real-world images. In fact, we usually identify a meaningful image with the existence of controlled correlation among its

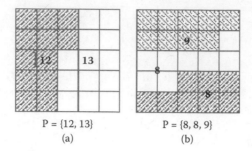

$P = \{12, 13\}$ $P = \{8, 8, 9\}$
(a)   (b)

FIGURE 5.5
Illustrating the various forms of partition $\mathcal{P}$.

pixels. As a result, we generalize the above i.i.d. case to incorporate correlations inside the $K \times K$ window.

We define the multiset $\mathcal{P} = \{P_1, \ldots, P_i, \ldots, P_m\}$, where $P_1 + \cdots + P_m = K^2$, as a *partition* of the K^2 variables in the window into m components. In addition, we assume that all variables within the ith component is correlated with each other with correlation coefficient ρ_i, and variables among different components are mutually uncorrelated with each other. Some examples of $\mathcal{P}$ in a 5×5 window are given in Figure 5.5. For example, we can describe the region around an edge pixel by the partition $\mathcal{P} = \{P_1, P_2\}$, where $P_1 \approx P_2$ (Figure 5.5a). In this general case, the variance σ'^2 after $K \times K$ averaging is given by

$$\sigma'^2 = \frac{1}{K^4} \mathbf{u}^T R \mathbf{u} = \frac{\sigma^2}{\kappa^2} \tag{5.64}$$

In this case, $\mathbf{R}$ is a *block-diagonal* matrix with the following structure

$$\mathbf{R} = diag[\mathbf{R}_1, \ldots, \mathbf{R}_m] \tag{5.65}$$

Each submatrix $\mathbf{R}_i, i = 1, \ldots, m$ is of dimension $P_i \times P_i$ with the following structure

$$\mathbf{R}_i = \sigma^2 \begin{bmatrix} 1 & \rho_i & \cdots & \rho_i \\ \rho_i & \ddots & \ddots & \vdots \\ \vdots & \ddots & & \rho_i \\ \rho_i & \cdots & \rho_i & 1 \end{bmatrix} \tag{5.66}$$

If we carry out the matrix multiplication in Equation (5.64), the square of the quantity κ in the equation is evaluated to be

$$\kappa^2 = \frac{K^4}{K^2 + \sum_{P_i \in \mathcal{P}} \rho_i (P_i^2 - P_i)} \tag{5.67}$$

Assuming that $0 \leq \rho_i \leq 1$ for all i, which implies positive correlation among pixels within a single component, the value of κ is maximized when $\rho_i = 0$, $\forall i$, giving $\kappa = K$ in Equation (5.67), which corresponds to the previous case of i.i.d. variables.

On the other hand, if we assume $\rho_i = 1$ for all i within a single element partition $\mathcal{P} = \{K^2\}$ and substituting into Equation (5.67), we have

$$\kappa^2 = \frac{K^4}{K^2 + (K^4 - K^2)} = 1 \qquad (5.68)$$

which implies $\kappa = 1$. This corresponds to the case where all the gray-level values within the window are highly correlated, or in other words, to smooth regions in the image. In general, the value of κ is between 1 and K. Thus it serves as an indicator of the degree of correlation within the $K \times K$ window. Larger values of κ indicate a low level of correlation among the pixels in the window, which is usually the case for textured and edge regions, while smaller values of κ usually correspond to smooth regions as indicated above. To provide an intuitive grasp of the values of κ corresponding to various features in the image, we carry out the calculation prescribed in Equation (5.67) for a 5×5 averaging window.

For $K = 5$, the minimum and maximum values of κ are 1 and 5, respectively. For a positive within-component correlation coefficient ρ_i, the value of κ is constrained within the interval [1, 5]. Referring to Figure 5.5a that describes image edge regions with the partition $\mathcal{P} = \{12, 13\}$, and further assumes that $\rho_i = \rho = 0.9$ for all components, we have, after substituting the corresponding quantities into Equation (5.67)

$$\kappa^2 = \frac{5^4}{5^2 + 0.9[(12^2 - 12) + (13^2 - 13)]} \approx 2.20 \qquad (5.69)$$

or $\kappa \approx 1.48$. This value of κ, which we designate as κ_2, serves to characterize all edge-like features in the image if a 5-by-5 averaging window is used. On the other hand, if we consider more complex features with the number of components $m > 2$, which usually correspond to textures in the images, we should expect the value of κ to be within the interval [1.48, 5]. This is confirmed by evaluating κ for the partition $\mathcal{P} = \{8, 8, 9\}$ as illustrated in Figure 5.5b, again assuming $\rho_i = 0.9$ for all i,

$$\kappa^2 = \frac{5^4}{5^2 + 0.9[2(8^2 - 8) + (9^2 - 9)]} \approx 3.28 \qquad (5.70)$$

or $\kappa \approx 1.81$, which we designate as κ_3. As a result, the value of κ indicates to a certain extent the qualitative attributes of a local image region, that is, whether it is more likely to be a textured region or an edge region, rather than just distinguishing the smooth background from the combined texture/edge regions as in the case of using the local standard deviation σ_{i_1, i_2} alone. We can therefore refer to this quantity κ as the edge-texture characterization measure.

TABLE 5.1

ETC Measure Values for Various
Different Partitions

κ_1	1.05	κ_6	2.54
κ_2	1.48	κ_7	2.72
κ_3	1.81	κ_8	2.91
κ_4	2.08	κ_9	3.07
κ_5	2.33	κ_{25}	5.00

Table 5.1 lists the values of the measure κ_1 to κ_9 and κ_{25}. We can notice that, in general, the value of κ increases with the number of correlated components within the pixel window, which confirms our previous observation. The case of 25 components corresponds to the case of i.i.d. random variables and results in the value $\kappa_{25} = 5$. It is also noted that the value of κ_1 is slightly larger than the case of fully correlated random variables in our previous calculation due to our present assumption of $\rho = 0.9$.

We can estimate the value of κ in a pixel neighborhood by the ratio $\hat{\sigma}/\hat{\sigma}'$ in accordance with Equation (5.64), where $\hat{\sigma}$ and $\hat{\sigma}'$ are the sample estimates of σ and σ', respectively.

$$\hat{\sigma}^2 = \frac{1}{|\mathcal{N}|} \sum_{(i,j)\in\mathcal{N}} (\tilde{f}_{i,j} - \overline{\tilde{f}})^2 \qquad (5.71)$$

$$\hat{\sigma}'^2 = \frac{1}{|\mathcal{N}|} \sum_{(i,j)\in\mathcal{N}} (\tilde{f}'_{i,j} - \overline{\tilde{f}'})^2 \qquad (5.72)$$

In the equations, $\mathcal{N}$ denotes a neighborhood set around the current pixel, $\tilde{f}_{i,j}$ denotes the gray-level value of pixel (i, j) in the set, and $\tilde{f}'_{i,j}$ is the corresponding smoothed gray-level value under $K \times K$ averaging. $\overline{\tilde{f}}$ and $\overline{\tilde{f}'}$ are, respectively, the mean of the gray-level values of the original and smoothed variables in the neighborhood set. In general, this empirically estimated κ is not restricted to the interval $[1, K]$ due to the use of the sample variances, but most of its values are restricted to the interval $[0, K]$.

5.10   ETC Fuzzy HMBNN for Adaptive Regularization

In this section, we propose a fuzzified version of the model-based neural network for adaptive regularization. A hierarchical architecture is again adopted for the current network, but we have included two neurons in each subnetwork instead of a single neuron. We referred to those two neurons as the edge neuron and the texture neuron, respectively. They in turn estimate *two* regularization parameters, namely, the edge parameter and the texture parameter, for each region with values optimal for regularizing the edge pixels and the textured pixels, respectively, within the region. Ideally, the parameter of the edge neuron should be updated using the information of the edge pixels only, and the texture neuron should be updated using the texture pixels only, which implies the necessity to distinguish between the edge and texture pixels. This is precisely the motivation for the formulation of the ETC measure, which achieves this very purpose. As the concepts of edge and texture are inherently fuzzy, we characterize this fact by defining two fuzzy sets, the EDGE fuzzy set and the TEXTURE fuzzy set, over the ETC measure domain. This is possible since the value of the ETC measure reflects the degree of resemblance of a particular local gray-level configuration to either textures or edges. More importantly, there exists definite intervals of measure values that we can claim that the corresponding gray-level configuration should be more like textures, and vice versa. As a result, the ETC measure domain serves as an ideal *universe of discourse* [37] for the definition of the EDGE and TEXTURE fuzzy sets. In view of the importance of the fuzzy-set concept in the current formulation, we will first have a review on this topic.

5.11   Theory of Fuzzy Sets

The inclusion of a member a in a set A is usually represented symbolically as $a \in A$. Alternatively, we can express this inclusion in terms of a *membership function* $\mu_A(a)$ as follows:

$$\mu_A(a) = \begin{cases} 1 & \text{if } a \in A \\ 0 & \text{if } a \notin A \end{cases} \tag{5.73}$$

The membership function takes values in the discrete set $\{0, 1\}$. Zadeh, in his 1965 paper [38], generalized the definition of the membership function such that it can take values in the real-valued interval $[0, 1]$. The generalized set corresponding to the new membership function is known as a *fuzzy set* [36–38,133–135] in contrast with the previous *crisp set*.

The implication of the new membership function is that, aside from the states of belonging wholly to a set or not belonging to a set at all, we can now allow an element to be a member of a fuzzy set "to a certain extent," with the exact degree of membership being expressed by the value of the corresponding membership function. Therefore, if the membership function value $\mu_A(a)$ of an element a is close to 1, we can say that a belongs "to a large degree" to the fuzzy set A, and we should expect that a possesses many of the properties that are characteristic of the set A. On the other hand, if the membership function value is small, we can expect that element a only bears a vague resemblance to a typical member of the set A.

This concept is particularly important for expressing how human beings characterize everyday concepts. Take, for example, the concept "tall." We can immediately visualize the meaning of the word. But in order for machine interpretation systems to recognize this concept, we must provide an exact

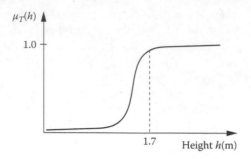

FIGURE 5.6
The fuzzy set representing the concept TALL.

definition. If restricted to the use of crisp sets, we may define this concept in terms of the set $T = \{h : h \geq 1.7 \text{ m}\}$ where h is the height variable. According to this definition, a value of the variable h such as $h = 1.699$ m will not be considered as belonging to the concept "tall," which is certainly unnatural from the human viewpoint. If we instead define "tall" to be a fuzzy set with the membership function shown in Figure 5.6, the value $h = 1.699$ m can still be interpreted as belonging "strongly" to the set, and thus conforms more closely to human interpretation.

The ordinary operations on crisp set can be generalized to the case of fuzzy set in terms of the membership function values as follows:

$$\mu_{A \cap B}(x) = \min\{\mu_A(x), \mu_B(x)\} \tag{5.74}$$

$$\mu_{A \cup B}(x) = \max\{\mu_A(x), \mu_B(x)\} \tag{5.75}$$

$$\mu_{A^c}(x) = 1 - \mu_A(x) \tag{5.76}$$

where $A \cap B$, $A \cup B$, and A^c are, respectively, the intersection of fuzzy sets A and B, the union of A and B, and the complement of A. These equations reduce to the ordinary intersection, union, and complement operations on crisp sets when the ranges of the membership functions $\mu_A(x)$, $\mu_B(x)$ are restricted to values in $\{0, 1\}$.

A *fuzzy inference relationship* is usually of the following form:

If $((x_1 \text{ has property } A_1) \otimes \ldots \otimes (x_n \text{ has property } A_n))$,

then $(y \text{ has property } B)$

where $x_1, \ldots, x_n$ and y are numerical variables, $A_1, \ldots, A_n$ and B are *linguistic* descriptions of the properties required of the corresponding numerical variables, and $\otimes$ denotes either the union or intersection operations. As described above, we can convert the linguistic descriptions $A_1, \ldots, A_n$ and B into fuzzy sets with membership functions $\mu_{A_1}(x_1), \ldots, \mu_{A_n}(x_n)$ and $\mu_B(y)$, where we have identified the names of the fuzzy sets with their corresponding linguistic descriptions. Formally, we can describe this inference operation as a mapping between fuzzy sets as follows:

$$B' = F(A_1, \ldots, A_n) \tag{5.77}$$

where $B' \subset B$ is a *fuzzy subset* of B. Fuzzy subsethood is formally described by the following condition

$$A \subset B \quad \text{if } \mu_A(x) \le \mu_B(x), \forall x \tag{5.78}$$

which reduces to the ordinary subsethood relationship between crisp sets if $\mu_A(x)$, $\mu_B(x)$ are allowed to take values only in $\{0, 1\}$.

The particular fuzzy subset B' chosen within the fuzzy set B (or equivalently, the particular form of the mapping F) depends on the degree to which the current value of each variable $x_i, i = 1, \ldots, n$ belongs to its respective fuzzy sets A_i. To summarize, the inference procedure accepts fuzzy sets as inputs and emits a single fuzzy set as output.

In practical systems, a *numerical* output that captures the essential characteristics of the output fuzzy set B' is usually required. This is usually done by specifying a *defuzzification* operation D on the fuzzy set B' to produce the numerical output y'

$$y' = D(B') \tag{5.79}$$

A common defuzzification operation is the following *centroid defuzzification* [37] operation

$$y' = \frac{\int_{-\infty}^{\infty} y \mu_{B'}(y) \, dy}{\int_{-\infty}^{\infty} \mu_{B'}(y) \, dy} \tag{5.80}$$

where we assign the centroid of the fuzzy membership function $\mu_{B'}(y)$ to the variable y'.

5.12 Edge-Texture Fuzzy Model Based on ETC Measure

We have previously defined the edge-texture characterization (ETC) measure κ that quantifies the degree of resemblance of a particular gray-level configuration to either textures or edges. In addition, we have established that, for values of κ within the interval $I_2 = [(\kappa_1 + \kappa_2/2), \kappa_3]$, we can consider the underlying gray-level configuration to be reasonably close to that of edges, and for $\kappa > \kappa_3$, we can conclude that the corresponding configuration has a closer resemblance to that of textures. However, if we consider the value $\kappa = \kappa_3 + \epsilon$, where ϵ is a small positive constant, we will classify the corresponding configuration as a texture configuration, but we can expect that it will still share many of the properties of an edge configuration due to the closeness of κ to an admissible edge value. In fact, it is difficult to define the concepts of "edge" and "textures" in terms of crisp sets in the ETC domain. In view of this, fuzzy set theory becomes a natural candidate for characterizing these concepts in terms of the ETC measure.

We therefore define two fuzzy sets, namely, the EDGE fuzzy set and the TEXTURE fuzzy set, on the ETC measure domain in terms of their membership functions $\mu_E(\kappa)$ and $\mu_T(\kappa)$, as follows:

$$\mu_E(\kappa) \equiv \frac{1}{1 + e^{\beta_E(\kappa - \kappa_E)}} \tag{5.81}$$

$$\mu_T(\kappa) \equiv \frac{1}{1 + e^{-\beta_T(\kappa - \kappa_T)}} \tag{5.82}$$

The two set membership functions are plotted in Figure 5.7. From the figure, it is seen that $\mu_E(\kappa)$ is a decreasing sigmoid function with the transition point at $\kappa = \kappa_E$, and $\mu_T(\kappa)$ is an increasing sigmoid function with the transition point at $\kappa = \kappa_T$. The parameters β_E and β_T control the steepness of transition of the two membership functions. In view of our previous observation that characterizes edges with values of κ within the interval I_2, we may have

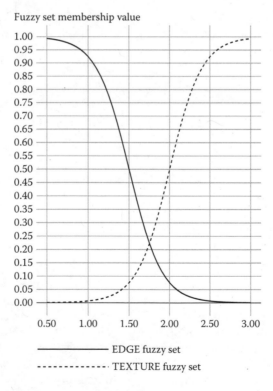

FIGURE 5.7
The Edge and Texture fuzzy membership functions.

expected the function $\mu_E(\kappa)$ to exhibit a peak around κ_2 and to taper off on both sides. Instead, we have chosen a decreasing sigmoid function with the transition point at $\kappa_E \approx \kappa_2$, that implicitly classifies the smooth regions with $\kappa < \kappa_2$ as belonging to the EDGE fuzzy set as well. This is due to our formulation of the current regularization algorithm in such a way that, whenever a certain pixel configuration has a larger membership value in the EDGE fuzzy set, larger values of regularization parameters will be applied to this configuration. As a result, it is reasonable to assign greater membership values to those configurations with $\kappa < \kappa_2$, which usually corresponds to weak edges or smooth regions, due to their less effective noise-masking capabilities compared with strong edges. In other words, we are interpreting the membership function $\mu_E(\kappa)$ as the indicator of the required amount of regularization for a certain pixel configuration rather than its true correspondence to a characteristic edge configuration.

On the other hand, the shape of the TEXTURE set membership function truly reflects the fact that we consider those gray-level configurations with $\kappa > \kappa_3$ to be essentially textures. However, instead of choosing $\kappa_T = \kappa_3$, which was shown to correspond to a gray-level configuration containing three uncorrelated components and may reasonably resemble textures, we instead choose the more conservative value of $\kappa_T = 2 > \kappa_3$.

From the two membership functions, we define the following *normalized ETC fuzzy coefficients* $\tilde{\mu}_E(\kappa)$ and $\tilde{\mu}_T(\kappa)$

$$\tilde{\mu}_E(\kappa) = \frac{\mu_E(\kappa)}{\mu(\kappa)} \tag{5.83}$$

$$\tilde{\mu}_T(\kappa) = \frac{\mu_T(\kappa)}{\mu(\kappa)} \tag{5.84}$$

where

$$\mu(\kappa) = \mu_E(\kappa) + \mu_T(\kappa) \tag{5.85}$$

5.13 Architecture of the Fuzzy HMBNN

Corresponding to the partition of the image into the combined edge/texture components $\mathcal{F}_{r_f}, r_f = 1, \ldots, R_f$ and the smooth regions $\mathcal{B}_{r_b}, r_b = 1, \ldots, R_b$, we assign individual subnetworks to each of these regions. For the smooth regions, we assign one neuron to each smooth region to estimate a single parameter $\lambda_{r_b}, r_b = 1, \ldots, R_b$ for each region. However, for the combined edge/textured regions, instead of employing only one neuron as in the previous case, we have assigned *two* neurons, which we designate as the edge neuron and the texture neuron with associated weight vectors $\tilde{\mathbf{p}}_{edge}(\lambda_{r_f}^{edge})$ and $\tilde{\mathbf{p}}_{tex}(\lambda_{r_f}^{tex})$, to each edge/texture subnetwork. The two weight vectors are, respectively, functions of the *edge-regularization parameter* $\lambda_{r_f}^{edge}$ and the *texture regularization parameter* $\lambda_{r_f}^{tex}$. As their names imply, we should design the training procedure in such a way that the parameter $\lambda_{r_f}^{edge}$ estimated through the edge neuron should be optimal for the regularization of edge-like entities, and $\lambda_{r_f}^{tex}$ estimated through the texture neuron should be optimal for textured regions.

Corresponding to these two weight vectors, we evaluate two estimates of the required pixel change, $\triangle \widetilde{f}_{i_1,i_2}^{edge}$ and $\triangle \widetilde{f}_{i_1,i_2}^{tex}$, for the same gray-level configuration $\widetilde{\mathbf{f}}_{i_1,i_2}$ as follows

$$\triangle \widetilde{f}_{i_1,i_2}^{edge} = \widetilde{\mathbf{p}}_{edge}(\lambda_{r_f}^{edge})^T \widetilde{\mathbf{f}}_{i_1,i_2} \tag{5.86}$$

$$\triangle \widetilde{f}_{i_1,i_2}^{tex} = \widetilde{\mathbf{p}}_{tex}(\lambda_{r_f}^{tex})^T \widetilde{\mathbf{f}}_{i_1,i_2} \tag{5.87}$$

where $(i_1, i_2) \in \mathcal{F}_{r_f}$. The quantities $\triangle \widetilde{f}_{i_1,i_2}^{edge}$ and $\triangle \widetilde{f}_{i_1,i_2}^{tex}$ are the required updates based on the assumptions that the underlying gray-level configuration $\widetilde{\mathbf{f}}_{i_1,i_2}$ corresponds to edge or textures, respectively.

Referring to the fuzzy formulation in the previous section, we can evaluate the ETC measure κ_{i_1,i_2} at pixel position (i_1, i_2) as a function of the pixel values in a local neighborhood of (i_1, i_2). From this we can calculate the normalized edge fuzzy coefficient value $\widetilde{\mu}_E(\kappa_{i_1,i_2})$ and the normalized texture fuzzy coefficient value $\widetilde{\mu}_T(\kappa_{i_1,i_2})$. As a result, a natural candidate for the final required pixel update value $\triangle \widetilde{f}_{i_1,i_2}$ can be evaluated as a convex combination of the quantities $\triangle \widetilde{f}_{i_1,i_2}^{edge}$ and $\triangle \widetilde{f}_{i_1,i_2}^{tex}$ with respect to the normalized fuzzy coefficient values

$$\triangle \widetilde{f}_{i_1,i_2} = \widetilde{\mu}_E(\kappa_{i_1,i_2})\triangle \widetilde{f}_{i_1,i_2}^{edge} + \widetilde{\mu}_T(\kappa_{i_1,i_2})\triangle \widetilde{f}_{i_1,i_2}^{tex} \tag{5.88}$$

From the above equation, it is seen that for regions around edges where $\widetilde{\mu}_E(\kappa_{i_1,i_2}) \approx 1$ and $\widetilde{\mu}_T(\kappa_{i_1,i_2}) \approx 0$, the final required gray-level update $\triangle \widetilde{f}_{i_1,i_2}$ is approximately equal to $\triangle \widetilde{f}_{i_1,i_2}^{edge}$, that is optimal for the restoration of the edges. On the other hand, in textured regions where $\widetilde{\mu}_T(\kappa_{i_1,i_2}) \approx 1$ and $\widetilde{\mu}_E(\kappa_{i_1,i_2}) \approx 0$, the required update $\triangle \widetilde{f}_{i_1,i_2}$ assumes the value $\triangle \widetilde{f}_{i_1,i_2}^{tex}$ which is optimal for textures, provided that proper training procedures are adopted for both of the neurons. Alternatively, this equation can be interpreted as the operation of a *two-layer* subnetwork, where $\triangle \widetilde{f}_{i_1,i_2}$ is the network output, $\widetilde{\mu}_E(\kappa_{i_1,i_2})$, $\widetilde{\mu}_T(\kappa_{i_1,i_2})$ are output weights and $\triangle \widetilde{f}_{i_1,i_2}^{edge}$ and $\triangle \widetilde{f}_{i_1,i_2}^{tex}$ are the hidden neuron outputs. The architecture of this two-layer subnetwork is shown in Figure 5.8.

5.13.1   Correspondence with the General HMBNN Architecture

Similar to our previous model-based neural network formulation, the current fuzzy HMBNN model also adopts the hidden-node hierarchical architecture, which is depicted in Figure 5.3. The architecture of the subnetworks for the regularization of the smooth regions remains the same. However, for the edge/texture regularization subnetworks, we have adopted a new

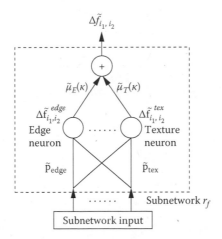

FIGURE 5.8
The architecture of the fuzzy HMBNN subnetwork.

architecture in the form of a two-layer network, with two neurons in the hidden layer as shown in Figure 5.8. Corresponding to the general form for evaluation of the subnetwork output, we can describe the operation of the edge/texture regularization subnetwork for region $\mathcal{F}_{r_f}$ as follows:

$$\phi(\widetilde{\mathbf{f}}_{i_1,i_2}, \widetilde{\mathbf{p}}_{r_f}) = \sum_{s_{r_f}=1}^{2} c_{s_{r_f}} \psi_{r_f}\left(\widetilde{\mathbf{f}}_{i_1,i_2}, \widetilde{\mathbf{p}}(\lambda_{r_f}^{s_{r_f}})\right) \tag{5.89}$$

where

$$\lambda_{r_f}^{1} \equiv \lambda_{r_f}^{edge} \tag{5.90}$$

$$\lambda_{r_f}^{2} \equiv \lambda_{r_f}^{tex} \tag{5.91}$$

$$\psi_{r_f}\left(\widetilde{\mathbf{f}}_{i_1,i_2}, \widetilde{\mathbf{p}}(\lambda_{r_f}^{s_{r_f}})\right) \equiv \widetilde{\mathbf{p}}(\lambda_{r_f}^{s_{r_f}})^T \widetilde{\mathbf{f}}_{i_1,i_2} \tag{5.92}$$

$$c_1 \equiv \widetilde{\mu}_E(\kappa_{i_1,i_2}) \tag{5.93}$$

$$c_2 \equiv \widetilde{\mu}_T(\kappa_{i_1,i_2}) \tag{5.94}$$

and $\widetilde{\mathbf{p}}_{r_f}$ can be considered as a concatenation of $\widetilde{\mathbf{p}}(\lambda_{r_f}^{s_{r_f}})$, $s_{r_f} = 1, 2$.

5.14 Estimation of the Desired Network Output

We have previously defined the desired network output $\triangle \widetilde{f}_{i_1,i_2}^{d}$ and the associated predicted gray-level value $\widetilde{f}_{i_1,i_2}^{d}$, where

$$\widetilde{f}_{i_1,i_2}^{d} = \widetilde{f}_{i_1,i_2} + \triangle \widetilde{f}_{i_1,i_2}^{d} \tag{5.95}$$

as follows: for smooth regions, we define the variable $\tilde{f}^d_{i_1,i_2}$ as the mean of the neighboring pixel values:

$$\tilde{f}^d_{i_1,i_2} = \overline{\tilde{f}}_{i_1,i_2} \tag{5.96}$$

$$\overline{\tilde{f}}_{i_1,i_2} = \frac{1}{N_p} \sum_{k=-L_p}^{L_p} \sum_{l=-L_p}^{L_p} \tilde{f}_{i_1+k,i_2+l} \tag{5.97}$$

where $N_p = (2L_p+1)^2$ is the size of the filter mask for averaging. For combined edge/textured regions, we consider the ranked pixel values in the prediction neighborhood set:

$$\mathcal{N}_p = \{ \tilde{f}^{(1)}_{i_1,i_2}, \ldots, \tilde{f}^{(n_p)}_{i_1,i_2}, \ldots, \tilde{f}^{(N_p)}_{i_1,i_2} \}$$

where $\tilde{f}^{(n_p)}_{i_1,i_2}$ is the n_pth order statistic of the pixels in the ordered prediction neighborhood set and corresponds to $\tilde{f}_{i_1+k,i_2+l}$ for appropriate k and l. The ranked gray-level values satisfy the following condition:

$$\tilde{f}^{(1)}_{i_1,i_2} \leq \cdots \leq \tilde{f}^{(n_p)}_{i_1,i_2} \leq \cdots \leq \tilde{f}^{(N_p)}_{i_1,i_2}$$

Adopting a prediction neighborhood size of $N_p = 9$, we define the predicted gray-level value $\tilde{f}^d_{i_1,i_2}$ for the combined edge/textured regions as follows:

$$\tilde{f}^d_{i_1,i_2} = \begin{cases} \tilde{f}^{(3)}_{i_1,i_2} & \tilde{f}_{i_1,i_2} < \overline{\tilde{f}}_{i_1,i_2} \\ \tilde{f}^{(7)}_{i_1,i_2} & \tilde{f}_{i_1,i_2} \geq \overline{\tilde{f}}_{i_1,i_2} \end{cases} \tag{5.98}$$

It was observed experimentally that this prediction scheme results in a noisy appearance for the restored edges at high noise levels due to its smaller noise-masking capability compared with textures. In view of this, we apply the following *edge-oriented* estimation scheme at high noise levels:

$$\tilde{f}^d_{i_1,i_2} = \begin{cases} \tilde{f}^{(4)}_{i_1,i_2} & \tilde{f}_{i_1,i_2} < \overline{\tilde{f}}_{i_1,i_2} \\ \tilde{f}^{(6)}_{i_1,i_2} & \tilde{f}_{i_1,i_2} \geq \overline{\tilde{f}}_{i_1,i_2} \end{cases} \tag{5.99}$$

This edge-oriented scheme results in a less noisy appearance for the edges, but the textured areas will appear blurred compared with the previous case. It would therefore be ideal if, at high noise levels, we can apply the texture-oriented estimation scheme to the textured areas only, and apply the edge-oriented estimation scheme to only the edges. This in turn requires the separation of the edges and textured areas, which is usually difficult in terms of conventional measures such as image gradient magnitudes or local variance due to the similar levels of gray-level activities for these two types of features. On the other hand, the previous fuzzy formulation in terms of the scalar ETC measure allows this very possibility.

In addition, Equations (5.98) and (5.99) predict the desired gray-level value $\tilde{f}^d_{i_1,i_2}$ in terms of a *single-order* statistic only. The estimation will usually be more meaningful if we can exploit information provided by the gray-level values of *all* the pixels in the prediction neighborhood set $\mathcal{N}_p$. In the next section, we investigate this possibility by extending the previous crisp estimation framework to a fuzzy estimation framework for $\tilde{f}^d_{i_1,i_2}$.

5.15 Fuzzy Prediction of Desired Gray-Level Value

In this section, we define two fuzzy sets, namely, the EDGE GRAY LEVEL ESTIMATOR (EG) fuzzy set and the TEXTURE GRAY LEVEL ESTIMATOR (TG) fuzzy set, over the domain of *gray-level values* in an image. This fuzzy formulation is independent of our previous ETC-fuzzy formulation where the EDGE and TEXTURE fuzzy sets are defined over the domain of the *ETC measure*. The purpose of this second fuzzy formulation is to allow the utilization of different prediction strategies for $\tilde{f}^d_{i_1,i_2}$ in the edge and textured regions, respectively, and the evaluation of this predicted gray-level value using all the gray-level values in the prediction neighborhood set $\mathcal{N}_p$ instead of a single crisp value.

5.15.1 Definition of the Fuzzy Estimator Membership Function

For this gray-level estimator fuzzy model, we define the two set membership functions, $\varphi_{EG}(\tilde{f}_{i_1,i_2})$ and $\varphi_{TG}(\tilde{f}_{i_1,i_2})$ in terms of Gaussian functions, as opposed to the use of sigmoid nonlinearity in the first model. Again denoting the gray-level value at location (i_1, i_2) as $\tilde{f}_{i_1,i_2}$ and the n_pth order statistic of the prediction neighborhood set $\mathcal{N}_p$ as $\tilde{f}^{(n_p)}_{i_1,i_2}$, and assuming that $\tilde{f}_{i_1,i_2} \geq \overline{\tilde{f}}_{i_1,i_2}$ without loss of generality, which corresponds to the second condition in Equations (5.98) and (5.99), we define the membership functions of the EG fuzzy set and the TG fuzzy set as follows:

$$\varphi_{EG}(\tilde{f}_{i_1,i_2}) = e^{-\xi_{EG}(\tilde{f}_{i_1,i_2} - \tilde{f}^{(6)}_{i_1,i_2})^2} \tag{5.100}$$

$$\varphi_{TG}(\tilde{f}_{i_1,i_2}) = e^{-\xi_{TG}(\tilde{f}_{i_1,i_2} - \tilde{f}^{(7)}_{i_1,i_2})^2} \tag{5.101}$$

For the condition $\tilde{f}_{i_1,i_2} < \overline{\tilde{f}}_{i_1,i_2}$, we will replace the centers of the EG and TG membership functions with $\tilde{f}^{(4)}_{i_1,i_2}$ and $\tilde{f}^{(3)}_{i_1,i_2}$, respectively, in accordance with Equations (5.98) and (5.99).

The two membership functions are depicted graphically in Figure 5.9. Under the condition $\tilde{f}_{i_1,i_2} \geq \overline{\tilde{f}}_{i_1,i_2}$, where we previously designate $\tilde{f}^{(6)}_{i_1,i_2}$ as the preferred gray-level estimator in the edge-oriented estimation scheme, we now generalize this concept by assigning a membership value of 1 for $\tilde{f}^{(6)}_{i_1,i_2}$. However, instead of adopting this value exclusively, we have also assigned nonzero membership function values for gray-level values close to $\tilde{f}^{(6)}_{i_1,i_2}$, thus

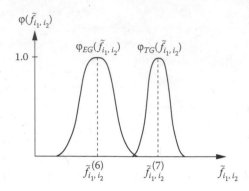

FIGURE 5.9
The fuzzy membership functions of the sets EG and TG.

expressing the notion that nearby values also have relevancy in determining the final predicted gray-level value. Similar assignments are adopted for the TG membership function with $\widetilde{f}^{(7)}_{i_1,i_2}$ as the center.

In the two equations, the two parameters ξ_{EG} and ξ_{TG} control the width of the respective membership functions, thus quantifying the notion of "closeness" in terms of gray-level value differences for both fuzzy sets. In general, the values of these two parameters are different from each other, and the optimal values of both parameters may also differ for different image regions. In view of these, we have devised a training algorithm for adapting these parameters in different regions and this will be described in a later section. The requirement for adaptation is also the reason we have chosen ξ_{EG} and ξ_{TG} to be multiplicative factors instead of their usual appearances as denominators in the form of variances for Gaussian functions: in the former case, we have only to ensure that both parameters are greater than zero. In the latter case, there is the additional difficulty that the adapting variances can easily become too small to result in an excessively large exponent for the Gaussian functions.

5.15.2   Fuzzy Inference Procedure for Predicted Gray-Level Value

Given the membership functions $\varphi_{EG}(\widetilde{f}_{i_1,i_2})$ and $\varphi_{TG}(\widetilde{f}_{i_1,i_2})$ of the two fuzzy sets EG and TG, we can define a mapping F, which assigns a single fuzzy set G, the GRAY LEVEL ESTIMATOR fuzzy set, to each pair of fuzzy sets EG and TG,

$$G = F(EG, TG) \tag{5.102}$$

in the following way: since any fuzzy set is fully defined by specifying its membership function, we will define the set mapping F in terms of real-valued mappings between the parameters of the membership function $\varphi_G(\widetilde{f}_{i_1,i_2})$ and the parameters of the membership functions $\varphi_{EG}(\widetilde{f}_{i_1,i_2})$ and

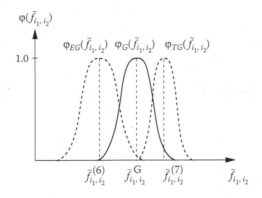

FIGURE 5.10
The mapping of the fuzzy sets EG and TG to G.

$\varphi_{TG}(\widetilde{f}_{i_1,i_2})$ as follows:

$$\varphi_G(\widetilde{f}_{i_1,i_2}) = e^{-\xi_G(\widetilde{f}_{i_1,i_2} - \widetilde{f}^G_{i_1,i_2})^2} \tag{5.103}$$

$$\widetilde{f}^G_{i_1,i_2} \equiv \widetilde{\mu}_E(\kappa_{i_1,i_2})\widetilde{f}^{(6)}_{i_1,i_2} + \widetilde{\mu}_T(\kappa_{i_1,i_2})\widetilde{f}^{(7)}_{i_1,i_2} \tag{5.104}$$

$$\xi_G \equiv \widetilde{\mu}_E(\kappa_{i_1,i_2})\xi_{EG} + \widetilde{\mu}_T(\kappa_{i_1,i_2})\xi_{TG} \tag{5.105}$$

under the condition $\widetilde{f}_{i_1,i_2} \geq \overline{\widetilde{f}}_{i_1,i_2}$. For $\widetilde{f}_{i_1,i_2} < \overline{\widetilde{f}}_{i_1,i_2}$, the terms $\widetilde{f}^{(6)}_{i_1,i_2}$ and $\widetilde{f}^{(7)}_{i_1,i_2}$ are replaced by $\widetilde{f}^{(4)}_{i_1,i_2}$ and $\widetilde{f}^{(3)}_{i_1,i_2}$, respectively.

The mapping operation is illustrated in Figure 5.10. From the figure, it is seen that the mapping performs a kind of interpolation between the two fuzzy sets EG and TG. The coefficients $\widetilde{\mu}_E(\kappa_{i_1,i_2})$ and $\widetilde{\mu}_T(\kappa_{i_1,i_2})$ refer to the previous normalized fuzzy coefficient values of the fuzzy ETC measure model, which is independent of the current gray-level estimator fuzzy model. If the current pixel belongs to an edge, the conditions $\widetilde{\mu}_E(\kappa) \approx 1$ and $\widetilde{\mu}_T(\kappa) \approx 0$ are approximately satisfied, the resulting final gray-level estimator fuzzy set G is approximately equal to the edge gray-level estimator fuzzy set EG, and the final prediction for the gray-level value, $\widetilde{f}^d_{i_1,i_2}$, will be based on the corresponding membership function of this fuzzy set. On the other hand, if the current pixel belongs to a textured area, the conditions $\widetilde{\mu}_E(\kappa_{i_1,i_2}) \approx 0$ and $\widetilde{\mu}_T(\kappa_{i_1,i_2}) \approx 1$ hold approximately, and the estimation of $\widetilde{f}^d_{i_1,i_2}$ will be mostly based on the membership function of the texture gray-level estimator fuzzy set TG. For all the other cases, the mapping operation results in a membership function for the set G with values of parameters intermediate between those of EG and TG.

5.15.3 Defuzzification of the Fuzzy Set G

The fuzzy inference procedure determines the final GRAY-LEVEL ESTIMATOR fuzzy set G with corresponding membership function $\varphi_G(\widetilde{f}_{i_1,i_2})$ from the membership functions $\varphi_{EG}(\widetilde{f}_{i_1,i_2})$ and $\varphi_{TG}(\widetilde{f}_{i_1,i_2})$. In order to provide a desired network output for the fuzzy HMBNN, we have to *defuzzify* the fuzzy set G to obtain a crisp prediction $\tilde{f}^d_{i_1,i_2}$ and the associated desired network output $\Delta \tilde{f}^d_{i_1,i_2}$. A common way to defuzzify a fuzzy set is to employ *centroid defuzzification* [37], where we assign the *centroid* of the fuzzy membership function to be the crisp value associated with the function. In the case of the fuzzy set G, we obtain

$$\tilde{f}^d_{i_1,i_2} \equiv \frac{\int_{-\infty}^{\infty} \tilde{f} \varphi_G(\tilde{f}) d\tilde{f}}{\int_{-\infty}^{\infty} \varphi_G(\tilde{f}) d\tilde{f}} = \tilde{f}^G_{i_1,i_2} \tag{5.106}$$

Due to the adoption of the Gaussian function for $\varphi_G(\tilde{f}_{i_1,i_2})$, this particular defuzzification strategy does not take into account the information provided by the width of the membership function. Specifically, rather than simply using $\tilde{f}^G_{i_1,i_2}$, which may not correspond to any of the pixel gray-level values in the prediction neighborhood set, as an estimator for the current pixel value, we would also like to incorporate those pixels in the neighborhood set with values "close" to $\tilde{f}^G_{i_1,i_2}$, in a sense depending on the exact value of the width parameter ξ_G, in the determination of the final estimate for the current gray-level value. The inclusion of these pixels from the neighborhood set will reduce incompatibility of the value of the estimator with those pixel values in its neighborhood. In accordance with these, we propose the following *discrete defuzzification procedure* for the fuzzy set G:

$$\tilde{f}^d_{i_1,i_2} \equiv \tilde{\varphi}_G(\tilde{f}^G_{i_1,i_2}) \tilde{f}^G_{i_1,i_2} + \sum_{n_p=1}^{9} \tilde{\varphi}_G(\tilde{f}^{(n_p)}_{i_1,i_2}) \tilde{f}^{(n_p)}_{i_1,i_2} \tag{5.107}$$

where $\tilde{\varphi}_G(\tilde{f}_{i_1,i_2})$ is the scaled membership function defined as follows:

$$\tilde{\varphi}_G(\tilde{f}_{i_1,i_2}) \equiv \frac{1}{C} \varphi_G(\tilde{f}_{i_1,i_2}) \tag{5.108}$$

$$C \equiv \varphi_G(\tilde{f}^G_{i_1,i_2}) + \sum_{n_p=1}^{9} \varphi_G(\tilde{f}^{(n_p)}_{i_1,i_2}) \tag{5.109}$$

From this expression, it is seen that the final crisp estimate $\tilde{f}^d_{i_1,i_2}$ does not necessarily equal the centroid $\tilde{f}^G_{i_1,i_2}$ as in continuous defuzzification, but it also depends on the pixel values in the prediction neighborhood set $\mathcal{N}_p$. In particular, if there exists a pixel with value $\tilde{f}^{(n_p)}_{i_1,i_2}$ in $\mathcal{N}_p$ that is not exactly equal to $\tilde{f}^G_{i_1,i_2}$, but is nevertheless "close" to it in the form of a large membership value $\tilde{\varphi}_G(\tilde{f}^{(n_p)}_{i_1,i_2})$ in the fuzzy set G, then the crisp estimate $\tilde{f}^d_{i_1,i_2}$ will be substantially influenced by the presence of this pixel in $\mathcal{N}_p$ such that the final value will be a compromise between $\tilde{f}^G_{i_1,i_2}$ and $\tilde{f}^{(n_p)}_{i_1,i_2}$. This is to ensure that, apart from using $\tilde{f}^G_{i_1,i_2}$, we incorporate the information of all those pixels in the prediction neighborhood set that are close in values to $\tilde{f}^G_{i_1,i_2}$ to obtain a more reliable estimate of the current pixel value.

5.15.4 Regularization Parameter Update

Given the final crisp estimate $\widetilde{f}^d_{i_1,i_2}$ for the current pixel, we can then define the desired network output $\triangle \widetilde{f}^d_{i_1,i_2}$ for the fuzzy HMBNN:

$$\triangle \widetilde{f}^d_{i_1,i_2} = \widetilde{f}^d_{i_1,i_2} - \widetilde{f}_{i_1,i_2} \qquad (5.110)$$

Adopting the same cost function $\mathcal{E}_t$, where

$$\mathcal{E}_t = \frac{1}{2} \left(\triangle \widetilde{f}^d_{i_1,i_2} - \triangle \widetilde{f}_{i_1,i_2} \right)^2 \qquad (5.111)$$

we can update the value of the regularization parameter by applying stochastic gradient descent on this cost function:

$$\lambda_r(t+1) = \lambda_r(t) - \eta \frac{\partial \mathcal{E}_t}{\partial \lambda_r} \qquad (5.112)$$

However, unlike the previous case, the fuzzy regularization network has two neurons, namely, the texture neuron and the edge neuron, associated with a single subnetwork for the combined edge/textured region, whereas there is only one neuron associated with a subnetwork for our previous network. We must therefore derive two update equations for the two regularization parameters $\lambda^{tex}_{r_f}(t)$ and $\lambda^{edge}_{r_f}(t)$ associated with the respective neurons. In addition, each update equation is to be designed in such a way that the resulting value of the regularization parameter would be appropriate for the particular types of regions assigned to each neuron. For example, we may expect that the parameter $\lambda^{tex}_{r_f}(t)$ of the texture neuron would be smaller than the parameter $\lambda^{edge}_{r_f}(t)$ of the edge neuron due to the better noise-masking capability of textures.

In view of these considerations, we can formulate the two update equations by regarding the current subnetwork as a *two-layer network*. More precisely, we consider the form of Equation (5.88), which is repeated here for convenience:

$$\triangle \widetilde{f}_{i_1,i_2} = \widetilde{\mu}_E(\kappa_{i_1,i_2}) \triangle \widetilde{f}^{edge}_{i_1,i_2} + \widetilde{\mu}_T(\kappa_{i_1,i_2}) \triangle \widetilde{f}^{tex}_{i_1,i_2}$$

where $\widetilde{\mu}_E(\kappa_{i_1,i_2})$, $\widetilde{\mu}_T(\kappa_{i_1,i_2})$ can be regarded as the *output weights* of the subnetwork, and $\triangle \widetilde{f}^{edge}_{i_1,i_2}$, $\triangle \widetilde{f}^{tex}_{i_1,i_2}$ can be considered as the outputs of the hidden edge and texture neurons. As a result, we can derive the update equations for both $\lambda^{edge}_{r_f}$ and $\lambda^{tex}_{r_f}$ using the *generalized delta rule* for multilayer neural networks.

For the parameter $\lambda_{r_f}^{edge}$ associated with the edge neuron, the update equation is derived as follows:

$$\lambda_{r_f}^{edge}(t+1) = \lambda_{r_f}^{edge}(t) - \eta\,\frac{\partial \mathcal{E}_t}{\partial \lambda_{r_f}^{edge}}$$

$$= \lambda_{r_f}^{edge}(t) - \eta\,\frac{\partial \mathcal{E}_t}{\partial \triangle \widetilde{f}_{i_1,i_2}^{edge}}\,\frac{\partial \triangle \widetilde{f}_{i_1,i_2}^{edge}}{\partial \lambda_{r_f}^{edge}}$$

$$= \lambda_{r_f}^{edge}(t) + \eta(\triangle \widetilde{f}_{i_1,i_2}^{d} - \triangle \widetilde{f}_{i_1,i_2})\frac{\partial \triangle \widetilde{f}_{i_1,i_2}}{\partial \triangle \widetilde{f}_{i_1,i_2}^{edge}}\,\frac{\partial \triangle \widetilde{f}_{i_1,i_2}^{edge}}{\partial \lambda_{r_f}^{edge}}$$

$$= \lambda_{r_f}^{edge}(t) + \eta(\triangle \widetilde{f}_{i_1,i_2}^{d} - \triangle \widetilde{f}_{i_1,i_2})\widetilde{\mu}_E(\kappa_{i_1,i_2})\frac{\partial \triangle \widetilde{f}_{i_1,i_2}^{edge}}{\partial \lambda_{r_f}^{edge}} \qquad (5.113)$$

Similarly, for $\lambda_{r_f}^{tex}$, we have the following update equation:

$$\lambda_{r_f}^{tex}(t+1) = \lambda_{r_f}^{tex}(t) + \eta(\triangle \widetilde{f}_{i_1,i_2}^{d} - \triangle \widetilde{f}_{i_1,i_2})\widetilde{\mu}_T(\kappa_{i_1,i_2})\frac{\partial \triangle \widetilde{f}_{i_1,i_2}^{tex}}{\partial \lambda_{r_f}^{tex}} \qquad (5.114)$$

From the equations, it is seen that if the current pixel belongs to a textured region, the conditions $\widetilde{\mu}_T(\kappa_{i_1,i_2}) \approx 1$ and $\widetilde{\mu}_E(\kappa_{i_1,i_2}) \approx 0$ are approximately satisfied, and we are essentially updating only the parameter of the texture neuron. On the other hand, if the pixel belongs to an edge, we have the conditions $\widetilde{\mu}_E(\kappa_{i_1,i_2}) \approx 1$ and $\widetilde{\mu}_T(\kappa_{i_1,i_2}) \approx 0$, and we are updating the edge neuron almost exclusively.

5.15.5 Update of the Estimator Fuzzy Set Width Parameters

Previously, it is seen that the width parameters of the gray-level estimator fuzzy sets, ξ_{EG} and ξ_{TE}, determine to what degree the various pixels in the prediction neighborhood set participate in estimating the final predicted gray-level value. In other words, the parameters establish the notion of "closeness" in gray-level values for the various edge/textured regions, which may differ from one such region to another.

In general, we have no a priori knowledge of the values of the width parameters required for each combined edge/textured region. In view of this, we propose a simple learning strategy that allows the parameters of each such region to be determined adaptively. Recall that, for the fuzzy sets EG and TG, we assign a membership value of 1 to the order-statistics $\widetilde{f}_{i_1,i_2}^{(6)}$ and $\widetilde{f}_{i_1,i_2}^{(7)}$, respectively (assuming $\widetilde{f}_{i_1,i_2} > \overline{f}_{i_1,i_2}$), indicating that these two values are highly relevant in the estimation of the current gray-level value. Since the shape of the Gaussian membership function can be fixed by two sample points, we can determine the width parameter by choosing a gray-level value in the set of order statistics that is the *least* relevant in estimating the final gray-level value, and assign a small membership value $\epsilon \ll 1$

to this gray-level value, thus completely specifying the entire membership function.

A suitable candidate for this particular gray-level value is the median, or alternatively, the fifth-order statistic $\widetilde{f}^{(5)}_{i_1,i_2}$. If we adopt this value as our estimate for the final gray-level value, we are effectively aiming at a median filtered version of the image. Although median filter is known to preserve edges while eliminating impulse noises for the case of images without blur [86], we can expect that, for blurred or partially restored images, the median will be close to the mean gray-level value in a local neighborhood, and thus would constitute an unsuitable estimate for the final gray-level value in the vicinity of edges and textures. We can thus assign a small membership value to the median gray-level value to completely determine the final form of the Gaussian membership function.

More precisely, we define the following error function C_ξ for adapting the two width parameters ξ_{EG} and ξ_{TG} as follows:

$$C_\xi = \frac{1}{2}\left(\epsilon - \varphi_G\left(\widetilde{f}^{(5)}_{i_1,i_2}\right)\right)^2 \tag{5.115}$$

which expresses the requirement that the value of $\varphi_G(\widetilde{f}_{i_1,i_2})$ should be small when $\widetilde{f}_{i_1,i_2}$ is close to the median value. The gradient descent update equations for the two parameters ξ_{EG} and ξ_{TG} with respect to C_ξ can then be derived by applying the chain rule according to Equations (5.103) and (5.105). For the parameter ξ_{EG},

$$\xi_{EG}(t+1) = \xi_{EG}(t) - \eta_\xi \frac{\partial C_\xi}{\partial \xi_{EG}}$$

$$= \xi_{EG}(t) - \eta_\xi \frac{\partial C_\xi}{\partial \varphi_G}\frac{\partial \varphi_G}{\partial \xi_G}\frac{\partial \xi_G}{\partial \xi_{EG}} \tag{5.116}$$

Similarly, the update equation for ξ_{TG} is

$$\xi_{TG}(t+1) = \xi_{TG}(t) - \eta_\xi \widetilde{\mu}_T(\kappa_{i_1,i_2})\varphi_G\left(\widetilde{f}^{(5)}_{i_1,i_2}\right)\left(\epsilon - \varphi_G\left(\widetilde{f}^{(5)}_{i_1,i_2}\right)\right)\left(\widetilde{f}^{(5)}_{i_1,i_2} - \widetilde{f}^{G}_{i_1,i_2}\right) \tag{5.117}$$

From the form of the two update equations, it is seen that, similar to the case for the update of the regularization parameters, when the current pixel belongs to a textured region, where $\widetilde{\mu}_T(\kappa_{i_1,i_2}) \approx 1$ and $\widetilde{\mu}_E(\kappa_{i_1,i_2}) \approx 0$, substantial update is performed on ξ_{TG} while there is essentially no update on ξ_{EG}, which is reasonable in view of the necessity to update the shape of the TG membership function using information from the textured pixels only. On the other hand, for edge pixels where $\widetilde{\mu}_E(\kappa_{i_1,i_2}) \approx 1$ and $\widetilde{\mu}_T(\kappa_{i_1,i_2}) \approx 0$, only the parameter ξ_{EG} will be substantially modified.

5.16   Experimental Results

The fuzzy modular MBNN was applied to three images, including the Lena image, and two images showing a flower and an eagle that are shown in Figure 5.11. For comparison purpose, the nonadaptive restoration results using the Wiener filter and Hopfield restoration network [136] proposed in [46,95] are included. In addition, we have also included results using conventional adaptive regularization approaches described in [27,28], where the original isotropic Euclidean norm in the restoration cost function is replaced

(a)

(b)

(c)

FIGURE 5.11
Original images: (a) Lena; (b) eagle; (c) flower.

by a weighted norm as follows:

$$E = \frac{1}{2}\|\mathbf{y} - \mathbf{H}\hat{\mathbf{f}}\|^2_{\mathbf{A}(\hat{\mathbf{f}})} + \frac{1}{2}\lambda(\hat{\mathbf{f}})\|\mathbf{D}\hat{\mathbf{f}}\|^2_{\mathbf{B}(\hat{\mathbf{f}})} \qquad (5.118)$$

The diagonal weighting matrix $\mathbf{B}(\hat{\mathbf{f}})$ allows the selective emphasis of either the data or regularization term in specific image pixel subsets to achieve spatially adaptive image restoration. Specifically, $\mathbf{B}(\hat{\mathbf{x}})$ is defined such that the associated matrix entries b_{ii}, $i = 1, \ldots, N_I$ are decreasing functions of the local variances, which in turn allows enforcement of the smoothness constraint in smooth regions for noise suppression, and relaxation of this constraint in high-variance regions for feature enhancement. In the experiments, the following form of $\mathbf{B}$, proposed in [28], was adopted:

$$b_{ii}(\hat{\sigma}_i) = \frac{1}{1 + \xi \max(0, \hat{\sigma}_i^2 - \sigma_n^2)} \qquad (5.119)$$

with $\hat{\sigma}_i^2$ representing the local variance at the ith pixel, σ_n^2 denoting the additive noise variance, and ξ is a tunable parameter. In order to provide a certain degree of adaptivity with respect to varying noise levels for this approach, we specify ξ to be equal to $1/\sigma_n^2$. The role of the weighting matrix $\mathbf{A}(\hat{\mathbf{f}})$ is complementary to that of $\mathbf{B}(\hat{\mathbf{f}})$ and is defined as $\mathbf{I} - \mathbf{B}(\hat{\mathbf{f}})$.

The global regularization parameter in Equation (5.118) is defined as the following function of the partially restored image $\hat{\mathbf{f}}$ as proposed in [27,102]

$$\lambda(\hat{\mathbf{f}}) = \frac{\|\mathbf{y} - \mathbf{H}\hat{\mathbf{f}}\|^2_{\mathbf{A}(\hat{\mathbf{f}})}}{2\|\mathbf{y}\|^2 - \|\mathbf{D}\hat{\mathbf{f}}\|^2_{\mathbf{B}(\hat{\mathbf{f}})}} \qquad (5.120)$$

In Figure 5.12, a 5×5 point spread function implementing a uniform blur is applied to the image Lena. In addition, Gaussian noise at the level of 30 dB BSNR (blurred signal-to-noise ratio) [23] is added to the image. The degraded image is shown in Figure 5.12a. The restoration result using the current fuzzy MBNN restoration algorithm is shown in Figure 5.12e. Figure 5.12b shows the result using the Wiener filter, where we can notice the blurring and severe ringing at the boundaries. Figure 5.12c shows the result using the nonadaptive Hopfield network-restoration approach proposed by Zhou et al. [46], where the global regularization parameter λ is determined as $\lambda = 1/\text{BSNR}$ according to [22]. We can notice the noisy appearance of the image due to the nonadaptive regularization approach. Figure 5.12d shows the result using the parameter assignment function [Equation (5.119)] with $\xi = 1/\sigma_n^2$. In addition, the global parameter $\lambda(\hat{\mathbf{x}})$ is updated iteratively according to Equation (5.120). Although Figure 5.12d is an improvement on Figure 5.12b and c, the ringing in the vicinity of the edges is still noticeable.

The above results can be compared with the restoration result using the current approach in Figure 5.12e. We can notice that noise suppression around the edges is achieved without compromising the details of the textured area.

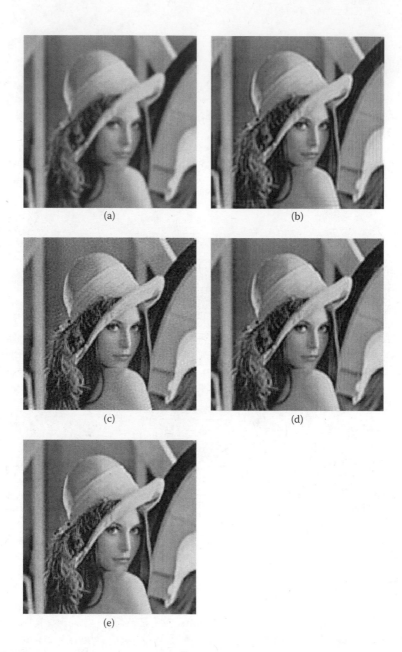

FIGURE 5.12
Restoration of the Lena image (5×5 uniform blur, 30 dB BSNR): (a) Degraded image; (b)–(d) restored images using (b) Wiener filter, (c) Hopfield *NN* ($\lambda = 1/BSNR$), (d) adaptive restoration approach using Equation (5.119) ($\xi = 1/\sigma_n^2$), (e) Fuzzy MBNN.

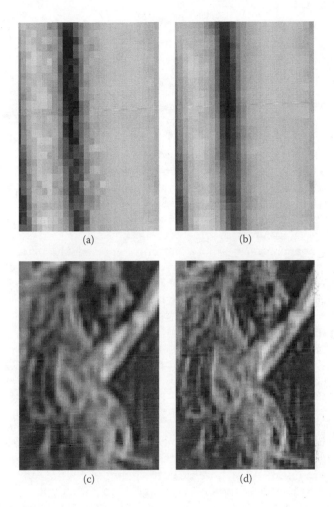

FIGURE 5.13
(a) Edges under texture-oriented regularization. (b) Edges under edge-oriented regularization. (c) Textures under edge-oriented regularization. (d) Textures under texture-oriented regularization.

The importance of adopting different regularization parameters for edges and textures is illustrated in Figure 5.13. In Figure 5.13a, we magnify the lower left portion of the Lena image to show the effect of adopting the texture-oriented regularization [Equation (5.98)] to edges. The noisy appearance of the restored edge is readily noticed when this is compared with the same region restored using edge-oriented regularization in Figure 5.13b. On the other hand, if we apply an edge-oriented regularization approach to textures as shown in Figure 5.13c, we can notice the blurred appearance that can be compared with the same region restored using the texture-oriented regularization approach in Figure 5.13d.

Figures 5.14a to e show the restoration results under 5×5 uniform blur with 20 dB BSNR, which represents a more severe degradation for the image. Comparing with the degraded image shown in Figure 5.14a, we can notice that the Wiener filter restoration result in Figure 5.14b does not result in an appreciable enhancement of image details. This is due to the increased emphasis of the Wiener filter on noise smoothing rather than feature enhancement at higher noise levels. Similarly, we can notice the noisy appearances of the non-adaptive Hopfield network result in Figure 5.14c and the spatially adaptive regularization result in Figure 5.14d. We can adopt alternative values for ξ in Equation (5.119) for noise reduction, but there are in general no effective criteria for choosing the optimal parameter value. As a result, to illustrate the best performance under Equation (5.119), the parameter ξ is adjusted in such a way that the root mean square (RMSE) between the restored and original image is minimized, which is only possible with the availability of the original image and does not genuinely represent the usual case where only the blurred image is available. In other words, we can expect the RMSE in practical restoration attempts to be higher. The restored image using this parameter-selection approach is shown in Figure 5.14e, with the associated RMSE value shown in Table 5.2. We can see that even the minimization of RMSE through a suitable choice of ξ does not necessarily result in a restored image with good visual quality, as seen in the blurred appearance of the restored image. This can be compared with the result using the current approach in Figure 5.14f, where it is seen that the visual quality is satisfactory even at this more serious level of degradation. This is achieved by the possibility of distinguishing between various feature types through the ETC measure and the corresponding assignment of different λ_r values to these feature classes.

We have also applied the current algorithm to the eagle and flower images. We apply the uniform PSF with 30 dB BSNR additive Gaussian noise to the eagle image in Figure 5.15a, and the corresponding restored image is shown in Figure 5.15c. Similarly, the flower image is degraded with 5×5 uniform blurring and 20 dB Gaussian additive noise in Figure 5.15b, and the restored image is shown in Figure 5.15d. In general, conclusions similar to those for the previous cases can be drawn from these results with regard to the restoration qualities in both the edge and textured regions. Comparison of the restoration results for these two images in terms of the root mean square measure is also shown in Table 5.2.

The spatial distribution of regularization parameter values for the Lena image is shown in Figure 5.16. In these images, light gray-level values correspond to large regularization parameter values λ_r, and dark areas correspond to small values of λ_r. The λ distribution corresponding to 5×5 times uniform blur at 30 dB BSNR are shown in Figure 5.16a, while those corresponding to the same PSF at 20 dB BSNR are shown in Figure 5.16b. It is seen that, for smooth regions, large λ_r values are generated for greater noise suppression, while small values are automatically adopted for high-variance regions to achieve feature enhancement. In addition, it is seen that the λ maps for the 20-dB case exhibit lighter gray values than those for the 30-dB case, which

FIGURE 5.14
Restoration of Lena image (5×5 uniform blur, 20-dB BSNR): (a) Degraded image; (b)–(f) restored images using (b) Wiener filter, (c) Hopfield *NN* ($\lambda = 1/BSNR$), (d) adaptive restoration using Equation (5.119)($\xi = 1/\sigma_n^2$), (e) adaptive restoration using Equation (5.119) (optimized ξ), (f) fuzzy MBNN.

TABLE 5.2

RMSE Values of the Restoration Results Using Different Restoration Algorithms

Image,BSNR	Blur.	Wien.	N.adap.	(119)	(119)(opt.ξ)	HMBNN
Lena, 30 dB	13.03	11.23	8.65	8.46	8.45	7.09
Lena, 20 dB	13.79	11.61	10.56	12.62	11.43	9.69
Eagle, 30 dB	9.88	9.61	10.28	7.81	7.62	6.89
Eagle, 20 dB	11.98	9.75	11.57	13.03	11.07	8.99
Flower, 30 dB	8.73	7.96	7.70	6.15	5.90	4.98
Flower, 20 dB	10.06	8.12	8.47	11.00	8.65	6.92

(a)   (b)

(c)   (d)

FIGURE 5.15
Further restoration results using fuzzy MBNN: (a) degraded eagle image (30 dB BSNR); (b) degraded flower image (20 dB BSNR); (c) restored eagle image; (d) restored flower image.

FIGURE 5.16
Distribution of local λ values: Lena (a) 30 dB, (b) 20 dB.

is reasonable considering the greater need of noise suppression in this case. Thus it is seen that, instead of attempting to assign regularization parameters directly, the optimal values of which may depend on the particular noise levels, it is more desirable to adopt a learning approach where the specification of a single gray-level target value for the fuzzy NN allows the emergence of different local regularization parameters in response to different noise levels.

In Figure 5.17, we show the thresholded distribution map of ETC values for the Lena image. Specifically, in Figure 5.17b, those pixels with their ETC coefficient values $\tilde{\mu}_T(\kappa_{r,i}) > \tilde{\mu}_E(\kappa_{r,i})$ are labeled. We should expect that, according to our previous discussion, these pixels will approximately correspond to the textures and be regularized essentially by the smaller λ_r^{tex}. This can

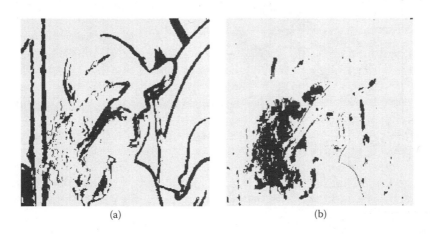

FIGURE 5.17
Lena: (a) edge pixels, (b) texture pixels.

be confirmed from the figures where it is seen that the labeled pixels are mostly clustered around the feathers on the hat. In Figure 5.17a, we have labeled those image pixels with $\widetilde{\mu}_E(\kappa_{r,i}) \geq \widetilde{\mu}_T(\kappa_{r,i})$. They will essentially be regularized with the larger parameter λ_r^{edge} and thus we expect they should approximately correspond to the edges, which is again confirmed from the figures.

In addition to the visual comparisons, we have included objective comparisons in terms of the RMSE measure, which is defined as follows:

$$RMSE = \left(\frac{1}{N_I} \|\mathbf{f} - \hat{\mathbf{f}}\|^2 \right)^{\frac{1}{2}} \tag{5.121}$$

where $\mathbf{f}$ and $\hat{\mathbf{f}}$ represent the original image and restored image, respectively, with the pixels arranged in lexicographic order, and N_I is the number of pixels in the image. The RMSE values of the restored images using the various algorithms are listed in Table 5.2. The improvement resulting from the adoption of the current approach is indicated by the corresponding small RMSE values at various noise levels. In particular, we can compare the RMSE values of the current approach with those using the adaptive strategy [Equation (5.119)] with optimized ξ with respect to the *original image* in Column 5. Although in some cases (especially under low noise levels), these values are only slightly higher than those using the current approach, we can expect that in practical cases where the original image is not available, the corresponding RMSE values will be higher than those shown. In addition, at higher noise levels, the RMSE values using the adaptive approach [Equation (5.119)] is significantly higher than those using the current approach.

5.17   Summary

An alternative formulation of the problem of adaptive regularization in image restoration was proposed in the form of a *model-based neural network with hierarchical architecture* (HMBNN) operating on nonoverlapping regions on the image. Based on the principle of adopting small parameter values for the textured regions for detail emphasis, while using large values for ringing and noise suppression in the smooth regions, we have developed an alternative viewpoint of adaptive regularization that centers on the concept of *regularization parameters as model-based neuronal weights*. From this we have derived a stochastic gradient descent algorithm for optimizing the parameter value in each region. In addition, incremental redefinition of the various regions using the nearest neighbor principle is incorporated to reverse the effects of initial inaccurate segmentation.

We have generalized the previous HMBNN framework for adaptive regularization to incorporate fuzzy information for edge/texture discrimination. Adopting the new ETC measure, we propose an edge/texture fuzzy model that expresses the degree to which a local image neighborhood resembles either edge or texture in terms of two fuzzy membership values. Correspondingly, we modify our previous network architecture to incorporate two neurons, namely, the edge neuron and the texture neuron, within each subnetwork. Each of the two neurons estimate an independent regularization parameter from the local image neighborhood and evaluate the corresponding required gray-level update value as a function of its own parameter. The two gray-level update values are then combined using fuzzy inference to produce the final required update in a way that takes into account the degree of edge/texture resemblance of the local neighborhood. In general, the algorithm is designed such that less regularization is applied to the textured areas due to their better noise-masking capability, and more regularization is applied to the edges where the noises are comparatively more visible.

The generalized algorithm is applied to a number of images under various conditions of degradations. The better visual quality of the restored images under the current algorithm can be appreciated by comparing the result with those produced using a number of conventional restoration algorithms. The current fuzzy HMBNN paradigm thus refines the previous notion of simply applying less regularization for the combined edge/textured regions to allow the possibility of using different levels of regularization to accommodate the different noise-masking capabilities of the various regional components.

6

Adaptive Regularization Using Evolutionary Computation

6.1 Introduction

In this chapter, we propose an alternative solution to the problem of adaptive regularization by adopting a new *global* cost measure. It was previously seen that the newly formulated ETC measure is capable of distinguishing between the textures and edges in an image. It is further observed in this chapter that the distribution function of this measure value in a typical image assumes a characteristic shape, which is comparatively invariant across a large class of images, and can thus be considered a *signature* for the images. This is in contrast with the distribution function of other quantities such as the gray-level values that varies widely from image to image, as can be confirmed by observing the gray-level histograms of different images.

It is also observed that the corresponding ETC distribution function [or equivalently the ETC probability density function (ETC-pdf)] of degraded images is usually very different from that of the nondegraded images. In other words, for optimal restoration results, we can assign the regularization parameters in such a way that the ETC distribution function of the restored image once again assumes the shape that is typical for nondegraded images, or more formally, we have to minimize the distance between the ETC-pdf of the restored image and the characteristic ETC-pdf of nondegraded images.

In practice, we can only approximate the ETC-pdf by the histogram of the measure values in the image, which cannot be expressed in closed form with respect to the regularization parameters, and thus the conventional gradient-based optimization algorithms are not applicable. We have therefore adopted an artificial evolutionary optimization approach where evolutionary programming (EP), belonging to the class of algorithms known as evolutionary computational algorithms, is used to search for the optimal set of regularization parameters with respect to the ETC-pdf criterion. One of the advantages of this class of algorithms is their independence from the availability of gradient information, which is therefore uniquely suited to our current optimization problem. In addition, these algorithms employ multiple search points in a population instead of a sequence of single search points as in conventional optimization algorithms, thus allowing many regions of the parameter space to be explored simultaneously. The characteristics of this class of algorithms are described in the following section.

6.2 Introduction to Evolutionary Computation

Evolutionary programming [40,137] belongs to the class of optimization algorithms known as evolutionary computational algorithms [39,40,138–140] that mimic the process of natural evolution to search for an optimizer of a cost function. There are three mainstreams of research activities in this field, which include research in genetic algorithm (GA) introduced by Holland [141], Goldberg [61] and Mitchell [142], evolutionary programming (EP) by Fogel [40,137], and evolutionary strategy (ES) by Schwefel [143,144]. The defining characteristics of this class of algorithms include its maintenance of a diversity of potential optimizers in a *population* and allowing highly effective optimizers to emerge through the processes of mutation, recombination, competition, and selection. The implicit parallelism resulting from the use of multiple search points instead of a single search point in conventional optimization algorithms allows many regions of the search space to be explored simultaneously. Together with the stochastic nature of the algorithms that allow search points to spontaneously escape from nonglobal optima, and the independence of the optimization process from gradient information, the instances of local minima are usually reduced, unlike the case with the gradient-based algorithms. In addition, this independence from gradient information allows the incorporation of highly irregular functions as fitness criteria for the evolutionary process, unlike the case with gradient-based algorithms where only differentiable cost functions are allowed. It is not even necessary for the cost function to have a closed form, as in those cases where the function values are obtained only through simulation. We will give a brief introduction of the above members of this class of algorithms in the following sections.

6.2.1 Genetic Algorithm

Genetic algorithm (GA) [61,141,142] is the most widely used among the three evolutionary computational algorithms. The distinguishing feature of GA includes its representation of the potential optimizers in the population as *binary strings*. Assuming the original optimizers are real-valued vectors $\mathbf{z} \in \mathbf{R}^N$, an encoding operation $\mathcal{G}$ is applied to each of these vectors to form the binary strings $\mathbf{g} = \mathcal{G}(\mathbf{z}) \in \mathbf{B}^H$, where $\mathbf{B} = \{0, 1\}$. In other words, the various evolutionary operations are carried out in the space of *genotypes*.

New individuals in the population are created by the operations of *crossover* and *mutation*. In GA, crossover is the predominant operation, while mutation only serves as an infrequent background operation. In crossover, two binary strings $\mathbf{g}_{p_1}$, $\mathbf{g}_{p_2}$ are randomly selected from the population. A random position $h \in \{1, \ldots, H\}$ is selected along the length of the binary string. After this, the substring to the left of h in p_1 is joined to the right substring of p_2, and similarly for the left substring of p_2 and the right substring of p_1, thus mimicking the biological crossover operation on the chromosomes. On the other hand, the mutation operator toggles the status of each bit for a certain binary string with a probability π_m, which is a very small value in the case of GA. The main purpose of mutation is to introduce new variants of genotypes into the population.

After the processes of crossover and mutation, the fitness of each binary string, $f(\mathbf{g}_p)$, $p = 1, \ldots, \mu$, is evaluated, where μ is the number of optimizers in the population, and f represents the *fitness function*. The function f usually reflects the requirement of the optimization problem at hand. In the case of a maximization task, we can usually equate the objective function of the problem with the fitness function. In the case of a problem that requires minimization, we can simply set the fitness function to be the negative of the current cost function.

After the evaluation of the individual fitness values, the optimizers in the population undergo a proportional selection process: each optimizer is to be included in the next generation with probability $\pi(\mathbf{g}_{p'})$, which reflects the relative fitness of individual $\mathbf{g}_{p'}$

$$\pi(\mathbf{g}_{p'}) = \frac{f(\mathbf{g}_{p'})}{\sum_{p=1}^{\mu} f(\mathbf{g}_p)} \tag{6.1}$$

In other words, an individual is more likely to survive into the next generation if it possesses a high fitness value. As the algorithm proceeds, the population will eventually consist of those optimizers with appreciable fitness values.

6.2.2 Evolutionary Strategy

As opposed to genetic algorithm, evolutionary strategy (ES) [143,144] represents an alternative evolutionary approach where the various adaptation operations are carried out in the space of *phenotypes*: instead of first encoding the individual optimizers $\mathbf{z} \in \mathbf{R}^N$ into binary strings, the *recombination* and *mutation* operations are carried out directly on the real-valued vectors as described below.

For the recombination operation, two optimizers $\mathbf{z}_{p_1}$, $\mathbf{z}_{p_2}$ are randomly selected from the population, and a new optimizer $\mathbf{z}$ is generated from these two according to the recombination operator C.

$$\mathbf{z} = C(\mathbf{z}_{p_1}, \mathbf{z}_{p_2}) \tag{6.2}$$

The simplest form for C is the linear combination operation

$$\mathbf{z} = \alpha \mathbf{z}_{p_1} + (1 - \alpha)\mathbf{z}_{p_2} \tag{6.3}$$

where $\alpha < 1$ is the combination coefficient, although other combination operations, as described in [39], are also possible.

The mutation operation in ES randomly perturbs each component of the optimizer vector z_j to form a new optimizer $\mathbf{z}'$ with components z'_j as follows:

$$z'_j = z_j + N(0, \sigma_j^m) \tag{6.4}$$

where $N(0, \sigma)$ is a Gaussian random variable with mean 0 and standard deviation σ. In the terminology of ES, the parameter σ_j^m associated with the component z_j is usually referred to as a mutation *strategy parameter*. The values of the mutation strategy parameters determine whether the current optimization process more resembles a global search, as when the σ_j^ms assume large values, which is desirable at the initial stages when many regions of the parameter space are simultaneously explored, or a local search that is more appropriate toward the final stages when the mutation strategy parameters assume small values to restrict the search within promising localities. The mutation strategy parameters themselves are usually adapted according to the following log-normal equation:

$$\sigma'^m_j = \sigma_j^m \exp(\tau' N(0, 1) + \tau N_j(0, 1)) \tag{6.5}$$

where τ', τ are predetermined constants, and $N(0, 1)$, $N_j(0, 1)$ are Gaussian random variables with mean 0 and standard deviation 1. In the case of $N_j(0, 1)$, the random variable is resampled for each new component j. The log-normal adaptation equation is adopted to preserve the positivity of the mutation strategy parameters σ_j^m.

The recombination and mutation operations are performed γ times to form γ new individuals from the original μ individuals in the population. In the case of $(\mu + \gamma)$ selection strategy [39], the fitness values $f(\mathbf{z}_p)$ of each individual in the $(\mu + \gamma)$ parent/descendant combination are evaluated, and those μ optimizers in this combination with the greatest fitness values are incorporated into the population in the next generation. In the (μ, γ) selection strategy [39], only the fitness values of the newly generated γ descendants are evaluated and the fittest of those incorporated into the next generation. The selection process is deterministic and depends solely on the fitness values of the individuals.

6.2.3 Evolutionary Programming

Evolutionary Programming (EP) [40,137] shares many common features with evolutionary strategy in that the primary adaptation operations are also carried out in the space of phenotypes. In addition, there are a number of important similarities:

- The individual components z_j of each optimizer $\mathbf{z} \in \mathbf{R}^N$ are also perturbed according to Equation (6.4), with the mutation strategy parameters σ_j^m similarly defined for the current component. In some variants of EP, the individual components are perturbed by an amount proportional to the square root of the objective function value [40].

- The individual mutation strategy parameters σ_j^m themselves are also subject to adaptations. In some EP variants, the log-normal relationship given in Equation (6.5) is also adopted for this purpose.

Despite these similarities, there are a number of important differences between EP and ES that clearly distinguishes the former from the latter:

- In EP, mutation is the only adaptation operation applied to the individuals in the population. *No* recombination operations are carried out.

- Instead of using a deterministic selection strategy as in ES, where γ descendant optimizers are created from μ parent optimizers, and the members of the new population selected from the resulting combination according to a ranking of their respective fitness values, EP uses a stochastic selection strategy as follows: for each optimizer in the $(\mu + \gamma)$ parent/descendant combination, we randomly select Q other optimizers in the same combination and compare their fitness values with the current optimizer. The current optimizer is included in the population in the next generation if its fitness value is greater than those of the Q "opponents" selected. In addition, the number of descendants γ is usually set equal to the number of parents μ. In other words, this *Q-tournament selection strategy* can be regarded as a probabilistic version of the $(\mu + \mu)$ selection strategy in ES.

The tournament selection strategy has the advantage that, compared with the deterministic selection strategy, even optimizers that are ranked in the lower half of the $(\mu + \mu)$ parent and descendant combination have a positive probability of being included in the population in the next generation. This is especially useful for nonstationary fitness functions when certain optimizers in the population which at first seem nonpromising turn out to be highly relevant due to the constant evolving nature of the fitness landscape. These optimizers may probably be already excluded in the initial stages of optimization if a deterministic strategy is adopted, but in the case of a tournament selection strategy, it is still possible for these optimizers to be included if there are indications that they are slightly distinguished from the truly nonpromising optimizers through the result of the tournament competition.

This is also the reason for our choice of EP for our adaptive regularization problem: in our algorithm, we have generated a population of *regularization strategies*, which are vectors of regularization and segmentation parameters, as our potential optimizers. The optimal regularization strategy is selected from the population at each generation according to criteria to be described in a later section, and this is used to restore the image for a single iteration. This partially restored image is then used as the basis for the evolution of regularization strategies in the next generation. In other words, the fitness landscape in the next generation depends on the particular optimizer selected in the previous generation, thus constituting a nonstationary optimization problem. The stochastic tournament selection strategy is therefore useful in retaining those regularization strategies that initially seem nonpromising but are actually highly relevant in later restoration stages.

6.3   ETC-pdf Image Model

In this chapter, we address the adaptive regularization problem by proposing a novel image model, the adoption of which in turn necessitates the use of powerful optimization algorithms such as those typical in the field of evolutionary computation. This model is observed to be capable of succinctly characterizing common properties of a large class of images, and thus we will regard any restored image that conforms to this image model as suitably regularized. In other words, this model can be regarded as a possible objective characterization of our usual notion of subjective quality. The model, which is specified as the probability distribution of the ETC measure introduced in Chapter 5, is approximated as a histogram, and the regularization parameters in the various image regions are chosen in such a way that the corresponding ETC histogram of the resulting restored image matches the model pdf closely, that is, minimizing the difference between the two distributions. It is obvious that the resulting error function, which involves differences of discrete distributions, is highly irregular and nondifferentiable, and thus necessitates the use of powerful optimization algorithms. In view of this, we have chosen evolutionary programming as the optimization algorithm to search for the minimizer of this cost function.

Denoting the probability density function of the ETC measure κ within a typical image as $p_\kappa(\kappa)$, we have plotted the ETC histograms, which are approximations of the ETC-pdf, for several images (shown in Figure 6.1). Notice that the histograms peak around $\kappa = 1$, indicating the predominance of smooth regions. As κ increases, values of the various histograms gradually decrease, with $p_\kappa(\kappa) \approx 0$ for $\kappa \approx K$, which is the size of the averaging window used ($K = 5$ in the current example). This indicates the smaller proportion of edges and textured regions. More importantly, it is seen that, although there are slight deviations between the various ETC histograms, they in general assume the typical form as shown in Figure 6.1. This is in contrast with the traditional gray-level histogram that is usually used to characterize the gray-level distribution in an image, and which can vary widely from image to image. This is illustrated in Figures 6.2a to d, where the gray-level histograms for the same images are shown. As a result, we can consider the ETC histogram as a form of *signature* for a large class of nondegraded images.

On the other hand, it is observed that the corresponding density functions for degraded images are usually very different from the standard density

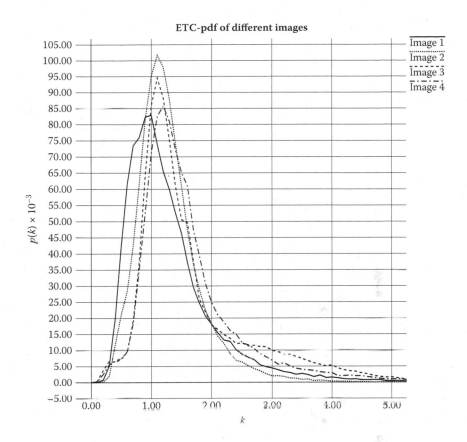

FIGURE 6.1

The ETC-pdf of different images.

function. Figure 6.3 illustrates this point by comparing the corresponding ETC-pdf of one of the images in Figure 6.1 and its blurred version. In the figure, the solid curve is the original ETC-pdf and the dotted curve is the ETC-pdf of the blurred image. It is seen that the rate of decrease is greater for the blurred image, indicating the higher degree of correlation among its pixels. Therefore, one possible regularization strategy to allow the restored image to more closely resemble the original image is to assign the parameters in such a way as to minimize the discrepancies between the ETC histogram of the restored image and that of the original.

Due to the similar shapes of the ETC-pdf in Figure 6.1, we can model the ETC-pdf using a combination of piecewise Gaussian and exponential functions. During restoration, we can adaptively assign the regularization parameters in such a way that the corresponding ETC-pdf in the restored image conforms closely to the model density function. In this work, we have

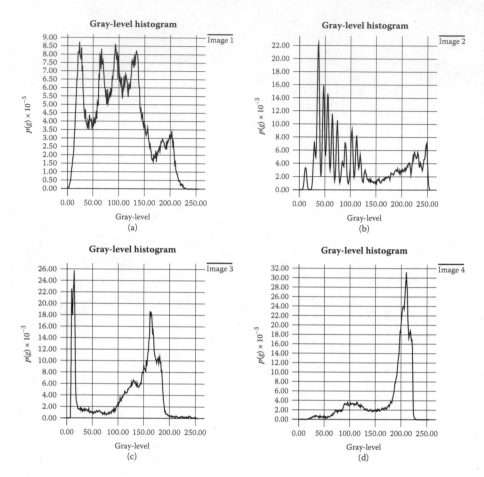

FIGURE 6.2
Gray-level histograms of different images: (a) image 1, (b) image 2, (c) image 3, (d) image 4.

adopted the following model $p_\kappa^M(\kappa)$ for the typical ETC-pdf, which can best characterize the density function of a large class of images.

$$p_\kappa^M(\kappa) = \begin{cases} ce^{-(\kappa-1)^2} & 0 \le \kappa \le 1 \\ ca_1^{\varepsilon(\kappa-1)} & 1 < \kappa \le \kappa_3 \\ ca_1^{\varepsilon(\kappa_3-1)}a_2^{\varepsilon(\kappa-\kappa_3)} & \kappa_3 < \kappa \le K \end{cases} \tag{6.6}$$

In Equation (6.6), we use a Gaussian function segment to model the density function when $\kappa \in [0, 1]$, and use two exponential function segments to model the tail distribution in the interval $[1, K]$, with the corresponding parameters a_1 and a_2 satisfying $a_1 < a_2$. The constant κ_3 in the equation is that value of κ that corresponds to an almost equipartition of the K^2 variables into three components. It has been shown for the case of 5×5 averaging window that $\kappa_3 \approx 1.8$. We could have modeled the density function in the interval $[1, K]$

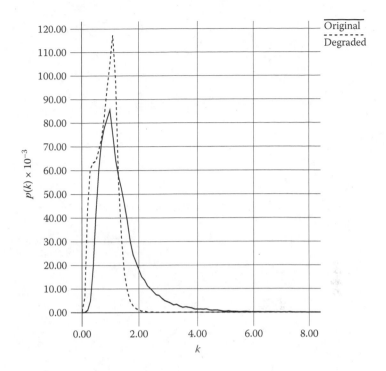

FIGURE 6.3
The ETC-pdf for a typical image and its blurred version. The solid curve is the ETC-pdf of the original image. The dotted curve is the ETC-pdf of the blurred image. Size of the averaging window is 5×5.

using a single exponential segment with parameter a, and in turn estimate this parameter from a histogram of κ using typical real-world images. Instead, we have used two separate exponential functions, with a_1 and a_2 chosen such that $a_1 < a_2$. The reason for doing this and choosing the particular transition point κ_3 is that: for $\kappa \in [1, \kappa_3]$, the neighborhood surrounding the current pixel consists of fewer than three components, which usually corresponds to a mixture of noises and smooth regions, and is particularly undesirable given the enhanced visibility of noises against the smooth background. One must therefore limit the probability of occurrence of such values of κ, which explains the adoption of $a_1 < a_2$ in the probability density model allowing a smaller probability of occurrences for $\kappa \in [1, \kappa_3]$. One may argue that this may adversely affect the probability of edge occurrence, which consists of two gross components with $\kappa = \kappa_2 \in [1, \kappa_3]$ as well, but edges usually occupy a small area in a typical image, which translates to a very small occurrence probability, so it is much more probable that those locations with κ in that interval correspond to a mixture of noises and smooth backgrounds, and the main effect of probability reduction in this interval is the elimination of this type of artifact. The variable ε controls the rates of decay for the two exponentials, and the constant c in the equation is a normalization factor such that

$$\int_0^K p_\kappa^M(\kappa)\, d\kappa = 1 \tag{6.7}$$

6.4 Adaptive Regularization Using Evolutionary Programming

As mentioned previously, the value of the regularization parameter in image restoration applications is usually determined by trial and error. Therefore, the purpose of establishing the model in Section 6.3 is to provide an objective way by which we can assign the parameters adaptively to result in the best subjective quality. Formally, we replace the constrained least square cost function for image restoration with the following cost function:

$$E = \frac{1}{2}\|\mathbf{y} - \mathbf{H}\hat{\mathbf{f}}\|_{\mathbf{A}}^2 + \frac{1}{2}\|\mathbf{D}\hat{\mathbf{f}}\|_{\mathbf{L}}^2 \tag{6.8}$$

where, instead of a single parameter λ, we employ a diagonal *weighting matrix* $\mathbf{L}$ defined as follows:

$$\mathbf{L} = diag[\lambda(\sigma_1), \dots, \lambda(\sigma_{N_I})] \tag{6.9}$$

where $\sigma_i, i = 1, \dots, N_I$ is the local standard deviation of the ith pixel in the image. This is similar to the cost function associated with the spatially adaptive iterative restoration algorithm that was adopted as a benchmark in Chapter 5. However, an important difference between the current formulation and the previous algorithm is that, while the previous algorithm performs adaptive regularization by separately adjusting a global parameter $\lambda(\hat{\mathbf{f}})$ and a weighting matrix $\mathbf{A}$, the current formulation combines these two into a single weighting matrix $\mathbf{L}$, the entries of which are determined through evolutionary programming. Similarly, the weighting matrix $\mathbf{A}$ is usually specified in such a way as to complement the effect of $\mathbf{L}$, or is sometimes simply replaced by the identity matrix $\mathbf{I}$, as we have chosen for the current algorithm.

Denote the ETC-pdf for the restored image $\hat{\mathbf{f}}$ as $p_\kappa(\kappa)$; our objective is to select the particular forms for $\lambda(\sigma_i)$ in the weighting matrix $\mathbf{L}$ in such a way as to minimize the following *weighted probability density error measure* of the ETC measure κ (ETC-pdf error measure).

$$E_{pdf}^\kappa = \int_0^K w(\kappa)\big(p_\kappa^M(\kappa) - p_\kappa(\kappa)\big)^2 d\kappa \tag{6.10}$$

where the weighting coefficients are defined as follows

$$w(\kappa) \equiv \frac{1}{\max\big(p_\kappa^M(\kappa), p_\kappa(\kappa)\big)^2} \tag{6.11}$$

to compensate for the generally smaller contribution of the tail region to the total probability. In practice, we replace the integral operation by a finite summation over a suitable discretization of $[0, K]$, and the density function $p_\kappa(\kappa)$ is approximated using the histogram of κ in the partially restored image.

$$E_{pdf}^\kappa = \sum_{r=1}^{d} w(r\triangle)\big(p_\kappa^M(r\triangle) - \hat{p}_\kappa(r\triangle)\big)^2 \tag{6.12}$$

where $\triangle$ is the width of the discretization interval, $\hat{p}_\kappa(r\triangle)$ is the estimated ETC-pdf of κ in terms of its histogram, and d is the number of discretization intervals. Since the histogram records the relative occurrence frequencies of the various values of κ based on the previous discretization, it involves the counting of discrete quantities. As a result, the overall error function given in Equation (6.12) is obviously nondifferentiable with respect to the regularization parameters, and the evolutionary programming approach provides a viable option to minimize this error function.

Evolutionary programming is a population-based optimization algorithm in which the individual optimizers in the population compete against each other with respect to the minimization of a cost function, or equivalently, the maximization of a fitness function [138]. In the current case, we have already specified the fitness function, which is Equation (6.12). In addition, we have to specify the form of the individual optimizers in the population as well. For the adaptive regularization problem, we consider the following *regularization profile* $\lambda(\sigma_i)$ defined on the local standard deviation σ_i around the ith pixel

$$\lambda(\sigma_i) = \frac{\lambda_{max} - \lambda_{min}}{1 + e^{\beta(\sigma_i - \alpha)}} + \lambda_{min} \tag{6.13}$$

Equation (6.13) defines a decreasing sigmoidal function on the local standard deviation range of the image, which is consistent with our previous view that large λ is required at low variance pixels to suppress noise and small λ is required at high variance pixels to enhance the features there. There are four parameters in Equation (6.13) that determine the overall λ assignment strategy: λ_{min} and λ_{max} represent the minimum and maximum parameter values used, respectively, α represents the offset of the sigmoidal transition from the origin, thus implicitly defining the standard deviation threshold that separates the small variance from the large variance region, and β controls the steepness of the sigmoidal transition. The various parameters of a typical regularization profile are illustrated in Figure 6.4.

Concatenating these four parameters together with the *mutation strategy parameters* $\sigma_{\lambda_{min}}^m$, $\sigma_{\lambda_{max}}^m$, σ_α^m, σ_β^m (not to be confused with the local standard deviation σ_i of an image) into an 8-tuple, we define the following *regularization strategy* S_p as the pth potential optimizer in the population.

$$S_p \equiv \big(\lambda_{min,p}, \lambda_{max,p}, \alpha_p, \beta_p, \sigma_{\lambda_{min},p}^m, \sigma_{\lambda_{max},p}^m, \sigma_{\alpha,p}^m, \sigma_{\beta,p}^m\big) \tag{6.14}$$

Employing the usual operations of evolutionary computational algorithms, which in the case of evolutionary programming is restricted to the mutation

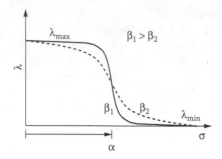

FIGURE 6.4
Illustrating the various parameters of a typical regularization profile $\mathcal{S}_p$.

operator [40], we generate a population P consisting of μ instances of $\mathcal{S}_p$ in the first generation, and we apply mutation to each of these μ parents to generate μ descendants in each subsequent generation according to the following equations

$$\lambda'_{\min, p} = \lambda_{\min, p} + N(0, \sigma^m_{\lambda_{\min, p}}) \tag{6.15}$$

$$\lambda'_{\max, p} = \lambda_{\max, p} + N(0, \sigma^m_{\lambda_{\max, p}}) \tag{6.16}$$

$$\alpha'_p = \alpha_p + N(0, \sigma^m_{\alpha, p}) \tag{6.17}$$

$$\beta'_p = \beta_p + N(0, \sigma^m_{\beta, p}) \tag{6.18}$$

where $N(0, \sigma)$ denotes a Gaussian random variable with zero mean and standard deviation σ. The variables $\lambda'_{\min, p}$, $\lambda'_{\max, p}$, α'_p, and β'_p are the components of the descendant strategy $\mathcal{S}'_p$. The mutation strategy parameters are updated according to the following *log-normal adaptation rule*.

$$\sigma'^m_{\lambda_{\min}, p} = \sigma^m_{\lambda_{\min}, p} \exp(N(0, \tau_1) + N_j(0, \tau_2)) \tag{6.19}$$

$$\sigma'^m_{\lambda_{\max}, p} = \sigma^m_{\lambda_{\max}, p} \exp(N(0, \tau_1) + N_j(0, \tau_2)) \tag{6.20}$$

$$\sigma'^m_{\alpha, p} = \sigma^m_{\alpha, p} \exp(N(0, \tau_1) + N_j(0, \tau_2)) \tag{6.21}$$

$$\sigma'^m_{\beta, p} = \sigma^m_{\beta, p} \exp(N(0, \tau_1) + N_j(0, \tau_2)) \tag{6.22}$$

where $N(0, \tau_1)$ is held fixed across all mutation parameters, while $N_j(0, \tau_2)$ is generated anew for each parameter. The values for τ_1 and τ_2 are $(\sqrt{2N})^{-1}$ and $(\sqrt{2\sqrt{N}})^{-1}$, respectively, as suggested in [39], where $2N$ is the dimension of the regularization strategy $\mathcal{S}_p$.

For each regularization strategy $\mathcal{S}_p$ in the population, we can in principle use it to restore a blurred image, and then build up the histogram $\hat{p}_\kappa(\kappa)$ by evaluating κ at each pixel and counting their relative occurrence frequencies. Finally, we can evaluate the ETC-pdf error $E^\kappa_{pdf}(\mathcal{S}_p)$ as a function of the strategy $\mathcal{S}_p$ by comparing the difference of the normalized histogram and the model ETC-pdf, and use this value as our basis for competition and selection. For the selection stage, we adopt tournament competition as our selection mechanism, where we generate a subset $T(\mathcal{S}_p) \subset P \setminus \{\mathcal{S}_p\}$ for each strategy $\mathcal{S}_p$ with $|T| = Q$ by randomly sampling the population Q times. We can then form the set $\mathcal{W}(\mathcal{S}_p)$ as below

$$\mathcal{W}(\mathcal{S}_p) = \left\{ \mathcal{S}_q \in T : E^{\kappa}_{pdf}(\mathcal{S}_q) > E^{\kappa}_{pdf}(\mathcal{S}_p) \right\} \tag{6.23}$$

which contains all those strategies $\mathcal{S}_q$ in T with their corresponding ETC-pdf error greater than that of $\mathcal{S}_p$. We then define the *win count* $w_c(\mathcal{S}_p)$ as the cardinality of $\mathcal{W}(\mathcal{S}_p)$

$$w_c(\mathcal{S}_p) = |\mathcal{W}(\mathcal{S}_p)| \tag{6.24}$$

Denoting the population at the current generation as $P(t)$, we order the regularization strategies $\mathcal{S}_p$ in $P(t)$ according to decreasing values of their associated win counts $w_c(\mathcal{S}_p)$, and choose the first μ individuals in the list to be incorporated into the next generation $P(t+1)$.

6.4.1 Competition under Approximate Fitness Criterion

As we have mentioned, we can in principle perform a restoration for each of the 2μ regularization strategies in $P(t)$ to evaluate its fitness, but each restoration is costly in terms of the amount of computations, since it has to be implemented iteratively, and is especially the case when the number of individuals in the population is large. We therefore resort to an approximate competition and selection process where each individual strategy in the population is used to restore only a part of the image. Specifically, we associate each strategy $\mathcal{S}_p$, $p = 1, \ldots, 2\mu$ in the population $P(t)$ with a subset $\mathcal{R}_p \subset \mathcal{X}$, where $\mathcal{X}$ denotes the set of points in an $N_y \times N_x$ image lattice

$$\mathcal{X} = \{(i_1, i_2) : 1 \le i_1 \le N_y, 1 \le i_2 \le N_x\} \tag{6.25}$$

and where the regions $\mathcal{R}_p$ form a partition of $\mathcal{X}$

$$\mathcal{R}_{p_1} \cap \mathcal{R}_{p_2} = \emptyset \quad p_1 \neq p_2 \tag{6.26}$$

$$\bigcup_{p=1}^{2\mu} \mathcal{R}_p = \mathcal{X} \tag{6.27}$$

In this way, we can define the quantity $\hat{E}^{\kappa}_{pdf}(\mathcal{S}_p, \mathcal{R}_p)$, which is both a function of $\mathcal{S}_p$ and $\mathcal{R}_p$, as the evaluation of $E^{\kappa}_{pdf}(\mathcal{S}_p)$ restricted to the subset $\mathcal{R}_p$ only. This serves as an estimate of the exact $E^{\kappa}_{pdf}(\mathcal{S}_p)$, which is evaluated over the entire image lattice $\mathcal{X}$ and is therefore a function of $\mathcal{S}_p$ only. In order to approximate the original error function closely, we must choose the subsets $\mathcal{R}_p$ in such a way that each of them captures the essential features of the distribution of the ETC measure in the complete image. This implies that each of these subsets should be composed of disjoint regions scattered almost uniformly throughout the lattice $\mathcal{X}$ and that both the ways of aggregating these regions into a single $\mathcal{R}_p$ and the ways of assigning these subsets to each individual in the population should be randomized in each generation. In view of these, we partition the image lattice $\mathcal{X}$ into nonoverlapping square blocks B_s of size $n \times n$

$$\mathcal{B} = \{B_s : s \in \mathcal{I}\} \tag{6.28}$$

where

$$B_{s_1} \cap B_{s_2} = \emptyset \quad s_1 \neq s_2 \tag{6.29}$$

$$\bigcup_{s \in \mathcal{I}} B_s = \mathcal{X} \tag{6.30}$$

In this equation, $\mathcal{I}$ is the index set $\{1, \ldots, S\}$ and S is the number of $n \times n$ square blocks in the image lattice. Assuming $2\mu < S$ and furthermore that $u \equiv S/2\mu$ is an integer, that is, 2μ is a factor of S, we can construct the following *random partition* $\mathcal{P}_{\mathcal{I}} = \{\mathcal{I}_p : p = 1, \ldots, 2\mu\}$ of $\mathcal{I}$ where

$$\mathcal{I}_{p_1} \cap \mathcal{I}_{p_2} = \emptyset \quad p_1 \neq p_2 \tag{6.31}$$

$$\bigcup_{p=1}^{2\mu} \mathcal{I}_p = \mathcal{I} \tag{6.32}$$

and $|\mathcal{I}_p| = u$ as follows:

- Randomly sample the index set $\mathcal{I}$ u times without replacement to form $\mathcal{I}_1$.
- For $p = 2, \ldots, 2\mu$, randomly sample the set $\mathcal{I} \setminus \bigcup_{q=1}^{p-1} \mathcal{I}_q$ u times to form $\mathcal{I}_p$.

Finally, we define the subsets $\mathcal{R}_p$ corresponding to each individual strategy $\mathcal{S}_p$ as follows.

$$\mathcal{R}_p = \bigcup_{s \in \mathcal{I}_p} B_s \tag{6.33}$$

We can then evaluate $\hat{E}_{pdf}^{\kappa}(\mathcal{S}_p, \mathcal{R}_p)$ for each strategy $\mathcal{S}_p$ with its corresponding region $\mathcal{R}_p$ and use it in place of the exact error $E_{pdf}^{\kappa}(\mathcal{S}_p)$ for the purpose of carrying out the tournament competition and ranking operation in Section 6.4. The process is illustrated in Figure 6.5. In this case, for the purpose of illustration, we have used a population with only four individuals. The image is divided into 12 subblocks and we have assigned 3 subblocks to each individual in the population. This assignment is valid only for one update iteration only, and a randomly generated alternative assignment is adopted for the next iteration to ensure adequate representation of the entire image through these assigned blocks.

1	4	3	2
3	4	1	3
4	1	2	2

FIGURE 6.5
Illustrating the assignment of regions $\mathcal{R}_p$ to individual regularization strategy $\mathcal{S}_p$ for a 4-member population.

6.4.2 Choice of Optimal Regularization Strategy

After the tournament competition and ranking operations, we can in principle choose the optimal strategy $\mathcal{S}^*$ from the population $P(t)$ and use it to restore the image for a single iteration. We can define optimality here in several ways. The obvious definition is to choose that strategy $\mathcal{S}^*$ with the minimum value of $\hat{E}^\kappa_{pdf}$

$$\mathcal{S}^* = \arg\min_p \hat{E}^\kappa_{pdf}(\mathcal{S}_p, \mathcal{R}_p) \qquad (6.34)$$

Alternatively, we can form a subset of those elite strategies with their win index $w_c(\mathcal{S}_p)$ equal to the maximum possible value, that is, the tournament size Q:

$$\mathcal{E}_Q = \{\mathcal{S}_p : w_c(\mathcal{S}_p) = Q\} \subset P(t) \qquad (6.35)$$

and then choose our strategy $\mathcal{S}^*$ by uniformly sampling from this subset.

The above selection schemes are suitable for the competition and selection stage of conventional evolutionary computational algorithms where the fitness value of each individual directly reflects its inherent optimality. In the current approximate competition scheme; however, we have replaced the exact fitness value E^κ_{pdf} with the estimate $\hat{E}^\kappa_{pdf}$, which depends on both the inherent fitness of the strategy $\mathcal{S}_p$ and its particular assigned region $\mathcal{R}_p$. Therefore, there may exist cases where a nonoptimal strategy will acquire a high fitness score due to a chance combination of image blocks $\mathcal{B}_s$ in forming its assigned region. To prevent this, a more sophisticated selection scheme involving the elite individuals in the population is required.

Recall that the estimated error $\hat{E}^\kappa_{pdf}$ is both a function of the inherent fitness of the strategy $\mathcal{S}_p$ and its assigned region $\mathcal{R}_p$, and a particular combination of image blocks $\mathcal{B}_s$ may result in a low estimated error even though the inherent fitness of $\mathcal{S}_p$ is not very high. However, we should expect that, for a strategy with low inherent fitness, it will quickly encounter a combination of image blocks $\mathcal{B}_s$ leading to a very large estimated error and become displaced from the population. On the other hand, for a strategy with high inherent fitness, we should expect that the corresponding estimated error will be low for a variety of image block combinations in each generation, and it will most probably survive into the next generation. In view of this, a proper way to estimate an optimal regularization strategy from the current population should involve the *survival time* t_p of each individual strategy $\mathcal{S}_p$ that is defined as follows:

$$t_p \equiv t - t_p^i \tag{6.36}$$

such that

$$\mathcal{S}_p \in \bigcap_{t'=t_p^i}^{t} P(t') \tag{6.37}$$

In other words, t_p^i is the generation where the strategy $\mathcal{S}_p$ first appears in the population, and t is the current generation. The survival time t_p is defined as the difference between these two indices. In general, it is reasonable to assume that a regularization strategy with a long survival time t_p is more likely to possess high inherent fitness, but whether a strategy with a short survival time possesses high inherent fitness is yet to be confirmed in later generations. As a result, it is more reasonable to choose the optimal regularization strategy based on those in the population with long survival times. Rearranging the corresponding survival time t_p of each individual in ascending order,

$$t_{(1)}, \ldots, t_{(p)}, \ldots, t_{(2\mu)} \tag{6.38}$$

such that

$$t_{(1)} \le \cdots \le t_{(p)} \le \cdots \le t_{(2\mu)} \tag{6.39}$$

and $t_{(p)}$ denotes the pth order statistic of the survival time sequence, we expect those values $t_{(p)}$ with large index p will most likely correspond to a strategy $\mathcal{S}_{(p)}$ with high inherent fitness. Choosing $p_0 > 1$ and regarding each strategy as a vector in $\mathbf{R}^8$, we define the following *combined regularization strategy*

$$\mathcal{S}_{p_0}^* = \frac{1}{2\mu - p_0 + 1} \sum_{p=p_0}^{2\mu} \mathcal{S}_{(p)} \tag{6.40}$$

To ensure the inclusion of only those individuals with high inherent fitness in the above averaging operation, the value p_0 is usually chosen such that $p_0 \gg 1$.

In addition, if we possess *a priori* knowledge regarding inherent properties of desirable regularization strategies characterized by the *constraint set* $\mathcal{C}_S$, we can modify the previous averaging procedure to include this knowledge as follows

$$S_{p_0}^* = \frac{\sum_{p=p_0}^{2\mu} I_{\{S_{(p)} \in \mathcal{C}_S\}} S_{(p)}}{\sum_{p=p_0}^{2\mu} I_{\{S_{(p)} \in \mathcal{C}_S\}}} \tag{6.41}$$

where $I_{\{S_{(p)} \in \mathcal{C}_S\}}$ is the indicator function of $\mathcal{C}_S$.

The optimal strategy $S_{p_0}^*$ is constructed based on the estimated ETC-pdf error $\hat{E}_{pdf}^{\kappa}$, and we have to evaluate its performance based on the true error measure E_{pdf}^{κ} eventually, which requires using $S_{p_0}^*$ to restore the whole image for one iteration and then collecting the statistics of the measure κ on the resulting image to form an estimation of $p_{\kappa}(\kappa)$. In the current algorithm, this has to be performed once only, again highlighting the advantage of using $\hat{E}_{pdf}^{\kappa}$, which requires performing restoration over an image subset only for each individual strategy in the population, instead of using E_{pdf}^{κ} in the competition process, which requires performing restoration for an entire image. In this way, the time required for fitness evaluation is independent of the number of individuals in the population.

Denoting the current optimal strategy as $S^*(t)$ and the last strategy used to update the image as $S^*(t-1)$, we should decide between using $S_{p_0}^*$ or $S^*(t-1)$ as $S^*(t)$, or we should simply leave the image unrestored for the present iteration. Denoting the *null strategy* as S_ϕ, which amounts to leaving the present image unrestored, the basis of this decision should be the true ETC-pdf error value $E_{pdf}^{\kappa}(S)$ where $S = S_{p_0}^*, S^*(t-1)$ or S_ϕ. Using $\hat{f}_r(S)$ to indicate the resulting restored image through the action of S, $\hat{f}(t)$ as the updated image, and $\hat{f}(t-1)$ as the pre-updated image, we adopt the following decision rule for choosing $S^*(t)$

- If $E_{pdf}^{\kappa}(S_{p_0}^*) = \min\{E_{pdf}^{\kappa}(S_{p_0}^*), E_{pdf}^{\kappa}(S^*(t-1)), E_{pdf}^{\kappa}(S_\phi)\}$
  1. $S^*(t) = S_{p_0}^*$
  2. $\hat{f}(t) = \hat{f}_r(S_{p_0}^*)$
- If $E_{pdf}^{\kappa}(S^*(t-1)) = \min\{(E_{pdf}^{\kappa}(S_{p_0}^*), E_{pdf}^{\kappa}(S^*(t-1)), E_{pdf}^{\kappa}(S_\phi)\}$
  1. $S^*(t) = S^*(t-1)$
  2. $\hat{f}(t) = \hat{f}_r(S^*(t-1))$
- If $E_{pdf}^{\kappa}(S_\phi) = \min\{(E_{pdf}^{\kappa}(S_{p_0}^*), E_{pdf}^{\kappa}(S^*(t-1)), E_{pdf}^{\kappa}(S_\phi)\}$
  1. $S^*(t) = S_\phi$
  2. $\hat{f}(t) = \hat{f}_r(S_\phi) = x(t-1)$

6.5 Experimental Results

The current algorithm was applied to the same set of images in Chapter 5. The parameters of the ETC-pdf image model were chosen as follows: $a_1 = 0.83, a_2 = 0.85$ and $\kappa_3 = 1.8$ using a 5×5 averaging window. The value

FIGURE 6.6
Restoration of the flower image (5×5 uniform blur, 30 dB BSNR). (a) Blurred image, (b) non-adaptive approach ($\lambda = 1/\text{BSNR}$), (c) HMBNN, (d) EP.

$\Delta = 0.1$ is adopted as the bin width. For the EP algorithm we have chosen $\mu = 16$, a subblock size of 32×32 for B_s, and a tournament size of $Q = 10$ in the selection stage.

In Figure 6.6, we apply the algorithm to the flower image. The degraded image is shown in Figure 6.6a, and the restored image using the EP algorithm is shown in Figure 6.6d. For comparison purposes, we also include the same image restored using alternative algorithms in Figures 6.6b and c. Figure 6.6b shows the restored image using the nonadaptive approach, where a single $\lambda = 1/\text{BSNR}$ is adopted for the whole image. We can notice the noisy appearance of the resulting restoration when compared with Figure 6.6d. We have also included the HMBNN restoration results in Figure 6.6c, and we can notice that the qualities of the two restored images are comparable.

In Figure 6.7, we apply the algorithm to the flower image degraded by a 5×5 uniform PSF at an increased noise level of 20 dB BSNR. The degraded image

FIGURE 6.7
Restoration of the flower image (5 × 5 uniform blur, 20 dB BSNR). (a) Blurred image, (b) non-adaptive approach ($\lambda = 1/BSNR$), (c) HMBNN, (d) EP.

is shown in Figure 6.7a. The increased noise level manifests itself in the more noisy appearance of the nonadaptive restoration result in Figure 6.7b. This can be compared with our result in Figure 6.7d, where the adoption of the ETC-pdf error criterion allows the redetermination of the relevant parameters through EP. We have also included the HMBNN result in Figure 6.7c. Again, the two restored results are comparable, although we can notice a slightly more noisy appearance in the EP result.

We also apply the EP restoration algorithm to the image in Figure 6.8. Figures 6.8a and b show the degraded images using a 5 × 5 uniform PSF with 30 dB and 20 dB noise added, respectively, and the EP restoration results in Figures 6.8c and d.

The λ assignment maps for the flower image under 5 × 5 uniform blur are shown in Figure 6.9. Figure 6.9a shows the assignment map under 30 dB additive noise, and Figure 6.9b shows the corresponding map under 20 dB noise.

FIGURE 6.8
Restoration of the Lena image (5 × 5 uniform blur). (a)-(b) Degraded images: (a) 30 dB BSNR,
(b) 20 dB BSNR, (c)-(d) restoration results using EP: (c) 30 dB BSNR, (d) 20 dB BSNR.

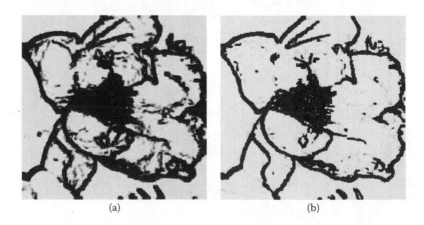

FIGURE 6.9
Distribution of λ values for the flower image: (a) 30 dB BSNR, (b) 20 dB BSNR.

In the two maps, the darker gray values correspond to small λ values, and the brighter values correspond to large λ values. In general, the regularization strategy discovered by the artificial evolutionary process assigns small parameter values to edge/textured regions. These smaller values in turn help to bring out more fine details in these regions in the accompanying restoration phase. On the other hand, large λ values are assigned to the smooth regions for noise suppression. In addition, the maps for the same image are different under different levels of additive noise: at low noise levels, the area over which large λ values are assigned is small compared with the corresponding maps under high noise levels. This implies that edge/texture enhancement takes precedence over noise suppression at low noise levels. On the other hand, for higher noise levels, most of the areas are assigned large λ values, and only the very strong edges and textured regions are assigned moderately smaller λ values. We can thus conclude that at low noise levels, the primary purpose is edge/texture enhancement, whereas for higher noise levels it is noise removal.

This can also be confirmed from Figure 6.10, where we show plots of the regularization profile corresponding to the λ-maps in Figure 6.10 for the flower image (the logarithmic plots are shown due to the large dynamic ranges of the parameters). It can be seen that the profile corresponding to 20 dB noise is shifted to the right with respect to the 30 dB profile, which implies a larger value of the threshold parameter α_p and resulting in a larger image area being classified as smooth regions. This agrees with our previous conclusion that noise suppression takes precedence over edge/texture enhancement when the noise level is high.

It is also seen from the figure that pixels with high local standard deviation, which possibly correspond to significant image features, are assigned different values of λ_i at different noise levels. At lower noise levels, the λ_i values assigned at large σ_i are seen to be smaller than the corresponding values at higher noise levels. This is reasonable due to the possibility of excessive noise amplification at higher noise levels, which in turn requires higher values of λ_i for additional noise suppression.

We have also applied this algorithm to the eagle image. Figure 6.11a shows the degraded images for eagle under 5×5 uniform blur at 30 dB BSNR and Figure 6.11b shows the result using the current EP approach. We can more readily appreciate the importance of adaptive processing from these additional results.

We list the RMSE of the restored images using the current algorithm, together with those restored using the HMBNN algorithm, in Table 6.1. Each RMSE value for the EP algorithm is evaluated as the mean of five independent runs (the values in the parentheses indicate the standard deviation among the different runs). It is seen that, in all the cases, the mean RMSE values of the EP algorithm are comparable to the HMBNN error values.

We also include the standard deviation associated with each RMSE value for the EP results over five independent runs. We can see that the values of these standard variations are within acceptable ranges. It was also observed that

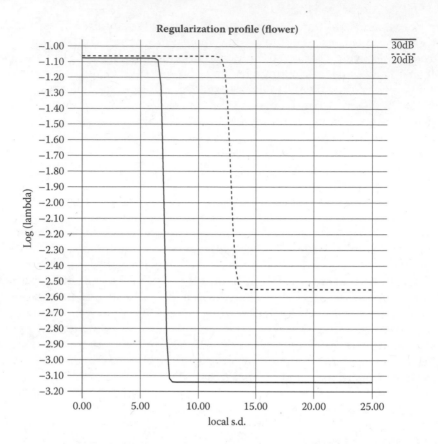

FIGURE 6.10
Regularization profiles under different noise levels for the flower image.

there are no noticeable differences between the appearances of the restored images in different trials. As a result, we can conclude that these variations are acceptable for the current problem.

6.6 Other Evolutionary Approaches for Image Restoration

An alternative evolutionary computational approach to adaptive image regularization based on a neural network model, the hierarchical cluster model (HCM) is proposed in [145]. An HCM is a hierarchical neural network that coordinates parallel, distributed subnetworks in a recursive fashion with its cluster structures closely matching the homogeneous regions of an image. In addition, its sparse synaptic connections are effective in reducing the computational cost of restoration. In the proposed algorithm, the degraded image is

(a)   (b)

FIGURE 6.11
Further restoration results (5×5 uniform blur, 30 dB BSNR): (a) blurred eagle image, (b) restored image using EP.

first segmented into different regions based on its local statistics. They are further partitioned into separate clusters and mapped onto a three-level HCM. Each image cluster, equipped with an optimized regularization parameter λ, performs restoration using the model-based neuron updating rules.

The main difference between this algorithm and our previous approaches is that an evolutionary strategy scheme is adopted to optimize the λ values by minimizing the HCM energy function. Specifically, the scheme progressively selects the well-evolved individuals that consist of a set of partially restored images with their associated cluster structures, segmentation maps, and the optimized λ values. Once the restoration is completed, the final empirical relationship between the optimized λ values and a specific local perception measure can be determined and reused in the restoration of other unseen, degraded images, which will remove the computational overhead of evolutionary optimization. The specific features of this approach are described below.

TABLE 6.1

RMSE Values of the Restoration Results Using Different Restoration Algorithms

Image, Noise Level	Blurred	Nonadaptive	HMBNN	EP
Lena, 30 dB	13.03	8.65	7.09	7.35(0.0611)
Lena, 20 dB	13.79	10.56	9.69	9.78(0.112)
Eagle, 30 dB	9.88	10.28	6.89	7.13(0.0600)
Eagle, 20 dB	11.98	11.57	8.99	8.97(0.0364)
Flower, 30 dB	8.73	7.70	4.98	5.45(0.0310)
Flower, 20 dB	10.06	8.47	6.92	7.06(0.0767)

6.6.1   Hierarchical Cluster Model

The hierarchical cluster model (HCM) is a nested neural network that consists of parallel, distributed subnetworks or clusters. It models the organization of the neocortex in human brain where functional groups of neurons organize themselves dynamically into multidimensional subnetworks. HCM can be constructed in a bottom-up manner. Individual neurons form the trivial level 0 clusters of the network. The neurons sharing similar functionality and characteristics coalesce into a number of level 1 clusters. Likewise, those level 1 clusters with homogeneous characteristics in turn coalesce into level 2 clusters. This process continues repeatedly to form a hierarchical neural network with multidimensional distributed clusters.

A three-level HCM is employed in the context of image regularization. There exists a close correspondence between the chosen HCM with image formation. The individual neuron or level 0 cluster of HCM corresponds to each image pixel. Similarly, level 1 clusters correspond to homogeneous image regions whereas level 2 clusters correspond to dynamic boundary adjustments between regions. The algorithm is composed of the following two stages.

6.6.2   Image Segmentation and Cluster Formation

In this first stage, a degraded image is segmented into different homogeneous clusters and restored with adaptive λ values to achieve optimum visual result. A three-level HCM is constructed from an image in two steps: an image is first segmented into different regions based on its local statistics and each individual region is then further partitioned into a set of clusters.

6.6.3   Evolutionary Strategy Optimization

A (μ, τ, ν, ρ) evolutionary scheme is adopted to optimize the regularization parameter values. The scheme uses μ parents, ν offspring with τ as the upper limit of life span, and ρ as the number of ancestors for each descendant. Since the regularization parameter associated with each cluster is closely related to the perception measure such as cluster average local variance, the evolutionary scheme continually explores the optimum relationship between the local λ value and the cluster average local variance by minimizing a suitable error measure. In this algorithm, a parameterized logarithmic function is adopted as a model for the nonlinear transformation in each cluster with the associated parameters of the function determined through the evolutionary scheme. In addition, the energy function of the HCM is adopted as the fitness function, the value of which is optimized through the two operations of mutation and tournament selection.

As a result, compared with the current algorithm, this alternative evolutionary restoration algorithm offers the possibility of refined local image characterization by adopting multiple sets of regularization profiles at the expense of increased computational requirements (due to the need to adopt EC to search for the optimal profile for each region), and the necessity to perform a prior segmentation of the image.

6.7 Summary

We have proposed an alternative solution to the problem of adaptive regularization in image restoration in the form of an artificial evolutionary algorithm. We first characterize an image by a model discrete probability density function (pdf) of the ETC measure κ, which reflects the degree of correlation around each image pixel, and effectively characterizes smooth regions, textures, and edges. An optimally regularized image is thus defined as the one with its corresponding ETC-pdf closest to this model ETC-pdf. In other words, during the restoration process, we have to minimize the difference between the ETC-pdf of the restored image, which is usually approximated by the ETC histogram, and the model ETC-pdf. The discrete nature of the ETC histogram and the nondifferentiability of the resulting cost function necessitates the use of evolutionary programming (EP) as our optimization algorithm to minimize the error function. The population-based approach of evolutionary programming provides an efficient method to search for potential optimizers of highly irregular and nondifferentiable cost function as the current *ETC-pdf error measure*. In addition, the current problem is also nonstationary, as the optimal regularization strategy in the current iteration of pixel updates is not necessarily the same as the next iteration. The maintenance of a diversity of potential optimizers in the evolutionary approach increases the probability of finding alternative optimizers for the changing cost function. Most significantly, the very adoption of evolutionary programming has allowed us to broaden the range of cost functions in image processing that may be more relevant to the current application, instead of being restricted to differentiable cost functions.

7

Blind Image Deconvolution

7.1 Introduction

We will again restrict the presentation of this chapter to the popular linear image degradation model:

$$g = H\hat{f} + n \qquad (7.1)$$

where g, f and n are the lexicographically ordered degraded image, original image and additive white Gaussian noise (AWGN), respectively [21,146]. H is the linear distortion operator determined by the point spread function (PSF), h. Blind image deconvolution is an inverse problem of rendering the best estimates, $\hat{f}$ and $\hat{h}$, to the original image and the blur based on the degradation model. It is a difficult ill-posed problem as the uniqueness and stability of the solution is not guaranteed [24]. Classical restorations require complete knowledge of the blur to be known prior to restoration [21,23,146] as discussed in the previous chapters. However, it is often too costly, cumbersome or, in some cases, impossible to determine the exact blur a priori. These could be due to various practical constraints such as the difficulty of characterizing air turbulence in aerial imaging, or the potential health hazard of employing a stronger incident beam to improve the image quality in x-ray imaging. In these circumstances, blind image deconvolutions are essential in recovering visual clarity from the degraded images.

Traditional blind methods such as *a priori* blur identifications formulate blind restoration into two disjoint processes where the blur is first estimated, followed by classical restoration based on the identified blur [147,148]. The methods are inflexible as they require the parametric structures of the blur to be known exactly, and are tailored specifically for the targeted blur type. In addition, they are ineffectual in identifying certain blurs such as a Gaussian mask, which does not exhibit prominent frequency nulls.

The success of linear regression theory in digital communication and signal processing motivates various researchers to consider a degraded image as an autoregressive moving average (ARMA) process. The original image is modeled as an autoregressive (AR) process, with the blur representing the moving average (MA) coefficients. Under this assumption, blind image deconvolution is transformed into an ARMA parameter estimation problem. Maximum likelihood (ML) [149] and generalized cross validation (GCV) [150] are two popular techniques used to perform the estimation. ML is commonly used in conjunction with expectation maximization (EM) to determine the parameters by maximizing their log-likelihood probabilities from the solution space. The side effect is that it suffers from insensitivity toward changes in individual parameters due to its huge input argument list. GCV evaluates the parameters by minimizing the weighted sum of the predictive errors. Its main disadvantage lies in the repetitive process of cross validations, which constitutes significant computational cost. Both methods require small AR and MA support dimensions, albeit at the cost of diminished modeling effectiveness. The problem is further complicated by the ARMA image stationarity constraint that is inconsistent with some real-life images consisting of inhomogeneous smooth, textured, and edge regions.

The conceptual simplicity of iterative image and blur estimations has attracted researchers to propose various iterative blind deconvolution schemes. These include iterative blind deconvolution (IBD) [151], simulated annealing (SA) [152], and nonnegativity and support constraints recursive inverse filtering (NAS-RIF) [153]. IBD alternates between spatial and frequency domains, imposing constraints onto the image and blur estimates repeatedly. In contrast, SA employs the standard annealing procedure to minimize a multimodal cost function. NAS-RIF extends the recursive filtering of the blurred image to optimize a convex cost function. These iterative algorithms require the image objects to have known support dimensions, and to be located in a uniform background, and thus impose restrictions on many applications. Other difficulties such as sensitivity to initial conditions, slow convergence, and interdomain dependency further complicate the iterative schemes.

Recent investigations into image regularization have been extended to address blind image deconvolution [154,155]. The problem is formulated into two symmetrical processes of image restoration and blur identification. However, the implicit symmetrical assumption conflicts with several observations. First, most blur functions have predominantly low-frequency contents. In contrast, the frequency spectrums of typical images vary from low in the smooth regions, to medium and high in the textured and edge regions. Second, studies show that practical blurs satisfy up to a certain degree of parametric structure. In comparison, we have little or no prior knowledge about the imaging scenes. Third, the methods ignore different characteristics of the image and blur domains, leading to poor incorporation of their unique properties. Recent work alleviates these difficulties by integrating the unique properties of image and blur domains into a recursive scheme based on soft blur identification and a hierarchical neural network [156]. It assigns different emphases to image restoration and blur identification according to their characteristics, thereby providing a priority-based subspace deconvolution.

This chapter presents two recent approaches to adaptive blind image deconvolution. The first is based on computational reinforced learning in an attractor-embedded solution space, and the second is a recursive scheme based on soft-decision blur identification and a hierarchical neural network.

7.1.1 Computational Reinforced Learning

The computational reinforced learning (CRL) method, an extended evolutionary strategy that integrates the merits of priority-based subspace deconvolution, is developed to generate the improved blurs and images progressively. Evolutionary algorithms (EA) are often used to solve difficult optimization problems where traditional path-oriented and volume-oriented methods fail [157,158]. Evolutionary strategies (ES), a mainstream of EA, model the evolutionary principles at the level of individuals or phenotypes. An important feature of this method lies in its capability to alleviate the common difficulties encountered by other blind algorithms, namely, interdomain dependency, poor convergence, and local minima trapping. It allows the most tangible performance measure to be adopted, rather than being restricted by the criteria's tractability as in other schemes. We embrace this feature by introducing a novel entropy-based performance measure that is both effective and intuitive in assessing the fitness of the solutions, as well as estimating the blur support dimensionality.

As blind deconvolution exhibits the property of reducibility, careful incorporation of image and blur information is instrumental in achieving good results. In accordance with the notion, a mutation attractor space is constructed by incorporating the blur knowledge domain into the algorithm. A maximum a posteriori estimator is developed to predict these attractors, and their relevance is evaluated within the evolutionary framework. A novel reinforced mutation scheme that integrates classical mutation and reinforced learning is derived. The technique combines self-adaptation of the attractors and intelligent reinforced learning to improve the algorithmic convergence, thereby reducing the computational cost significantly. The new evolutionary scheme features a multithreaded restoration that is robust toward divergence and poor local minima trappings.

Unlike traditional blind methods where a hard decision on the blur structure has to be made prior to restoration, this approach offers a continual relevance-feedback learning throughout image deconvolution. This addresses the formulation dilemma encountered by other methods, namely, integrating the information of well-known blur structures without compromising its overall flexibility. As image restoration is a high-dimensional problem, fast convergence is particularly desirable. Therefore, a stochastic initialization procedure is introduced to speed the initial searching. A probability neural network is devised to perform landscape surveying and blur dimensionality estimation. The underlying principle is that the solution subspace with better performance should, accordingly, be given more emphasis.

7.1.2   Blur Identification by Recursive Soft Decision

Similar to the CRL method, the soft-decision method offers a continual soft-decision blur adaptation throughout the restoration. It incorporates the knowledge of standard blur structure while preserving the flexibility of the blind restoration scheme. A new cost function that comprises the data fidelity measure, image and blur-domain regularization terms, and a novel soft-decision blur estimation error is introduced. A blur estimator is devised to determine various soft parametric blur estimates. The significance of the estimates is evaluated by its proximity to the currently computed blur.

In this method, we formulate blind image restoration into a two-stage subspace optimization. The overall cost function is projected and optimized with respect to the image and blur domains recursively. A nested neural network, called the hierarchical cluster model (HCM) [103,159] is employed to provide an adaptive, perception-based restoration by minimizing the image-domain cost function [160]. Its sparse synaptic connections are effective in reducing the computational cost of restoration. A blur compensation scheme is developed to rectify the statistical averaging effect that arises from the ambiguous blur identifications in the edge and texture regions. This method does not require assumptions such as image stationarity as in ARMA modeling, or known support objects in uniform backgrounds as in deterministic constraint restoration. The approach also addresses the asymmetry between image restoration and blur identification by employing the HCM to provide a perception-based restoration and conjugate gradient optimization to identify the blur.

Conjugate gradient optimization is adopted, wherever appropriate, in image restoration and blur identification. It is chosen ahead of other gradient-based approaches such as quasi-Newton and steepest descent techniques as it offers a good compromise in terms of robustness, low computational complexity, and small storage requirement.

7.2   Computational Reinforced Learning

7.2.1   Formulation of Blind Image Deconvolution as an Evolutionary Strategy

Classical image restorations usually involve the minimization of a quadratic cost function:

$$J(\hat{\mathbf{f}}) = \frac{1}{2}\hat{\mathbf{f}}^T\mathbf{Q}\hat{\mathbf{f}} + \mathbf{r}^T\hat{\mathbf{f}} \tag{7.2}$$

As the matrix $\mathbf{Q}$ is symmetric positive definite, the minimization of $J(\hat{\mathbf{f}})$ will lead to the restored image, $\hat{\mathbf{f}}$. In comparison, blind image deconvolution is an inverse problem of inferring the best estimates, $\hat{\mathbf{f}}$ and $\hat{\mathbf{h}}$, to the original image and the blur. Its formulation involves the development and optimization of a multimodal cost function $J(\hat{\mathbf{f}}, \hat{\mathbf{h}}|\mathbf{g})$.

As the image and blur domains exhibit distinctive spatial and frequency profiles, we project and minimize the cost function $J(\hat{\mathbf{f}}, \hat{\mathbf{h}}|\mathbf{g})$ with respect to each domain iteratively. The scheme can be summarized by

1. Initialize $\hat{\mathbf{h}}$
2. Recursively minimize the ith iterative subspace cost functions:

$$\min J_i(\hat{\mathbf{f}}|\mathbf{g}, \hat{\mathbf{h}}) = \min \mathbf{P}_{\hat{\mathbf{f}}}\{J_i(\hat{\mathbf{f}}, \hat{\mathbf{h}}|\mathbf{g})\} \qquad (7.3)$$

and

$$\min J_i(\hat{\mathbf{h}}|\mathbf{g}, \hat{\mathbf{f}}) = \min \mathbf{P}_{\hat{\mathbf{h}}}\{J_i(\hat{\mathbf{f}}, \hat{\mathbf{h}}|\mathbf{g})\} \qquad (7.4)$$

3. Stop when the termination criterion is satisfied

$J_i(\hat{\mathbf{f}}|\mathbf{g}, \hat{\mathbf{h}})$ and $J_i(\hat{\mathbf{h}}|\mathbf{g}, \hat{\mathbf{f}})$ are the ith iterative subspace cost functions, and $\mathbf{P}_{\hat{\mathbf{f}}}$ and $\mathbf{P}_{\hat{\mathbf{h}}}$ are the projection operators with respect to the image and blur domains. The advantages of these approaches include the simplicity of constraint and algorithmic formulation. The cost function $J(\hat{\mathbf{f}}, \hat{\mathbf{h}}|\mathbf{g})$ is usually chosen to be quadratic with respect to the image and blur domains, thereby ensuring convergence in their respective subspaces. However, the iterative projections create interdependency between $\hat{\mathbf{f}}$ and $\hat{\mathbf{h}}$, giving rise to potential poor convergence and local minima trapping. This often leads to inadequate image restorations and blur identifications.

To address these difficulties, we extend the alternating minimization procedure into an (μ, κ, ν, ρ) evolutionary strategy. The scheme uses μ parents, ν offspring with κ as the upper limit of life span, and ρ as the number of ancestors for each descendant. The mathematical representation of the new scheme can be summarized by

1. Initialize Φ_0
2. For ith generation, determine the dynamic image and blur solution spaces, Ω_i and Φ_i:

$$\Omega_i = \left\{ \hat{\mathbf{f}}_i | \hat{\mathbf{f}}_i = \arg \min_{\hat{\mathbf{f}}} J(\hat{\mathbf{f}}|\mathbf{g}, \hat{\mathbf{h}}_i \in \Phi_i) \right\} \qquad (7.5)$$

$$\Phi_i = \{ \hat{\mathbf{h}}_i | \hat{\mathbf{h}}_i = \mathbf{M} \circ \mathbf{R} \circ \mathbf{S} \circ \mathbf{F}(\hat{\mathbf{f}}_{i-1} | \hat{\mathbf{f}}_{i-1} \in \Omega_{i-1}) \} \qquad (7.6)$$

3. Stop when convergence or the maximum number of generation is reached

The blur solution space Φ_i is generated based on concatenation of the performance evaluation operator $\mathbf{F}$, the candidate selection operator $\mathbf{S}$, the niche-space recombination operator $\mathbf{R}$ and the reinforced mutation operator $\mathbf{M}$. The new technique preserves the algorithmic simplicity of the projection-based deconvolutions by performing image restoration in Equation (7.5) and blur identification in Equation (7.6). Moreover, it exploits the virtue of the evolutionary strategies to alleviate interdomain dependency and poor convergence, thereby enhancing the robustness of the deconvolution scheme.

Probability-Based Stochastic Initialization

The main challenge of applying evolutionary algorithms lies in the significant computational cost associated usually with the evolution of good solutions and characteristics. In the context of blind image deconvolution, the inherent high dimensionality puts further demands on computational efficiency. One of the key issues in blind image deconvolution is the determination of blur dimensionality. Most blind methods either assume the blur supports are known *a priori*, or impose some highly restrictive assumptions to estimate them. The combination of high image dimensionality and uncertainty in blur supports indicates the importance of intelligent initialization. In view of this, we introduce a new probability neural network shown in Figure 7.1 to provide a performance-feedback initialization.

The network consists of three layers, namely, input, intermediary, and output layers. The input layer comprises neurons representing the blur knowledge base $x_i, i = 1, \ldots, K$, and a single neuron representing the stochastic perturbation, x_s. The knowledge base consists of K established blur structures that can be adjusted to incorporate any information or previous experience. The intermediary layer comprises neurons $y_j, j = 1, \ldots, D$ representing D different blur dimensionalities. The output neuron z initializes the blur estimates by combining the performances of x_i and y_j. The neurons x_i and x_s, are instantiated dynamically to $M \times N$ dimensional vectors when connected to y_j of $M \times N$ blur support.

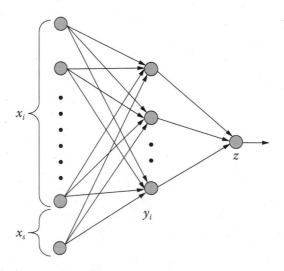

FIGURE 7.1
Probability neural network for stochastic initialization.

The linear summation at $\mathbf{y}_j$ is given by

$$\mathbf{y}_j = \sum_i w_{ji}\mathbf{x}_i + w_{js}\mathbf{x}_s \quad j = 1, \ldots, D \tag{7.7}$$

where w_{ji} is the connection weight between $\mathbf{y}_j$ and $\mathbf{x}_i$, and w_{js} is the connection weight between $\mathbf{y}_j$ and $\mathbf{x}_s$. The output node $\mathbf{z}$ combines $\mathbf{y}_j$, $j = 1, \ldots, D$ to give

$$\mathbf{z} = \sum_j \delta_{jz}\mathbf{y}_j \tag{7.8}$$

where δ_{jz} is the connection weight between $\mathbf{y}_j$ and $\mathbf{z}$. The network evaluates the performances of $\mathbf{x}_i$, and adjusts the weights w_{ji} and w_{js} to reflect their degree of relevance accordingly. The blur dimensionalities are determined from the nearest neighbor of the 2-D mixture Gaussian distribution. The systematic landscape surveying and support estimation provide an efficient approach to achieve intelligent initialization.

Network Weight Estimation

The probability-based weight estimation is formulated in accordance with the notion that subspace with good solutions should be given more emphasis. The weight estimation algorithms for w_{ji} and w_{js} are summarized as follows:

1. Initialize $\mathbf{x}_i$, $i = 1, \ldots, K$ to blur knowledge base with random parameters, and instantiate them with different dimensionalities of $\mathbf{y}_j$, $j = 1, \ldots, D$ to form the *a priori* subspace, $\mathbf{X}_{ji}$.

2. For each $\hat{\mathbf{h}}_{ji} \in \mathbf{X}_{ji}$, restore the image $\hat{\mathbf{f}}_{ji}$,

$$\hat{\mathbf{f}}_{ji} = \arg\min_{\hat{\mathbf{f}}} J\left(\hat{\mathbf{f}}|\mathbf{g}, \hat{\mathbf{h}}_{ji} \in \mathbf{X}_{ji}\right) . \tag{7.9}$$

and evaluate their fidelity measure E_{ji},

$$E_{ji} = \frac{1}{2}\left\|\mathbf{g} - \hat{\mathbf{H}}_{ji}\hat{\mathbf{f}}_{ji}\right\|^2 \tag{7.10}$$

where $\hat{\mathbf{H}}_{ji}$ is the block-Toeplitz matrix formed by $\hat{\mathbf{h}}_{ji}$.

3. Compute the knowledge and stochastic probabilities, P_{ji} and P_{js}:

$$P_{ji} = e^{-\chi \frac{E_{ji}}{\sum_i E_{ji}}} \quad i = 1, \ldots, K \quad j = 1, \ldots, D \tag{7.11}$$

and

$$P_{js} = e^{-\chi \frac{\min\{E_{ji}, \overline{\Delta E}_{ji}\}}{\sum_i E_{ji}}} \quad j = 1, \ldots, D \tag{7.12}$$

where $\overline{\Delta E}_{ji}$ is the average difference between E_{ji} and χ is the scaling coefficient.

4. Estimate the probability-based connection weights:

$$w_{ji} = \frac{P_{ji}}{\sum_i P_{ji} + P_{js}}, \quad w_{js} = \frac{P_{js}}{\sum_i P_{ji} + P_{js}} \quad i = 1, \ldots, K \quad j = 1, \ldots, D$$

$$(7.13)$$

The scheme incorporates the knowledge base, instantiates them with random parameters, evaluates their performances, and integrates the feedback into the connection weights. The fidelity error E_{ji} is used as a performance indicator. The probabilities P_{ji} and P_{js} indicate the extent of prior knowledge and stochastic perturbation that should be assigned in the initialization scheme. The weights w_{ji} and w_{js} are the normalized probabilities for P_{ji} and P_{js}, respectively. If the instantiated ϖth blur structure $\mathbf{x}_\varpi$ dominates the others, its fidelity error $E_{j\varpi}$ will be small, leading to a similar range of values for $P_{j\varpi}$ and P_{js}. This combination induces a higher emphasis surrounding the subspace of $\mathbf{x}_\varpi$. On the other hand, if none of the knowledge-based blur instantiations shows good performance, E_{ji} tend to cluster together, leading to a small $\overline{\Delta E}_{ji}$ value. The resulting small P_{ji} and large P_{js} values imply a random initialization where the knowledge base is irrelevant. The asymptotic contribution of the stochastic perturbation ranges from 50% in the first case to 100% in the second case, thereby ensuring the randomness of the initial search space.

The main idea of the new support estimation scheme is centered on modeling the 2-D joint probability distribution of the blur supports. We model the blur dimensionality $\mathbf{s} = (s_x, s_y)$ as a 2-D mixture Gaussian distribution centered at selected mean $\mathbf{s}_j = (s_{xj}, s_{yj})$, $j = 1, \ldots, D$. This is analogous to function extrapolation in the regression neural network. If a selected dimension $\mathbf{s}_j$ shows good performance, its probability distribution is strengthened, resulting in more emphasis surrounding its support dimensionality. A novel entropy-based performance measure is introduced to assess the likelihood of $\mathbf{s}_j$, and their performances are consequently encoded into the connection weights. The support dimensionality $\mathbf{s}$ is determined by firing the neuron that corresponds to the nearest neighbor. The algorithm of the weight estimation for δ_{jz} is summarized below:

1. Initialize $\mathbf{x}_i$ to random blur and $\mathbf{y}_j$ to different dimensions to form the support subspace,

$$\mathbf{S} = \{\mathbf{s_j} | \mathbf{s_j} = (s_{xj}, s_{yj})\} \quad j = 1, \ldots, D \qquad (7.14)$$

2. For each $\hat{\mathbf{h}}_j \in \mathbf{S}$, restore the image $\hat{\mathbf{f}}_j$, and evaluate the support estimation measure E_j:

$$\hat{\mathbf{f}}_j = \arg \min_{\hat{\mathbf{f}}} J(\hat{\mathbf{f}} | \mathbf{g}, \hat{\mathbf{h}}_j \in \mathbf{S}) \qquad (7.15)$$

$$E_j = \frac{1}{2} \|\mathbf{g} - \hat{\mathbf{H}}_j \hat{\mathbf{f}}_j\|^2 + \frac{1}{2} \mathbf{u}^T \mathbf{W}_u \mathbf{u} + \frac{1}{2} \mathbf{v}^T \mathbf{W}_v \mathbf{v} \qquad (7.16)$$

where **u** and **v** are the entropy and inverse-entropy vector, and $\mathbf{W}_u$ and $\mathbf{W}_v$ are the corresponding weight matrices.

3. Compute the support probability, P_j:

$$P_j = e^{-\chi \frac{E_j}{\sum_j E_j}} \tag{7.17}$$

4. Model the support estimate **s** with a 2-D mixture Gaussian distribution centered at $\mathbf{s}_j$:

$$\mathbf{s} = \frac{\sum_j P_j \mathbf{N}(\mathbf{s}_j, \sigma_j^2)}{\sum_j P_j} \tag{7.18}$$

where $\mathbf{N}(\mathbf{s}_j, \sigma_j^2)$ is a 2-D Gaussian distribution with mean $\mathbf{s}_j = (s_{xj}, s_{yj})$ and variance $\sigma_j^2 = (\sigma_{xj}^2, \sigma_{yj}^2)$. The two elements variance σ_j^2 is given by

$$\sigma_j = \mathbf{d} \frac{E_j}{\sum_j E_j} \tag{7.19}$$

where **d** is the mean Euclidean separation between $\mathbf{s}_j$

5. Determine the firing neuron z:

$$z = \arg \min_{-j} \|\mathbf{s} - \mathbf{s}_j\| \tag{7.20}$$

6. Compute the support weight δ_{jz}:

$$\delta_{jz} = \begin{cases} 1 & \text{if } z = j \\ 0 & \text{if } z \neq j \end{cases} \tag{7.21}$$

where δ_{jz} is the Kronecker delta function.

The performance measure E_j employs a complementary pair of entropy-based criteria to estimate blur dimensionality. Figure 7.2 provides an illustration of the blur support estimation. The original "Lena" image in Figure 7.2a was degraded by a 5×5 Gaussian mask with a standard deviation of 2.0 to form Figure 7.2b. If the selected dimension $\mathbf{s}_j$ is smaller than the actual support, hardly any textured or edge details are restored as shown in Figure 7.2c, leading to a costly inverse-entropy term. Conversely, if $\mathbf{s}_j$ is larger than the actual support, extreme ringing and noise amplification dominates the scene as in Figure 7.2d, leading to a significant entropy term. Their compromise induces a higher probability in achieving good blur support initialization, exemplified by Figure 7.2e. It is clear that the restored image based on the random mask with good support dimension in Figure 7.2e outperforms those with undersized or oversized estimations in Figures 7.2c and d, respectively. The ringing artifact in Figure 7.2e that arises from random initialization will

FIGURE 7.2
Illustrations of blur dimensionality estimation. (a) Original Lena image. (b) Image degraded by
5 × 5 Gaussian blur with a standard deviation of 2.0. (c) Restored image using an undersized 3 ×
3 random mask. (d) Restored image using an oversized 7 × 7 random mask. (e) Restored image
using a 5 × 5 random mask.

improve throughout the blind deconvolution process in later stages. The formulation of the performance measure will be explained thoroughly below.

7.2.2 Knowledge-Based Reinforced Mutation

Dynamic Mutation Attractors

Studies show that most real-life blurs satisfy up to a certain degree of parametric structures. These include motion blur, out-of-focus blur, uniform blur, pillbox uniform blur, Gaussian blur, and sinc-square blurs, among others. Traditional blind methods require a hard decision on whether or not the blurs satisfy a parametric structure prior to problem formulation. This is elusive as the blur structures are often unknown *a priori*. To address this difficulty, we integrate the blur information into a solution space with embedded dynamic mutation attractors. This encourages a soft learning approach where relevant information is incorporated into the blur identification continually. A maximum a posteriori (MAP) estimator is devised to determine these attractors:

$$\vec{h} = \arg \max_{\tilde{h} \in \tilde{H}} L(\tilde{h}|\hat{h}) \tag{7.22}$$

$$= \arg \max_{\tilde{h} \in \tilde{H}} \log p(\tilde{h}|\hat{h}) \tag{7.23}$$

where $\tilde{H}$ is the knowledge-based solution space, $L(\tilde{h}|\hat{h})$ is the log likelihood function and $p(\tilde{h}|\hat{h})$ is the conditional probability density function of $\tilde{h}$ given the observation $\hat{h}$. Assuming that $n = \tilde{h} - \hat{h}$ follows a multivariate Gaussian distribution, we can rewrite Equation (7.22) in terms of the covariance matrix, Σ_{nn}:

$$\vec{h} = \arg \max_{\tilde{h} \in \tilde{H}} \log p(\tilde{h}|\hat{h})$$

$$= \arg \max_{\tilde{h} \in \tilde{H}} \log \left\{ \frac{1}{(\sqrt{2\pi})^{MN} |\Sigma_{nn}|^{\frac{1}{2}}} e^{-\frac{1}{2}(\tilde{h}-\hat{h})^T \Sigma_{nn}^{-1}(\tilde{h}-\hat{h})} \right\} \tag{7.24}$$

$$= \arg \max_{\tilde{h} \in \tilde{H}} \left\{ -\frac{1}{2}MN\log(2\pi) - \frac{1}{2}\log|\Sigma_{nn}| - \frac{1}{2}(\tilde{h} - \hat{h})^T \Sigma_{nn}^{-1}(\tilde{h} - \hat{h}) \right\} \tag{7.25}$$

If the blur coefficients are uncorrelated, we can further simplify (7.24) by letting $\Sigma_{nn} = \sigma_n^2 I$:

$$\vec{h} = \arg \max_{\tilde{h} \in \tilde{H}} \left\{ -\frac{1}{2}MN\log(2\pi) - \frac{1}{2}\log(\sigma_n^2 MN) - \frac{1}{2\sigma_n^2}(\tilde{h} - \hat{h})^T (\tilde{h} - \hat{h}) \right\}$$

$$= \arg \min_{\tilde{h} \in \tilde{H}} \left\{ \frac{1}{2}MN\log(2\pi) + \frac{1}{2}\log(\sigma_n^2 MN) + \frac{1}{2\sigma_n^2}(\tilde{h} - \hat{h})^T (\tilde{h} - \hat{h}) \right\} \tag{7.26}$$

The MAP estimator determines the structure and support of the mutation attractors from $\tilde{H}$. The noise variance, σ_n^2 has little effect on the estimator especially for small or known blur dimension.

The construction of solution space $\tilde{H}$ involves the incorporation of the blur knowledge. The main feature of the technique includes its ability to integrate any prior experience or devise any targeted structures without compromising the flexibility of the scheme. The solution space $\tilde{H}$ can be expressed as:

$$\tilde{H} = \bigcup_i \tilde{H}_i = \bigcup_i \{\tilde{h}:\tilde{h} = \varphi_i(s_i, \theta_i)\} \tag{7.27}$$

where $\tilde{H}_i$ is the blur subspace, φ_i is the 2-D blur function, s_i is the support, and θ_i is the defining parameters for ith parametric structure. In this chapter, we will focus on 2-D blur structures for general purposed image deconvolution. Several well-known blurs such as uniform blur, Gaussian blur, and concentric linear blurs are chosen for the stochastic initialization and attractor space construction. The respective subspace of the uniform, Gaussian, and linear concentric blurs are given by $\tilde{H}_u$, $\tilde{H}_g$, and $\tilde{H}_l$:

$$\tilde{H}_u = \left\{ \tilde{h}:\tilde{h}(x, y) = \frac{1}{MN} \quad x = 1,\ldots,M \ \ y = 1,\ldots,N \right\} \tag{7.28}$$

$$\tilde{H}_g = \left\{ \tilde{h}:\tilde{h}(x, y) = Ce^{-\frac{x^2+y^2}{2\sigma^2}} \quad x = 1,\ldots,M \ \ y = 1,\ldots,N \right\} \tag{7.29}$$

$$\tilde{H}_l = \left\{ \tilde{h}:\tilde{h}(x, y) = p\sqrt{x^2 + y^2} + q \quad x = 1,\ldots,M \ \ y = 1,\ldots,N \right\} \tag{7.30}$$

where (M, N) is the blur support, σ and C are the standard deviation and normalizing constant of the Gaussian blur, and p, q are the gradient and central peak value of the linear blur. Uniform blur is the 2-D extension of 1-D motion blur, and is characterized completely by its dimension. The Gaussain blur is widely observed in applications such as x-ray imaging, and is difficult to estimate using traditional blur identification approaches. The 2-D linear blur is implemented as the first-order estimation to the generic blur.

The construction of the solution space $\tilde{H}$ also includes constraints to preserve the average intensity of the original image by imposing the unity and nonnegativity constraints on $\tilde{H}$:

$$\sum_x \sum_y h(x, y) = 1 \tag{7.31}$$

$$h(x, y) \geq 0 \tag{7.32}$$

Reinforced Mutation in Attractor Space

Conventional evolutionary strategies employ mutation to provide a biased random walk through the solution space. The common disadvantage of this approach lies in their slow convergence, particularly for high-dimensional problems such as image restoration. To address this difficulty, a novel reinforced mutation scheme that integrates classical mutation and reinforced learning is introduced. The new technique offers a compromise between stochastic search through traditional mutation, and pattern acquisition through reinforced learning. If the evolved blur shows a close proximity to the attractor, suggesting a strong likelihood that the actual blur indeed satisfies a high degree of parametric structures, reinforced learning toward the attractors is emphasized. Otherwise, classical mutation is employed to explore the search space. The reinforced mutation paradigm integrates the blur knowledge domain to improve the convergence of the scheme, leading to significant reduction in the computational cost.

The reinforced mutation operator $\mathbf{M}$ involves the functional mapping of $\mathbf{M}:\mathfrak{R}^{M \times N} \rightarrow \mathfrak{R}^{M \times N}$ given by

$$\hat{\mathbf{h}}_k^{i+1} = \mathbf{M}(\hat{\mathbf{h}}_k^i) \quad k = 1, \ldots, \mu \tag{7.33}$$

where $\hat{\mathbf{h}}_k^i$ and $\hat{\mathbf{h}}_k^{i+1}$ are the kth blur estimate for ith and $(i+1)$th generations. The functional mapping equation can be decoupled into two reinforced mutation equations described by

$$\hat{\mathbf{h}}_k^{i+1} = \hat{\mathbf{h}}_k^i + (1-\alpha)\Delta\hat{\mathbf{h}}_k^i - \alpha \, \frac{\partial L\left(\hat{\mathbf{h}}_k^i, \tilde{\mathbf{h}}(\theta_k^i)\right)}{\partial \hat{\mathbf{h}}_k^i} \tag{7.34}$$

$$\vec{\theta}_k^i = \tilde{\theta}_k^i + a^i \Delta\tilde{\theta}_k^i - \beta \, \frac{\partial L\left(\tilde{\theta}_k^i, \theta_k^{i-1}\right)}{\partial \tilde{\theta}_k^i} \tag{7.35}$$

The first decoupling equation provides the fundamental reinforced learning process, where $\Delta\hat{\mathbf{h}}_k^i$ is the stochastic perturbation, $\tilde{\mathbf{h}}(\theta_k^i)$ is the predictor for the dynamic attractor, α and $L(\hat{\mathbf{h}}_k^i, \tilde{\mathbf{h}}(\theta_k^i))$ are the blur-domain learning rate and cost function. The second decoupling equation functions as the parametric vector estimation that characterizes the attractors. θ_k^i is the reinforced parametric vector, $\tilde{\theta}_k^i$ is the MAP parametric vector, $\Delta\tilde{\theta}_k^i$ is the stochastic perturbation, $\Delta\theta_k^{i-1}$ is the best vector obtained up to $(i-1)$th generation, β and $L(\tilde{\theta}_k^i, \theta_k^{i-1})$ are the parametric-domain learning rate and cost function.

The new scheme estimates the dynamic attractors and their defining parametric vectors, assesses their relevance with respect to the computed blurs and provides pattern acquisition through reinforced mutation based on the relevance of the estimates. If $\hat{\mathbf{h}}_k^i$ evolves within the vicinity of the attractor $\tilde{\mathbf{h}}(\theta_k^i)$, the likelihood that it actually assumes the parametric structure of $\tilde{\mathbf{h}}(\theta_k^i)$ is significant. In these cases, pattern acquisition toward the predictor is encouraged, and stochastic search is weakened. The reverse scenario applies if the evolved blurs are located farther away from the mutation attractors.

The classical mutation $\Delta\hat{\mathbf{h}}_k^i$ in Equation (7.34) can be expressed in terms of its strategic vectors γ_k^{i-1} by

$$\Delta\hat{\mathbf{h}}_k^i = \mathbf{N}(0, \gamma_k^{i-1}) = \left(\gamma_{k,1}^{i-1} e^{z_0 + z_1}, \ldots, \gamma_{k,MN}^{i-1} e^{z_0 + z_{MN}}\right)^T \tag{7.36}$$

where the variables z_0 and z_n provide constant and variable perturbations to γ_k^{i-1} and are given by

$$z_0 = N\left(0, \tau_0^2\right) \tag{7.37}$$

$$z_n = N\left(0, \tau_n^2\right) \quad n = 1, \ldots, MN \tag{7.38}$$

The typical values of τ_0 and τ_n given by Equations (7.39) and (7.40) are adopted in this work:

$$\tau_0 = \frac{1}{\sqrt{2MN}} \tag{7.39}$$

$$\tau_n = \frac{1}{\sqrt{2\sqrt{2MN}}} \tag{7.40}$$

The corresponding mutation in the parametric domain $\Delta\tilde{\theta}_k^i$ can be expressed in the similar fashion:

$$\Delta\tilde{\theta}_k^i = N\left(0, \eta_k^{i-1}\right) = \left(\eta_{k,1}^{i-1} e^{z_0 + z_1}, \ldots, \eta_{k,R}^{i-1} e^{z_0 + z_R}\right)^T \tag{7.41}$$

where η_k^{i-1} is the strategic vector, and R is the dimensionality of $\Delta\tilde{\theta}_k^i$.

The relevance of the attractor $\tilde{h}(\vec{\theta_k^i})$ with respect to the evolved blur $\hat{h}_k^i$ is captured by the learning rate α:

$$\alpha = \alpha_0 e^{-\xi \left\| \tilde{h}(\tilde{\theta}_k^i) - \hat{h}_k^i \right\|^2} \tag{7.42}$$

where ξ is the attractor field strength, and α_0 is the upper acquisition threshold. The field strength coefficient ξ defines the range of vicinity and the extent of learning, while the threshold α_0 is used to ensure that a minimum degree of stochastic exploration is available throughout the process.

The predictive scheme for the attractors in Equation (7.35) incorporates the previous vector estimation into the steepest descent learning term. The stochastic mutation and experience-based reinforced learning form a complementary pair to explore the search space efficiently. The MAP vector $\tilde{\theta}_k^i$, estimated from the parametric space Θ, is given by

$$\tilde{\theta}_k^i = \arg\max_{\theta \in \Theta} \log p\left(\tilde{h}(\theta) | \hat{h}_k^i\right) \tag{7.43}$$

The experiences of the previous estimations are captured by the momentum term as

$$\theta_k^{i-1} = \min_{j=1}^{i-2} \left\{ E\left[\tilde{h}(\tilde{\theta}_k^j)\right] \right\} \tag{7.44}$$

where $E(.)$ is the performance measure given by Equation (7.16). The momentum term is integrated into the learning rate β, which functions as a relevance measure between $\tilde{\theta}_k^i$ and $\widehat{\theta_k^{i-1}}$. The coefficient β, expressed in terms of the performance measure and scaling factor ω is given by

$$\beta = 1 - e^{-\omega \left\| E(\tilde{\theta}_k^i) - E(\widehat{\theta_k^{i-1}}) \right\|^2} \tag{7.45}$$

If the performance of θ_k^{i-1} is superior to that of $\tilde{\theta}_k^i$, namely, $E(\theta_k^{i-1})$ is much smaller than $E(\tilde{\theta}_k^i)$, we should emphasize the momentum term to take advantage of the previous good estimations. On the other hand, if their performances are comparable, no obvious advantage is gained through the previous experiences. In these cases, we will reduce the reinforced learning and encourage the classical stochastic search instead.

An important criterion in the formulation of cost function $L(\hat{\mathbf{h}}_k^i, \tilde{\mathbf{h}}(\theta_k^i))$ and $L(\tilde{\theta}_k^i, \theta_k^{i-1})$ is their proximity interpretation in the multidimensional vector space. The Euclidean distance measure is adopted due to its simplicity and proximity-discerning capability. The cost functions $L(\hat{\mathbf{h}}_k^i, \tilde{\mathbf{h}}(\theta_k^i))$ and $L(\tilde{\theta}_k^i, \theta_k^{i-1})$ are given by

$$L\left(\hat{\mathbf{h}}_k^i, \tilde{\mathbf{h}}(\tilde{\theta}_k^i)\right) = \frac{1}{2}\left\|\tilde{\mathbf{h}}(\tilde{\theta}_k^i) - \hat{\mathbf{h}}_k^i\right\|^2 \tag{7.46}$$

$$L\left(\tilde{\theta}_k^i, \theta_k^{i-1}\right) = \frac{1}{2}\left\|\tilde{\theta}_k^i - \theta_k^{i-1}\right\|^2 \tag{7.47}$$

An interesting feature of the new approach is its ability to provide parametric blur abstraction. This can be achieved by providing a generic parametric model such as a second-order quadratic function, or targeted structure if there is any prior experience, to the blur knowledge domain. The idea of parametric abstraction is particularly important if we would like to generalize a class of imaging processes whose blurring characteristics are unknown. The new approach will render an abstraction if the actual blur satisfies up to a certain degree of parametric structure, subject to our modeling. Even if the actual blur differs from any of the parametric structures, the obtained abstraction may function as an alternative blur model to facilitate problem formulations and simplifications.

The new technique makes a trade-off between the convergence efficiency and the chances that the actual blur may lie near to but not at the parametric structure. Fortunately, this dilemma can be alleviated in two ways. First, the upper learning threshold α_0 ensures that a minimum degree of stochastic mutation is always available, thereby providing a certain degree of search space surrounding the attractor. Second, even if the parametric blur is obtained as an abstraction for the actual blur, their close proximity will guarantee a good image restoration. In addition, the computational cost of achieving such a satisfactory deconvolution is reduced considerably.

7.2.3 Perception-Based Image Restoration

The image-domain solution space Ω_i given in Equation (7.5) is generated based on the evolved blur population Φ_i:

$$\Omega_i = \left\{ \hat{\mathbf{f}}_i | \hat{\mathbf{f}}_i = \arg \min_{\hat{\mathbf{f}}} J\left(\hat{\mathbf{f}}|\mathbf{g}, \hat{\mathbf{h}}_i \in \Phi_i\right) \right\} \tag{7.48}$$

The cost function $J\left(\hat{\mathbf{f}}|\mathbf{g}, \hat{\mathbf{h}}_i \in \Phi_i\right)$ is the convex set projection of $J\left(\hat{\mathbf{f}}, \hat{\mathbf{h}}|\mathbf{g}\right)$ onto the image subspace, and is given by

$$J\left(\hat{\mathbf{f}}|\mathbf{g}, \hat{\mathbf{h}}_i \in \Phi_i\right) = \frac{1}{2}\|\mathbf{g} - \hat{\mathbf{H}}_i \hat{\mathbf{f}}\|^2 + \frac{1}{2}\hat{\mathbf{f}}^{\mathrm{T}} \Lambda \mathbf{D}^{\mathrm{T}} \mathbf{D} \hat{\mathbf{f}} + C_f \tag{7.49}$$

where Λ is the image-domain regularization matrix, $\mathbf{D}$ is the Laplacian high-pass operator, and C_f is the projection constant. The cost function consists of a data-conformance and a model-conformance term. The data-conformance term measures the fidelity error of the restored image. However, the term alone is sensitive to singularities, and tends to form undesirable ringing and noise amplification across regions with abrupt visual activities. Therefore, the model-conformance term or regularization functional is introduced to provide a smoothing constraint on the restored image.

A main issue in the formulation of the cost function involves providing a fine balance between the fidelity term and regularization functional. We adopt an adaptive, perception-based regularization matrix Λ to control their relative contributions:

$$\Lambda = \mathrm{diag}(\lambda_1, \lambda_2, \ldots, \lambda_{PQ}) \tag{7.50}$$

where $P \times Q$ is the image dimension, and λ_i is the regularization coefficient for the ith image pixel, and is given as

$$\lambda_i = \max\left\{\lambda_l, -0.5[\lambda_h - \lambda_l]\log \sigma_i^2 + \lambda_h\right\} \tag{7.51}$$

The coefficients λ_l and λ_h are the lower and upper regularization values, and σ_i^2 is the local variance of ith pixel. λ_i is assigned as a decreasing logarithmic function of σ_i^2 because the textured and edge image pixels that exhibit high local variance require a small regularization parameter to preserve their fine details. The reverse holds true for the smooth regions that prefer noise and ringing suppression. The logarithmic function is adopted as it is used commonly to model the nonlinear transformation of human visual perception.

Constrained conjugate-gradient optimization is chosen to minimize Equation (7.49) due to its lower computational complexity and storage requirement as compared to quasi-Newton, and faster convergence with more robustness with respect to the steepest-descent technique. As the underlying Toeplitz distortion matrix is highly sparse, the convergence of the solution can be achieved very quickly. In practical applications, the algorithm is terminated when convergence or a predetermined number of iterations is reached.

The $(P \times Q)$-dimensional gradient of $J(\hat{\mathbf{f}}|\mathbf{g}, \hat{\mathbf{h}}_i \in \Phi_i)$ is denoted as

$$\mathbf{J}_{\hat{\mathbf{f}}} = \left[J_{\hat{f}_1} \cdots J_{\hat{f}_{PQ}} \right]^T \tag{7.52}$$

where $J_{\hat{f}_i}$ is the partial derivative of $J(\hat{\mathbf{f}}|\mathbf{g}, \hat{\mathbf{h}}_i \in \Phi_i)$ with respect to $\hat{f}_i$ and is given by

$$\begin{aligned} J_{\hat{f}_i} &= \frac{\partial J(\hat{\mathbf{f}}|\mathbf{g}, \hat{\mathbf{h}}_i \in \Phi_i)}{\partial \hat{f}_i} \\ &= -\{(g_i - h_i * f_i) * h_{-i}\} + \lambda_i \{(d_i * f_i) * d_{-i}\} \end{aligned} \tag{7.53}$$

where g_i, h_i, f_i, d_i, and λ_i are the corresponding scalar entries of $\mathbf{g}$, $\mathbf{h}$, $\mathbf{f}$, $\mathbf{D}$, and Λ, respectively.

The mathematical formulations of image restoration based on conjugate gradient optimization are summarized as follows. To simplify the notation, the indices of the ith blur estimates are assumed implicitly.

1. Initialize the conjugate vector, $\mathbf{q}$ with the gradient, $(\mathbf{J}_{\hat{\mathbf{f}}})_0$:

$$\mathbf{q}_0 = (\mathbf{J}_{\hat{\mathbf{f}}})_0 \tag{7.54}$$

2. Calculate the kth iteration's adaptive step size to update the image:

$$\eta_k = \frac{\|\mathbf{q}_k\|_\Gamma^2}{\|\mathbf{H}_k \mathbf{q}_k\|_\Gamma^2 + \mathbf{q}_k^T \Lambda \mathbf{D}^T \mathbf{D} \mathbf{q}_k} \tag{7.55}$$

where $\mathbf{H}_k$ is the Toeplitz matrix formed by $\mathbf{h}_k$ and Γ is the image-domain support.

3. Compute the updated image:

$$\hat{\mathbf{f}}_{k+1} = \hat{\mathbf{f}}_k + \eta_k \mathbf{q}_k \tag{7.56}$$

4. Calculate the kth iteration's adaptive step size to update the conjugate vector:

$$\rho_k = \frac{\|(\mathbf{J}_{\hat{\mathbf{f}}})_{k+1}\|_\Gamma^2}{\|(\mathbf{J}_{\hat{\mathbf{f}}})_k\|_\Gamma^2} \tag{7.57}$$

5. Update the new conjugate vector:

$$\mathbf{q}_{k+1} = -(\mathbf{J}_{\hat{\mathbf{f}}})_{k+1} + \rho_k \mathbf{q}_k \tag{7.58}$$

6. Impose image constraint to preserve the consistency of the display range:

$$0 \leq \hat{f}_{k+1}(i, j) \leq L \quad \forall (i, j) \in \Gamma \tag{7.59}$$

where $L = 2^b$ is the highest gray level, and b is the number of bits used in image representation.

7. Repeat steps (2) to (6) until convergence or a maximum number of iterations is reached.

7.2.4   Recombination Based on Niche-Space Residency

Many classical evolutionary schemes employ recombination to generate offspring by combining two or more randomly chosen candidates [40,157]. These techniques include global and local intermediary recombination, uniform crossovers, and γ-point crossovers. They often emphasize the randomness of recombination, with little consideration for the behaviors of the solutions. In this section, we introduce a new recombination technique based on subspace residency of the candidates. The underlying principle is that the solutions sharing a uniquely defined subspace should be reinforced, while those from different subspaces should be encouraged to recombine freely.

We introduce a new unique subspace called *niche-space* N_k, that describes the spatial association of the solutions. The niche-space of each solution is centered at $\mathbf{n}_\nu$ with the identifier ν given by

$$\nu = \min_s \left\| \hat{\mathbf{h}} - \arg \max_{\tilde{\mathbf{h}} \in \tilde{H}_s} \log \ p(\tilde{\mathbf{h}}|\hat{\mathbf{h}}) \right\| \tag{7.60}$$

$$\mathbf{n}_\nu = \arg \max_{\tilde{\mathbf{h}} \in \tilde{H}_\nu} \log \ p(\tilde{\mathbf{h}}|\hat{\mathbf{h}}) \tag{7.61}$$

The identifier ν represents the *a priori* blur type with $\mathbf{n}_\nu$ being the MAP estimate from its class. The niche-space is defined by

$$N_\nu = \left\{ \hat{\mathbf{h}} : d(\hat{\mathbf{h}}, \mathbf{n}_\nu) \leq \varepsilon \right\} \tag{7.62}$$

where ε is the niche-space radii, and $d(\hat{\mathbf{h}}, \mathbf{n}_\nu)$ expressed in term of the scaling coefficient χ, is given by

$$d(\hat{\mathbf{h}}, \mathbf{n}_\nu) = e^{-\chi \|\hat{\mathbf{h}} - \mathbf{n}_\nu\|^2} \tag{7.63}$$

The niche-space is a region of interest because it describes the subspace where the parametric blurs are likely to be found. Therefore, solutions originating from the same niche-space should be strengthened to enhance their existence. In contrast, classical recombination is employed to recombine solutions coming from different niche-spaces. The recombination operator $\mathbf{R}:\Re^{M \times N} \rightarrow \Re^{M \times N}$ involves the selection of random candidates, their niche-space determination, and recombination according to their residency.

The new recombination algorithm is summarized as follows:

For $n = 1$ to μ
   begin

- Generate two random candidates: $\mathbf{h}_i, \mathbf{h}_j, i, j \ \in 1, \ldots, \mu$
- If their support size differs, namely $\mathbf{s}_i \neq \mathbf{s}_j$, choose either of the candidates:

$$\mathbf{h}_r = \text{random}(\mathbf{h}_i, \mathbf{h}_j)$$

Else if $\mathbf{h}_i, \mathbf{h}_j \in N_\nu$, perform intra-niche-space recombination:

$$\mathbf{h}_r = R_{intra}(\mathbf{h}_i, \mathbf{h}_j)$$

Else, perform inter-niche-space recombination:

$$\mathbf{h}_r = R_{inter}\,(\mathbf{h}_i, \mathbf{h}_j)$$

end

The inter-niche-space recombination operator, R_{inter} employs global intermediary recombination to give

$$\mathbf{h_r} = \frac{1}{2}(\mathbf{h}_i + \mathbf{h}_j) \tag{7.64}$$

On the other hand, the intra-niche-space recombination operator, R_{intra} is computed using

$$\mathbf{h}_r = R_{intra}(\mathbf{h}_i, \mathbf{h}_j) = \frac{d\,(\mathbf{h}_i, \mathbf{n}_{vi})\,\mathbf{h}_i + d(\mathbf{h}_j, \mathbf{n}_{vj})\mathbf{h}_j}{d(\mathbf{h}_i, \mathbf{n}_{vi}) + d(\mathbf{h}_j, \mathbf{n}_{vj})} \tag{7.65}$$

If $\hat{\mathbf{h}}_i$ is close to the $\mathbf{n}_{vi}$, $d(\hat{\mathbf{h}}_i, \mathbf{n}_{vi})$ will have a large contribution, leading to a stronger $\hat{\mathbf{h}}_i$ emphasis in the weighted blur estimate. The reverse applies if $\hat{\mathbf{h}}_i$ and $\mathbf{n}_{vi}$ are farther apart. The new scheme is developed based on the principle that the evolved blurs that originate from a similar niche-space, namely, sharing similar *a priori* blur structure, should be enhanced. In comparison, solutions that reside in different niche-spaces are encouraged to recombine freely as in classical recombination.

7.2.5 Performance Evaluation and Selection

Algorithmic tractability has always been the core consideration in most optimization problems. This is because conventional methods usually require these preconditions to enable their formulations. For instance, the gradient-based optimization methods such as steepest descent and quasi-Newton techniques require the cost function to be differentiable. They fail to function properly if the preconditions are not satisfied. In addition, the chosen cost function may not be the best criterion that provides the most tangible interpretation to the problem. In these circumstances, evolutionary optimizations become an instrumental tool as they allow the most tangible performance measure to be adopted.

In view of this, an entropy-based objective function $F:\mathfrak{R}^{P \times Q + M \times N} \to \mathfrak{R}$ is introduced that functions as the restoration performance indicator:

$$F(\hat{\mathbf{h}}, \hat{\mathbf{f}}) = \frac{1}{2}\|\mathbf{g} - \hat{\mathbf{H}}\hat{\mathbf{f}}\|^2 + \frac{1}{2}\mathbf{u}^T\mathbf{W}_u\mathbf{u} + \frac{1}{2}\mathbf{v}^T\mathbf{W}_v\mathbf{v} \tag{7.66}$$

where $\mathbf{u}$ and $\mathbf{v}$ are the entropy and inverse-entropy vectors, and $\mathbf{W}_u$ and $\mathbf{W}_v$ are the corresponding weight matrices. The new objective function utilizes the intuitive local entropy statistics to assess the performance of the restored images. The entropy and inverse-entropy vectors are the lexicographically ordered visual activity measure for the local image neighborhood. They can be represented by

$$\mathbf{u} = \left[u_1 \cdots u_{PQ} \right]^T \tag{7.67}$$

$$\mathbf{v} = \left[v_1 \cdots v_{PQ} \right]^T \tag{7.68}$$

where u_i and v_i are the complementary entropy and inverse-entropy pair defined by

$$u_i = -\sum_j p_j \log p_j \tag{7.69}$$

$$v_i = \log n - \sum_j p_j \log p_j \tag{7.70}$$

The intensity probability histogram p_j is used to evaluate u_i and v_i, with n representing the number of classes in the histogram. The entropy u_i has larger values in high visual activity areas such as textured and edge regions. On the other hand, the inverse-entropy v_i has larger values in the low visual activity regions such as smooth backgrounds. The weight matrices $\mathbf{W}_u$ and $\mathbf{W}_v$ are given by

$$\mathbf{W_u} = \mathrm{diag}(w_{u1}, w_{u2}, \ldots, w_{uPQ}) \tag{7.71}$$

$$\mathbf{W_v} = \mathrm{diag}(w_{v1}, w_{v2}, \ldots, w_{vPQ}) \tag{7.72}$$

The weighted coefficients w_{ui} and w_{vi} present an adaptive cost-penalty effect, and is described by

$$w_{ui} = \alpha \left\{ 1 - \tanh \left(\frac{\sigma_i^2 - m}{m} \right) \right\} \tag{7.73}$$

$$w_{vi} = \beta \left\{ 1 + \tanh \left(\frac{\sigma_i^2 - m}{m} \right) \right\} \tag{7.74}$$

where α and β are the upper entropy and inverse-entropy emphasis thresholds, and m is the transition threshold between the smooth and textured regions. The shifted hyperbolic tangential function is adopted due to its negative symmetric property centered at the transition threshold m. The weighted matrices are used in conjunction with entropy and inverse-entropy vectors to ensure detail recovery in the textured regions, and ringing suppression in the smooth backgrounds.

The weight matrix $\mathbf{W}_u$ has large values in the smooth backgrounds and small values at the textured regions. This is combined with an entropy vector to ensure that ringing and noise amplification are penalized in smooth backgrounds. The reverse argument applies for the weight matrix $\mathbf{W}_v$. It has small weights at smooth backgrounds and large weights at textured regions. Together with the inverse-entropy vector, it encourages detail recovery in the textured regions by penalizing low visual activities in the respective areas. The incorporation of the superior entropy-based performance measure is made feasible owing to the virtue of evolutionary strategies, as its nondifferentiability will restrict its application in traditional optimization methods.

A combination of deterministic and stochastic selection $\mathbf{S}:\mathfrak{R} \to \mathfrak{R}^{P \times Q + M \times N}$ is adopted to choose the offspring for the future generation. In accordance with classical evolutionary strategies, deterministic selection is employed to propagate those candidates with superior performance into the future generations. To complement the deterministic process, stochastic selection such as tournament selection is adopted to improve the diversity and randomness of the solutions. The combination of deterministic and stochastic selections ensures the convergence of the algorithm by virtue of the convergence theorem given in [161].

The evolutionary process continues until convergence or the maximum number of iterations is reached. The termination criterion is satisfied when the relative reduction in the objective function falls below a predefined threshold ε:

$$|F(T) - F(T-1)| < \varepsilon |F(T)| \tag{7.75}$$

where $F(T)$ and $F(T-1)$ are the objective functions for the Tth and $(T-1)$th generations, respectively.

7.3 Soft-Decision Method

7.3.1 Recursive Subspace Optimization

As with the CRL method, we incorporate regularization in the soft-decision method. The new cost function that comprises a data fidelity measure, image and blur-domain regularization terms, and the soft-decision blur estimation error is given as

$$J(\hat{\mathbf{h}}, \hat{\mathbf{f}}) = \frac{1}{2}||\mathbf{g} - \hat{\mathbf{H}}\hat{\mathbf{f}}||^2 + \frac{1}{2}\hat{\mathbf{f}}^T \Lambda \mathbf{D}^T \mathbf{D}\hat{\mathbf{f}} + \frac{1}{2}\hat{\mathbf{h}}^T \Psi \mathbf{E}^T \mathbf{E}\hat{\mathbf{h}} + \frac{1}{2}\tilde{p}||\mathbf{w}^T(\hat{\mathbf{h}} - \tilde{\mathbf{h}})||^2 \tag{7.76}$$

where Λ and Ψ are the image and blur-domain regularization matrices, $\mathbf{D}$ and $\mathbf{E}$ are the Laplacian high-pass operators, $\tilde{p}$ is the soft-decision proximity measure, $\mathbf{w}$ is the weighted blur emphasis, and $\hat{\mathbf{h}}$ is the soft parametric blur estimate. The data fidelity term functions as a restoration criterion. However, the measure alone is sensitive to abrupt changes in visual activities, and tends to form undesirable ringing over smooth image backgrounds. Therefore, the image and blur-domain regularization terms are incorporated into the cost function to lend stability to the system. The image-domain regularization term encapsulates the regularization matrix Λ, to provide an adaptive image restoration. Likewise, the blur-domain regularization term includes the regularization matrix Ψ, to render piece-wise smoothness in the blur. The blur estimation error determines the proximity measure and modeling error of $\hat{\mathbf{h}}$ with respect to the soft estimate, $\tilde{\mathbf{h}}$ generated by the estimator. If there exists a high degree of proximity, the confidence that the current blur assumes a parametric structure increases, leading to a gradual adaptation of $\hat{\mathbf{h}}$ toward the soft estimate. Otherwise, the fidelity and regularization terms will dominate the restoration criteria and the soft estimation error will wither away.

Due to the distinctive characteristics and properties of $\hat{\mathbf{h}}$ and $\hat{\mathbf{f}}$, the optimization objectives and priorities for them should be handled differently. Therefore, it is both algorithmically and intuitively sensible to project $J(\hat{\mathbf{h}}, \hat{\mathbf{f}})$ into the image and blur subspaces to form the following cost functions:

$$J(\hat{\mathbf{f}}|\hat{\mathbf{h}}) = \frac{1}{2}||\mathbf{g} - \hat{\mathbf{H}}\hat{\mathbf{f}}||^2 + \frac{1}{2}\hat{\mathbf{f}}^T \Lambda \mathbf{D}^T \mathbf{D}\hat{\mathbf{f}} + C_f \tag{7.77}$$

$$J(\hat{\mathbf{h}}|\hat{\mathbf{f}}) = \frac{1}{2}||\mathbf{g} - \hat{\mathbf{H}}\hat{\mathbf{f}}||^2 + \frac{1}{2}\hat{\mathbf{h}}^T \Psi \mathbf{E}^T \mathbf{E}\hat{\mathbf{h}} + \frac{1}{2}\tilde{p}||\mathbf{w}^T(\hat{\mathbf{h}}-\tilde{\mathbf{h}})||^2 + C_h \tag{7.78}$$

where C_f and C_h are the projection constants of $J(\hat{\mathbf{h}}, \hat{\mathbf{f}})$ onto the image and blur domains, respectively.

The restored image, $\hat{\mathbf{f}}$, and identified blur, $\hat{\mathbf{h}}$, are estimated by minimizing the cost functions in their respective subspaces. An alternating minimization (AM) process is employed to optimize $J(\hat{\mathbf{f}}|\hat{\mathbf{h}})$ and $J(\hat{\mathbf{h}}|\hat{\mathbf{f}})$ recursively, thereby improving the $\hat{\mathbf{f}}$ and $\hat{\mathbf{h}}$ estimates progressively. The algorithmic simplicity of AM enables the scheme to project and minimize the image and blur cost functions one at a time.

The mathematical formulations of recursive subspace optmization are summarized as follows, with Γ and H denoting the solution spaces of $\hat{\mathbf{f}}$ and $\hat{\mathbf{h}}$, respectively:

1. Initialize $\hat{\mathbf{h}}_0$ to a random mask
2. For $(i + 1)$th recursion, solve for $\hat{\mathbf{f}}_{i+1}$ and $\hat{\mathbf{h}}_{i+1}$:

$$\hat{\mathbf{f}}_{i+1} = \arg\min_{\hat{\mathbf{f}}\in\Gamma} J(\hat{\mathbf{f}}|\hat{\mathbf{h}}_i) \tag{7.79}$$

$$\hat{\mathbf{h}}_{i+1} = \arg\min_{\hat{\mathbf{h}}\in H} J(\hat{\mathbf{h}}|\hat{\mathbf{f}}_{i+1}) \tag{7.80}$$

3. Stop when the convergence or the maximum number of iterations is reached

This algorithm differs from the symmetrical double regularization (SDR) approaches suggested by You and Kaveh [154], and Chan and Wong [155] in several ways. Their investigations assume that blind image deconvolution can be decomposed into two symmetrical processes of estimating the image and blur. This assumption conflicts with several observations. First, most PSFs exist in the form of low-pass filters. In contrast, typical images consist of smooth, texture, and edge regions, whose frequency spectrums vary considerably from low in the smooth regions to mediocre and high in the texture and edge regions. Second, image restoration is inherently a perception-based process as the quality of the restored image relies heavily upon the visual inspection of humans. Third, most real-life PSFs satisfy up to a certain degree of parametric structures. In comparison, we have little or no prior knowledge about most imaging scenes. These observations illustrate the importance of performing image restoration and blur identification in accordance with their priorities and characteristics. The new approach attempts to address these asymmetries by integrating parametric blur information into the scheme, and tailors image restoration and blur identification in accordance with their unique properties.

7.3.2 Hierarchical Neural Network for Image Restoration

Structure and Properties of Hierarchical Cluster Model

The formation of each degraded image pixel, g_{ij}, involves the filtering operation of a f_{ij} neighborhood window by a spatially finite PSF. As a result, only a finite local neighborhood window of g_{ij} is required to estimate each f_{ij}. This implies that image restoration is conceptually a collection of local parallel processes with a global coordination. The hierarchical cluster model (HCM) is a nested neural network consisting of parallel, distributed subnetworks or clusters. Its distributed nature underscores the unique properties of local pixel formation while its hierarchy provides an overall coordination. HCM can be constructed in a bottom-up manner. Individual neurons form the trivial level 0 clusters of the network. The neurons sharing similar functionality and characteristics coalesce into numerous level 1 clusters. Likewise, those level 1 clusters with homogeneous characteristics in turn coalesce into level 2 clusters. This process continues repeatedly to form a hierarchical neural network with multidimensional distributed clusters.

The sparse intercluster synaptic connections of HCM are effective in reducing the computational cost of restoration. As its cluster structures closely match the homogeneous image regions, HCM is uniquely placed to provide a perception-based restoration. The notion of clustering lends tolerance to the image stationarity constraint, as it divides a potentially inhomogeneous image into several quasi-stationary clusters. In addition, the modular structure encourages the developments of visually pleasing homogeneous regions.

Optimization of Image-Domain Cost Function as HCM
Energy Minimization

The formulation of $J(\hat{\mathbf{f}}|\hat{\mathbf{h}})$ optimization, as the HCM energy minimization based on a novel regularization matrix Λ, is developed in this section. The image-domain cost function, $J(\hat{\mathbf{f}}|\hat{\mathbf{h}})$ in Equation (7.77) can be expressed as

$$J(\hat{\mathbf{f}}|\hat{\mathbf{h}}) = \frac{1}{2}\hat{\mathbf{f}}^{\mathrm{T}}(\hat{\mathbf{H}}^{\mathrm{T}}\hat{\mathbf{H}} + \Lambda \mathbf{D}^{\mathrm{T}}\mathbf{D})\hat{\mathbf{f}} - (\mathbf{g}^{\mathrm{T}}\hat{\mathbf{H}})\hat{\mathbf{f}} + \frac{1}{2}\mathbf{g}^{\mathrm{T}}\mathbf{g} + C_f \qquad (7.81)$$

We can simplify Equation (7.81) by assigning $\mathbf{P} = \hat{\mathbf{H}}^T\hat{\mathbf{H}}, \mathbf{Q} = \mathbf{D}^T\mathbf{D}, \mathbf{r} = \hat{\mathbf{H}}^T\mathbf{g}$ and ignoring the constants $\frac{1}{2}\mathbf{g}^T\mathbf{g}$ and C_f:

$$J(\hat{\mathbf{f}}|\hat{\mathbf{h}}) = \frac{1}{2}\hat{\mathbf{f}}^T(\mathbf{P} + \Lambda\mathbf{Q})\hat{\mathbf{f}} - \mathbf{r}^T\hat{\mathbf{f}} \qquad (7.82)$$

The cost function can be decomposed into smooth S, texture T, and edge E partitions by introducing

$$\hat{\mathbf{f}} = \begin{pmatrix} \hat{\mathbf{fs}} \\ \hat{\mathbf{ft}} \\ \hat{\mathbf{fe}} \end{pmatrix},$$

$$\Lambda = \begin{bmatrix} \Lambda_s & \Phi & \Phi \\ \Phi & \Lambda_t & \Phi \\ \Phi & \Phi & \Lambda_e \end{bmatrix},$$

$$\mathbf{P} = \begin{bmatrix} \mathbf{P}_{ss} & \mathbf{P}_{st} & \mathbf{P}_{se} \\ \mathbf{P}_{ts} & \mathbf{P}_{tt} & \mathbf{P}_{te} \\ \mathbf{P}_{es} & \mathbf{P}_{et} & \mathbf{P}_{ee} \end{bmatrix},$$

$$\mathbf{Q} = \begin{bmatrix} \mathbf{Q}_{ss} & \mathbf{Q}_{st} & \mathbf{Q}_{se} \\ \mathbf{Q}_{ts} & \mathbf{Q}_{tt} & \mathbf{Q}_{te} \\ \mathbf{Q}_{es} & \mathbf{Q}_{et} & \mathbf{Q}_{ee} \end{bmatrix},$$

$$\mathbf{r} = \begin{pmatrix} \mathbf{r}_s \\ \mathbf{r}_t \\ \mathbf{r}_e \end{pmatrix} \qquad (7.83)$$

where $\hat{\mathbf{f}}_s, \hat{\mathbf{f}}_t, \hat{\mathbf{f}}_e$ are the lexicographically ordered pixels, and $\Lambda_s, \Lambda_t, \Lambda_e$ are the regularization submatrices for smooth, texture, and edge cluster types. The matrices $\mathbf{P}, \mathbf{Q},$ and vector $\mathbf{r}$ are partitioned likewise into their corresponding submatrices and subvector. It is observed from Equation (7.83) that there exists a cyclic symmetry between the smooth, texture and edge partitions. Substituting Equations (7.83) into (7.82) and simplifing the results using the implicit cyclic symmetry, we obtain

$$J(\hat{\mathbf{f}}|\hat{\mathbf{h}}) = \sum_\alpha \frac{1}{2}\hat{\mathbf{f}}_\alpha^T(\mathbf{P}_{\alpha\alpha} + \Lambda_\alpha\mathbf{Q}_{\alpha\alpha})\hat{\mathbf{f}}_\alpha + \sum_{\alpha,\beta:\,\alpha\neq\beta}\frac{1}{2}\hat{\mathbf{f}}_\alpha^T(\mathbf{P}_{\alpha\beta} + \Lambda_\alpha\mathbf{Q}_{\alpha\beta})\hat{\mathbf{f}}_\beta - \sum_\alpha \mathbf{r}_\alpha^T\hat{\mathbf{f}}_\alpha$$

$$(7.84)$$

where $\alpha, \beta \in \{S, T, E\}$ are the cluster types. The first and second summation terms in Equation (7.84) represent the contributions from similar and different cluster types, respectively.

Consider the decomposition of Equation (7.84) into the cluster level:

$$\hat{\mathbf{f}}_\alpha = \begin{pmatrix} \hat{\mathbf{f}}_1^\alpha \\ \vdots \\ \hat{\mathbf{f}}_\varphi^\alpha \end{pmatrix},$$

$$\Lambda_\alpha = \begin{bmatrix} \lambda_1^\alpha \mathbf{I} & & \Phi \\ & \ddots & \\ \Phi & & \lambda_\varphi^\alpha \mathbf{I} \end{bmatrix},$$

$$\mathbf{P}_{\alpha\beta} = \begin{bmatrix} \mathbf{P}_{11}^{\alpha\beta} & \cdots & \mathbf{P}_{1\gamma}^{\alpha\beta} \\ \vdots & & \vdots \\ \mathbf{P}_{\varphi 1}^{\alpha\beta} & \cdots & \mathbf{P}_{\varphi\gamma}^{\alpha\beta} \end{bmatrix},$$

$$\mathbf{Q}_{\alpha\beta} = \begin{bmatrix} \mathbf{Q}_{11}^{\alpha\beta} & \cdots & \mathbf{Q}_{1\gamma}^{\alpha\beta} \\ \vdots & & \vdots \\ \mathbf{Q}_{\varphi 1}^{\alpha\beta} & \cdots & \mathbf{Q}_{\varphi\gamma}^{\alpha\beta} \end{bmatrix},$$

$$r_\alpha = \begin{pmatrix} r_1^\alpha \\ \vdots \\ r_\varphi^\alpha \end{pmatrix} \tag{7.85}$$

where φ and γ are the number of clusters for partition types, α and β.

Substituting Equations (7.85) into (7.84) to express $J(\hat{\mathbf{f}}|\hat{\mathbf{h}})$ in terms of cluster levels, we obtain

$$J(\hat{\mathbf{f}}|\hat{\mathbf{h}})$$
$$= \left\{ \begin{array}{l} \sum_k \sum_\alpha \tfrac{1}{2}\hat{\mathbf{f}}_k^{\alpha^\mathsf{T}} (\mathbf{P}_{kk}^{\alpha\alpha} + \lambda_k^\alpha \mathbf{Q}_{kk}^{\alpha\alpha})\hat{\mathbf{f}}_k^\alpha + \sum_{k,l:\,k\neq l}\sum_\alpha \tfrac{1}{2}\hat{\mathbf{f}}_k^{\alpha^\mathsf{T}} (\mathbf{P}_{kl}^{\alpha\alpha} + \lambda_k^\alpha \mathbf{Q}_{kl}^{\alpha\alpha})\hat{\mathbf{f}}_l^\alpha + \\[2mm] \sum_{k,l:\,k\neq l}\sum_{\alpha,\beta:\,\alpha\neq\beta} \tfrac{1}{2}\hat{\mathbf{f}}_k^{\alpha^\mathsf{T}} (\mathbf{P}_{kl}^{\alpha\beta} + \lambda_k^\alpha \mathbf{Q}_{kl}^{\alpha\beta})\hat{\mathbf{f}}_l^\beta + \sum_k \sum_\alpha r_k^{\alpha^\mathsf{T}}\hat{\mathbf{f}}_k^\alpha \end{array} \right\} \tag{7.86}$$

where k, l are the cluster designations. The first term in Equation (7.86) refers to the intracluster contribution. The second and third terms correspond to intercluster contributions arising from similar and different cluster types.

The energy function of a three-level HCM is given by

$$E = -\frac{1}{2}\left\{ \sum_{k=1}^{K}\sum_{i=1}^{Pk}\sum_{j=1}^{Pk} w_{ik,jk}s_{ik}s_{jk} + \sum_{k\neq l}\gamma_{kl}\sum_{i=1}^{Pk}\sum_{j=1}^{Pl} w_{ik,jl}s_{ik}s_{jl} \right\} - \sum_{k=1}^{K}\sum_{i=1}^{Pk} b_{ik}s_{ik} \tag{7.87}$$

where $s_{ik}, s_{jl} \in \{0, \ldots, L-1\}$ are the state of the ith neuron in cluster k and the jth neuron in cluster l, $w_{ik,jk}$ is the intracluster connection weight between neurons s_{ik} and s_{jk}, $w_{ik,jl}$ is the intercluster connection weight between neurons s_{ik} and s_{jl}, γ_{kl} is the average strength of intercluster connections to intracluster connections, b_{ik} is the bias input to neuron s_{ik}, $L = 2^b$ is the number of gray levels, and b is the number of bits used in pixel representation. There are K first-level partitions with P_k and P_l representing the number of neurons in cluster k and cluster l, respectively. The first and second summation triplets in {.} represent the intracluster and intercluster contributions to the overall energy function.

A close inspection reveals the correspondence between the image-domain cost function in Equation (7.86) and the energy function of HCM in Equation (7.87). The HCM energy function is expressed in terms of generic clusters from smooth, texture, and edge partitions. Comparing these two expressions by mapping neuron s to pixel $\hat{f}$ and extending the generic clusters in Equation (7.87) to encapsulate the detailed clusters in Equation (7.86), we obtain

$$w_{ik,jk}^{\alpha\alpha} = - \left(p_{kk}^{\alpha\alpha} \right)_{ij} - \lambda_k^\alpha \left(q_{kk}^{\alpha\alpha} \right)_{ij} \qquad \alpha \in \{S, T, E\} \tag{7.88}$$

$$w_{ik,jl}^{\alpha\alpha} = - \left(p_{kl}^{\alpha\alpha} \right)_{ij} - \lambda_k^\alpha \left(q_{kl}^{\alpha\alpha} \right)_{ij} \qquad \alpha \in \{S, T, E\} \tag{7.89}$$

$$w_{ik,jl}^{\alpha\beta} = - \left(p_{kl}^{\alpha\beta} \right)_{ij} - \lambda_k^\alpha \left(q_{kl}^{\alpha\beta} \right)_{ij} \qquad \alpha, \beta \in \{S, T, E\} \tag{7.90}$$

$$b_{ik} = r_{ik} \tag{7.91}$$

$$\gamma_{kl} = 1 \tag{7.92}$$

where $w_{ik,jk}^{\alpha\alpha}$, $w_{ik,jl}^{\alpha\alpha}$, and $w_{ik,jl}^{\alpha\beta}$ are the intracluster connection weight of cluster type α, intercluster connection weight within cluster type α, and intercluster connection weight between cluster types α and β. $(p_{kk}^{\alpha\alpha})_{ij}, (q_{kk}^{\alpha\alpha})_{ij}, (p_{kl}^{\alpha\alpha})_{ij}, (q_{kl}^{\alpha\alpha})_{ij},$ $(p_{kl}^{\alpha\beta})_{ij}, (q_{kl}^{\alpha\beta})_{ij}$ are the ijth scalar entries of submatrices $\mathbf{P}_{kk}^{\alpha\alpha}, \mathbf{Q}_{kk}^{\alpha\alpha}, \mathbf{P}_{kl}^{\alpha\alpha}, \mathbf{Q}_{kl}^{\alpha\alpha}, \mathbf{P}_{kl}^{\alpha\beta},$ $\mathbf{Q}_{kl}^{\alpha\beta}$, respectively. We can express Equations (7.88) to (7.91) in terms of image degradation parameters by restoring $\mathbf{P} = \hat{\mathbf{H}}^T \hat{\mathbf{H}}$, $\mathbf{Q} = \mathbf{D}^T \mathbf{D}$, and $\mathbf{r} = \hat{\mathbf{H}}^T \mathbf{g}$ to give

$$w_{ik,jk}^{\alpha\alpha} = - \left[\sum_{n=1}^{N} \hat{h}_{ni} \hat{h}_{nj} \right]_{kk}^{\alpha\alpha} - \lambda_k^\alpha \left[\sum_{n=1}^{N} d_{ni} d_{nj} \right]_{kk}^{\alpha\alpha} \tag{7.93}$$

$$w_{ik,jl}^{\alpha\alpha} = - \left[\sum_{n=1}^{N} \hat{h}_{ni} \hat{h}_{nj} \right]_{kl}^{\alpha\alpha} - \lambda_k^\alpha \left[\sum_{n=1}^{N} d_{ni} d_{nj} \right]_{kl}^{\alpha\alpha} \tag{7.94}$$

$$w_{ik,jl}^{\alpha\beta} = - \left[\sum_{n=1}^{N} \hat{h}_{ni} \hat{h}_{nj} \right]_{kl}^{\alpha\beta} - \lambda_k^\alpha \left[\sum_{n=1}^{N} d_{ni} d_{nj} \right]_{kl}^{\alpha\beta} \tag{7.95}$$

$$b_{ik} = \left[\sum_{n=1}^{N} \hat{h}_{ni} g_n \right]_k \tag{7.96}$$

where N is the number of neurons in the network. The parameter γ_{kl} is unity as the space-invariance of PSF causes the intercluster synaptic strength to be as strong as the intracluster synaptic strength. The regularization parameter of the kth cluster for tth AM recursion, $\lambda_k(t)$ is given by

$$\lambda_k(t) = \max\{\lambda_l(t), -0.5[\lambda_h(t) - \lambda_l(t)]\log \sigma_k^2 + \lambda_h(t)\} \qquad (7.97)$$

where σ_k^2 is the average local variance of the kth cluster, and $\lambda_l(t)$, $\lambda_h(t)$ are the lower and upper regularization thresholds given by

$$\lambda_l(t) = [\lambda_l(0) - \lambda_l(\infty)]e^{-t} + \lambda_l(\infty) \qquad (7.98)$$
$$\lambda_h(t) = [\lambda_h(0) - \lambda_h(\infty)]e^{-t} + \lambda_h(\infty) \qquad (7.99)$$

$\lambda_l(0)$, $\lambda_h(0)$, $\lambda_l(\infty)$, $\lambda_h(\infty)$ are the lower and upper regularization values in the beginning and final AM recursion. $\lambda_k(t)$ is assigned as a decreasing logarithm function of σ_k^2 because the texture and edge clusters that have high local variance require a small regularization parameter to preserve the fine details. The reverse holds true for the smooth clusters. The logarithm function is adopted as it is commonly used to model the nonlinear transformation of human visual perception. In general, $\lambda_l(0)$ and $\lambda_h(0)$ are comparatively greater than $\lambda_l(\infty)$ and $\lambda_h(\infty)$. This enables the restoration scheme to recover more details progressively as we become more confident about the identified blur and the restored image.

Cluster Formation and Restoration

A three-level HCM is constructed from the blurred image in two stages. In the first stage, the image is segmented into smooth, texture, and edge regions. These regions are further partitioned into separate homogeneous clusters in the second stage. The mathematical representations for the cluster formation are

$$I = S \cup T \cup E \qquad (7.100)$$

$$S = \bigcup_i^x S_i \quad T = \bigcup_j^y T_j \quad E = \bigcup_k^z E_k \qquad (7.101)$$

where S_i, T_j and E_k are the ith smooth cluster, jth texture cluster, and kth edge cluster, respectively.

The blurred image is segmented into different regions based on its local statistics. The standard Sobel operator is used to extract the edges as they exhibit a distinctive change of gray levels compared to blurred texture regions. To further differentiate smooth backgrounds from texture regions, the blurred image is uniformly restored for an iteration using a random mask with estimated dimension of Section 7.2.1. We observe that smooth backgrounds in the partially restored image are mostly monotonous with low visual activities. Therefore, several visual measures such as contrast, entropy, and local variance are extracted from the image. An expert voting technique is employed to discriminate whether a pixel belongs to smooth or texture regions. These regions are further divided into nonneighboring clusters to preserve their homogeneity. Throughout HCM formation, small fragmented clusters that arise due to misclassification are continually incorporated into larger neighboring clusters by morphological operations. Our investigation shows that multivalue neuron modeling with steepest gradient descent is both robust and efficient in restoring the image. The neuron updating rule in 2-D image indices is given by

$$\Delta s(x, y) = -\frac{1}{w_{(x,y)}^{(x,y)}} \sum_{(x,y)\in\Re} w_{(x+m,y+n)}^{(x,y)} s(x+m, y+n) + b(x,y) \qquad (7.102)$$

$x = 1, \ldots, P$ $y = 1, \ldots, Q$ where $\Delta s(x, y)$ is the update increment, $w_{(x+m,y+n)}^{(x,y)}$ is the connection weight from neuron $s(x+m, y+n)$ to $s(x, y)$, $b(x, y)$ is the self-bias, and P, Q are the image dimensions. Equation (7.32) reveals that the update of a neuron or pixel value involves a weight kernel $\Re$, of dimension $(2M-1) \times (2N-1)$ for an $M \times N$ blur size. This implies a transformation from the initial high dimensional optimization problem to a smaller system with a time complexity of O($MNPQ$) per iteration, thereby reducing the computational cost of restoration. At the end of each iteration, the cluster boundaries are adjusted dynamically according to the mean and average local variance of each cluster. This renders a continual improvement in the segmentation map, its cluster structure and, most importantly, the quality of the restored image.

7.3.3   Soft Parametric Blur Estimator

We will use the same soft parametric estimator to model the computed blur with the best-fit parametric structure from a predefined solution space $\tilde{H}$. Therefore, the MAP estimator given in Equations (7.22) to (7.26) applies. Again the solution subspaces is the union of uniform, $\tilde{H}_u$, Gaussian, $\tilde{H}_g$ and concentric linear blur, $\tilde{H}_l$ given in Equations (7.28) to (7.30), respectively,

$$\tilde{H} = \bigcup_i \tilde{H}_i = \tilde{H}_u \cup \tilde{H}_g \cup \tilde{H}_l \qquad (7.103)$$

As the parametric blurs can be uniquely determined by their defining parameters, we derive a gradient-based approach to estimate these parameters, thereby further improving the processing speed. The simplified parametric estimators are given as

$$\sigma^2 = \mathrm{E}\left(\frac{\hat{h}r}{|\nabla\hat{h}|}\right)_{\mathrm{H}} \qquad (7.104)$$

where $\mathrm{E}(.)_{\mathrm{H}}$ is the expectation operator spanning over H, r is the radial distance of the $\hat{h}$ coefficient from the mask center, and $|\nabla\hat{h}|$ is the magnitude of gradient at $\hat{h}$.

The gradient p and central peak value q of the linear blur are estimated by

$$p = \mathrm{E}\left(|\nabla\hat{h}|\right)_{\mathrm{H}} \qquad (7.105)$$

Imposing the unity constraint given in Equation (7.31) on the linear blur, we obtain the q estimate:

$$q = \frac{E\left(|\nabla \hat{h}|\right)}{MN} \sum_x \sum_y \sqrt{x^2 + y^2} \qquad (7.106)$$

An important issue of the estimator is to evaluate the relevance of the parametric estimate to the computed blur. Therefore, a proximity measure is devised as the confidence criteria in our $\tilde{h}$ estimation. If the current $\hat{h}$ resembles one of the parametric structures in the solution space, a close proximity will be achieved that enhances our confidence in the $\tilde{h}$ estimation. The proximity measure is given as

$$\tilde{p} = \begin{cases} 0 & \text{if } \sum_i (\tilde{\mathbf{h}}_i - \hat{\mathbf{h}}_i)^2 > T \\ Ae^{-\xi(\tilde{\mathbf{h}}-\hat{\mathbf{h}})^{\mathsf{T}}(\tilde{\mathbf{h}}-\hat{\mathbf{h}})} & \text{if } \sum_i (\tilde{\mathbf{h}}_i - \hat{\mathbf{h}}_i)^2 \le T \end{cases} \qquad (7.107)$$

where A and ξ are the scaling factors, and T is the matching threshold.

The proximity measure corresponds to two scenarios. The first scenario describes those irregular $\hat{h}$ that do not resemble any parametric structures. As their chances of assuming a parametric structure are very slim, a zero value is appropriately assigned to these cases. On the other hand, if a satisfactory match occurs between $\hat{h}$ and $\tilde{h}$, a likelihood-based proximity measure is employed. We used a threshold T of 0.02 and ξ of 500 in our experiments. The scale factor, A, is determined by ensuring the contribution of the soft estimation term is an order less than the fidelity or regularization term for medium matches, and the same order for close matches. Experimental simulations show that it is unnecessary to evaluate the optimal A value as long as it has the right order of magnitude.

7.3.4 Blur Identification by Conjugate Gradient Optimization

The mathematical objective of blur identification can be obtained by combining Equations (7.78) and (7.80) to give

$$\hat{\mathbf{h}}_{i+1} = \arg\min_{\hat{\mathbf{h}}\in H}\left\{ \frac{1}{2}||\mathbf{g} - \hat{\mathbf{H}}\hat{f}||^2 + \frac{1}{2}\hat{\mathbf{h}}^{\mathsf{T}}\Psi\mathbf{E}^{\mathsf{T}}\mathbf{E}\hat{\mathbf{h}} + \frac{1}{2}\tilde{p}||\mathbf{w}^{\mathsf{T}}(\hat{\mathbf{h}}-\tilde{\mathbf{h}})||^2 \right\} \qquad (7.108)$$

The cost function consists of three criteria, namely, the data conformance measure, the blur-domain regularization term and the soft-decision blur estimation error. The data conformance and regularization terms ensure the fidelity and piecewise smoothness of the solutions. Their relative contribution is controlled by the regularization matrix, $\Psi = \text{diag}(\psi_1, \psi_2, \dots, \psi_{MN})$ with

$$\psi_i = [\psi_h - \psi_l]e^{-\sigma_i^2} + \psi_l \quad \forall i \in H \qquad (7.109)$$

where σ_i^2 is the local variance at $\hat{h}_i$, ψ_l and ψ_h are the lower and upper limits of the blur-domain regularization parameter.

The parametric blur information is incorporated into the cost function as the soft estimation error. Its functionality is analogous to Hebbian learning in neural networks where $\tilde{p}$ is the adaptive learning step size, and $||\mathbf{w}^T(\hat{\mathbf{h}}-\tilde{\mathbf{h}})||^2$ is the weighted learning error. The proximity measure, $\tilde{p}$, renders a compromise between the estimation error and the rest of the criteria in J $(\hat{\mathbf{h}}|\hat{\mathbf{f}})$. If the correlation between $\hat{\mathbf{h}}$ and the estimate $\tilde{\mathbf{h}}$ improves significantly, $\tilde{p}$ will become dominant and gradually locks $\hat{\mathbf{h}}$ into one of the parametric structures. Otherwise, the soft error will wither away and the combination of fidelity and regularization terms will determine the $\hat{\mathbf{h}}$ estimation. The flexibility of the approach lies in its ability to incorporate and adjust the relevance of the parametric structure throughout the restoration.

The raised cosine function, w, of the soft error assigns weighted emphasis to different regions of the blur. The inner coefficients of $\tilde{\mathbf{h}}$ and $\hat{\mathbf{h}}$ are emphasized because estimates of circumferential coefficients are prone to distortions by undesirable ringing in the restored image, particularly in the early stage of restoration. The raised cosine function given in terms of the roll-off factor, α, is

$$w(x, y) = \begin{cases} \frac{1}{2}\left\{1 + \cos\left[\frac{\pi}{\alpha}\left(\sqrt{\frac{x^2}{M^2} + \frac{y^2}{N^2}} - \frac{1-\alpha}{2}\right)\right]\right\} & \sqrt{\frac{x^2}{M^2} + \frac{y^2}{N^2}} \geq \frac{1-\alpha}{2} \\ 1 & \text{otherwise} \end{cases}$$

(7.110)

Conjugate gradient optimization is again chosen to minimize as given in Equation (7.112) for similar reasons in minimizing the image cost function in CRL. The convergence of an $(M \times N)$-dimensional blur can be achieved in a maximum of $M \times N$ steps. However, faster convergence is expected when the eigenvalues of the Hessian matrix are clustered. In practical applications, the algorithm is terminated when the convergence or the maximum number of iterations is reached. We extend the conjugate gradient approach in [154] to incorporate the soft estimation error in our blur identification. The $(M \times N)$-dimensional gradient of J $(\hat{\mathbf{h}}|\hat{\mathbf{f}})$ is denoted as

$$\mathbf{J}_{\hat{\mathbf{h}}} = \begin{bmatrix} J_{\hat{h}_1} \\ \vdots \\ J_{\hat{h}_{MN}} \end{bmatrix}$$

(7.111)

where $J_{\hat{h}_i}$ is the partial derivative of J $(\hat{\mathbf{h}}|\hat{\mathbf{f}})$ with respect to $\hat{h}_i$, $i \in H$ and is given by

$$J_{\hat{h}_i} = \frac{\partial J\,(\hat{\mathbf{h}}|\hat{\mathbf{f}})}{\partial \hat{h}_i}$$
$$= -\left\{(g_i - h_i * f_i) * f_{-i}\right\} + \mu_i\{(e_i * h_i) * e_{-i}\} + \tilde{p}w_i\left(\tilde{h}_i - \hat{h}_i\right) \quad (7.112)$$

where g_i, h_i, f_i, e_i, μ_i, and w_i are the corresponding scalar entries of $\mathbf{g}$, $\mathbf{h}$, $\mathbf{f}$, $\mathbf{E}$, Ψ, and $\mathbf{w}$, respectively.

The mathematical formulations of blur identification based on conjugate gradient optimization are summarized as follows:

1. Initialize the conjugate vector, $\mathbf{q}$, with the gradient, $(\mathbf{J_{\hat{h}}})_0$:

$$\mathbf{q}_0 = (\mathbf{J_{\hat{h}}})_0 \qquad (7.113)$$

2. Calculate the kth iteration's adaptive step size to update the blur:

$$\eta_k = \frac{\|\mathbf{q}_k\|_H^2}{\|\mathbf{Q}_k\hat{\mathbf{f}}\|_\Gamma^2 + \mathbf{q}_k^T \Psi \mathbf{E}^T \mathbf{E} \mathbf{q}_k + \tilde{p}\mathbf{q}_k^T \mathbf{W}\mathbf{q}_k} \qquad (7.114)$$

where $\mathbf{Q}_k$ is the Toeplitz matrix formed by $\mathbf{q}_k$ and $\mathbf{W} = \text{diag}(w_1, w_2, \dots, w_{MN})$

3. Compute the updated blur:

$$\hat{\mathbf{h}}_{k+1} = \hat{\mathbf{h}}_k + \eta_k \mathbf{q}_k \qquad (7.115)$$

4. Calculate the kth iteration's adaptive step size to update the conjugate vector:

$$\rho_k = \frac{\|(\mathbf{J_{\hat{h}}})_{k+1}\|_H^2}{\|(\mathbf{J_{\hat{h}}})_k\|_H^2} \qquad (7.116)$$

5. Update the new conjugate vector:

$$\mathbf{q}_{k+1} = -(\mathbf{J_{\hat{h}}})_{k+1} + \rho_k \mathbf{q}_k \qquad (7.117)$$

6. Repeat steps (2) to (5) until convergence or a maximum number of iterations is reached

To ensure the realizability of the blurs in most physical applications, we impose the unity and nonnegativity constraints in Equations (7.31) and (7.32) on the blur estimate. In addition, the blur is assumed to be centrosymmetric:

$$h(x, y) = h(-x, -y) \quad \forall (x, y) \in H \qquad (7.118)$$

These assumptions confine the solution space to those blurs that we are interested in. For example, the unity constraint in Equation (7.31) is a consequence of similar overall intensity levels between the degraded and original images. The centrosymmetric condition is usually satisfied by most physical PSFs. In general, these constraints improve the robustness and convergence of the scheme.

7.3.5   Blur Compensation

Typical images are comprised of smooth, texture, and edge regions. The monotonous smooth regions are ineffectual in recovering the blur. The task of blur estimation relies mainly on the information available in the texture and edge regions. In this section, we will investigate the effect of texture and edge orientation on blur estimation, and devise a compensation scheme to account for the necessary adjustment.

The scalar equivalence of Equation (7.1), ignoring the additive noise, is given by

$$g_i = f_i * h_i = \sum_{j \in H} f_{i-j} h_j \tag{7.119}$$

where H is the 2-D blur support. If a segment of f assumes an edge structure with F_b and F_e corresponding to the background and edge gray levels, respectively, we can rewrite Equation (7.119) as

$$G_e = F_e \sum_{j \in E} h_j + F_b \sum_{j \in B} h_j \tag{7.120}$$

where **E** and **B** are the sets of blur coefficients overlying on top of the edge and backgrounds, and G_e is the gray level of the corresponding edge in **g**. Imposing the unity constraint on Equation (7.120) and further simplifying it, we obtain

$$\sum_{j \in E} h_j = \frac{G_e - F_b}{F_e - F_b} \tag{7.121}$$

Equation (7.121) reveals the ambiguity of the solutions as the blur cannot be uniquely determined. As the blur coefficients along the edge are indistinguishable, their estimations from these edges will result in statistical averaging. For an image degraded by nonuniform blur that contains a dominant edge orientation, the averaging effect of blur coefficients is stronger along the dominant axis compared to the minor axis. The implication on nonuniform blur, such as the Gaussian mask, is that the central peak coefficients will be smeared out, the major axis coefficients boosted, and the minor axis coefficients weakened. It is worth noting that the extent of these effects depends upon the degree of dominance between the perpendicular major and minor axes. The more asymmetric these two axes are, the more prominent is the effect. In addition, these effects do not affect the uniform blur.

In view of the potential distortion to the estimated blur, a correlation-based approach is developed to compensate for the blur. As correlation is a good measure of edges and textures, its incorporation enables a systematic approach to the blur compensation. We introduce a new neighborhood correlation matrix, Z, that characterizes the correlation of neighborhood image pixels throughout the whole image:

$$
Z = \begin{bmatrix}
\zeta\left(-\frac{M-1}{2}, -\frac{N-1}{2}\right) & \cdots & \zeta\left(-\frac{M-1}{2}, \frac{N-1}{2}\right) \\
\vdots & \zeta_{(0,0)} & \vdots \\
\zeta\left(\frac{M-1}{2}, -\frac{N-1}{2}\right) & \cdots & \zeta\left(\frac{M-1}{2}, \frac{N-1}{2}\right)
\end{bmatrix}
\tag{7.122}
$$

where $\zeta_{(i,j)}$ is defined as

$$
\zeta_{(i,j)} = \frac{\sum\limits_{k} \mathbf{R}_{\mathrm{ff}}(k, k+i+jM)}{\frac{1}{MN}\sum\limits_{i}\sum\limits_{j}\sum\limits_{k} \mathbf{R}_{\mathrm{ff}}(k, k+i+jM)}
\tag{7.123}
$$

with $\mathbf{R}_{ff}(i, j)$ being the (i, j)th entry of the correlation matrix, $\mathbf{R}_{\mathrm{ff}} = \mathrm{E}[\hat{\mathbf{f}}\hat{\mathbf{f}}^{\mathrm{T}}]$.

The neighborhood correlation matrix provides useful information on the overall correlation structure of the image spanning over the blur support dimension. In particular, the higher-order correlation of Z along the vertical and horizontal directions indicates the extent of edges or hairy textures in the respective axes. Therefore, the information can be extracted and incorporated into the compensation scheme.

We present a modified 2-D Gaussian function, $c(x, y)$, to perform the compensation as below:

$$
\hat{h}_c(x,y) = c(x,y) \times \hat{h}(x,y) \quad \forall(x, y) \in H
\tag{7.124}
$$

where $\hat{h}_c(x, y)$ is the compensated blur, and $c(x, y)$ is given by

$$
c(x,y) = B e^{-\gamma S(\nabla E_h)\left(\frac{x^2}{2\sigma_x^2} + \frac{y^2}{2\sigma_y^2}\right)} \quad \forall(x, y) \in H
\tag{7.125}
$$

The factor γ functions as the compensation coefficient, ∇E_h is the gradient of blur radial energy profile, σ_x^2 and σ_y^2 are the horizontal and vertical variances, B is the normalizing constant, and $S(.)$ is the squash function. A modified Gaussian function is adopted because it is effective in rectifying the distortions, namely, strengthening the central peak coefficient, weakening the major axis coefficients and boosting the minor axis coefficients. The emphasis or deemphasis of the coefficients is controlled by the vertical and horizontal variances, σ_x^2 and σ_y^2:

$$\sigma_x^2 = \frac{\sum_i \sigma_i^2}{\frac{1}{M}\sum_i \left(\prod_j \zeta_{ij}\right)} \tag{7.126}$$

$$\sigma_y^2 = \frac{\sum_j \sigma_j^2}{\frac{1}{N}\sum_j \left(\prod_i \zeta_{ij}\right)} \tag{7.127}$$

where σ_i^2 and σ_j^2 are the intravariance of the ith row and jth column of matrix **Z**. Both the numerator and denominator in Equations (7.130) and (7.131) are characterizations of edges. If the image is dominated by vertical edges, the intravariance along the column is smaller and the product of neighborhood coefficients along the column is larger. The combined effects cause σ_y^2 to be smaller than σ_x^2, therefore deemphasizing the coefficients along the dominant vertical axis and emphasizing the coefficients along the horizontal axis. The reverse holds true when the image is dominated by horizontal edges.

The squash function, $S(\nabla E_h)$, takes the radial energy profile of the PSF as

$$S(\nabla E_h) = \tanh\left(\mathrm{E}[\|\nabla E_h\|]\right) = \mathrm{E}[\tanh\left(\|\nabla E_h\|\right)] \tag{7.128}$$

A hyperbolic tangent function is employed because its value ranges from zero for uniform blur, up to one for other blurs with decreasing radial energy profiles. This corresponds to negligible compensation for the uniform blur, and desirable correction for the others. The normalizing constant, B, is introduced to ensure the unity condition in Equation (7.31) is satisfied. The compensation factor, γ, can be evaluated by adjusting the extent of correction using

$$\frac{c(0, 0)}{c\left(\frac{M-1}{2}, \frac{N-1}{2}\right)} \leq \beta \tag{7.129}$$

where β is the upper limit for the ratio of central peak to corner coefficient. It is observed experimentally that for most images, a compensation of less than 10% is adequate, which translates to a β value of 1.3 approximately. Finally, the support dimension of the blur estimate is pruned if the relative energy of the circumferential coefficients falls below a threshold, namely,

$$E_c < \ell E_T \tag{7.130}$$

where E_c is the energy of the circumferential coefficients on two sides of the blur, E_T is the total energy of all the coefficients, and ℓ is the pruning threshold.

7.4 Simulation Examples

We demonstrate the performance of the two methods by simulation examples, and compare them with the SDR method. In the evolutionary strategy, the following parameters are used in the simulation: $\mu = 10$, $\kappa = \infty$, $\nu = 10$, and $\rho = 2$. Stochastic initialization was performed using the probability neural network. The blur population was recombined based on niche-space residency. Reinforced mutation with $\alpha_0 = 0.8$, $\xi = 150$, $a = 0.8$, and $\omega = 150$ was adopted. Image restoration based on conjugate gradient optimization with $\lambda_h = 10^{-2}$ and $\lambda_l = 10^{-4}$ was employed to generate the image-domain solution space. The performances of the restored images and blurring functions were evaluated using the entropy-based objective function with $\alpha = \beta = 0.01$. The combination of the deterministic and stochastic selection was used to propagate those solutions with good characteristics into the future generations. The process of blur recombination, reinforced mutation, image restoration, performance evaluation, and evolution selection was allowed to continue until convergence or a predetermined number of generations was reached.

In the soft-decision method, AM was used to restore the image and identify the blur recursively. The blur estimate was initialized to a random mask whose support size was estimated similar to CRL. The pruning procedure with an ℓ value of 0.15 gradually trimmed the blur support to the actual dimension when necessary. A three-level HCM was constructed using Equations (7.97) to (7.103) with $\lambda_l(0) = 0.005$, $\lambda_h(0) = 0.05$, $\lambda_l(\infty) = 0.0001$, $\lambda_h(\infty) = 0.01$, and the cluster restoration was performed to achieve perception-based restoration. On the other hand, conjugate gradient optimization was used to identify the blur by minimizing the blur-domain cost function. The blur-domain regularization parameters are taken as $\psi_l = 10^6$ and $\psi_h = 1.2 \times 10^6$. The AM continued until the convergence or the maximum number of recursions was reached.

The experimental results given in the following subsections cover the blind image deconvolution of various degraded images under different noise levels and blur structures. To provide an objective performance measure, the signal to noise ratio improvement is given as

$$\Delta\text{SNR} = 10 \log_{10} \frac{\|\mathbf{g} - \mathbf{f}\|^2}{\|\hat{\mathbf{f}} - \mathbf{f}\|^2} \tag{7.131}$$

where $\mathbf{g}$, $\mathbf{f}$, and $\hat{\mathbf{f}}$ represent the degraded, original and restored image, respectively. Nevertheless, the best criterion to assess the performance of the restoration schemes remains the human inspection of the restored images. This is consistent with the view that human vision is, fundamentally, a subjective perception.

(a)   (b)

(c)   (d)

FIGURE 7.3
Blind restoration of image degraded by Gaussian blur with quantization noise. (a) Image degraded by 5 × 5 Gaussian blur with quantization noise. (b) Restored image using CRL. (c) Restored image using RSD. (d) Restored image using SDR.

7.4.1 Identification of 2-D Gaussian Blur

The blind deconvolution of images degraded by Gaussian blur is presented in this subsection. Traditional blind methods are ineffectual in identifying Gaussian blur as it exhibits distinctive gradient changes throughout the mask, conflicting with the usual precondition of piecewise smoothness. In addition, it does not exhibit prominent frequency null, resulting in the failure of frequency-based blind deconvolutions. The original "Lena" image in Figure 7.2a has a dimension of 256 × 256 with 256 gray levels. It was degraded by a 5 × 5 Gaussian mask with a standard deviation of 2.0, coupled with some quantization noise to form Figure 7.3a. Applying the two algorithms on the degraded image, we obtained the restored images given in Figures 7.3b and c, respectively. It is observed that the two methods achieve almost perfect restoration by recovering the visual clarity and sharpness of the image. It suppresses the ringing and noise amplification in the smooth backgrounds, while preserving the fine details in textured and edge regions.

The restored image using the SDR approach proposed in [154] is given in Figure 7.3d. Comparing the results in Figures 7.3b and c, and that in Figure 7.3d, we observe that the SDR restored image does not recover fine details such as the edges and textured feather regions as well as the methods described in this chapter. Furthermore, ringing starts to develop in the smooth backgrounds near the shoulder and the hat regions. The visual observation is confirmed by the ΔSNR improvement as CRL and RSD methods offer an improvement of 6.1 and 3.9 dB, respectively, compared to the 3.5 dB obtained by the SDR approach. The significant ΔSNR improvement by the CRL method illustrates the advantages of incorporating dynamic attractor into the reinforced mutation scheme.

The corresponding blur estimates by CRL and RSD are given in Figures 7.4a and b, respectively. We observe that the blur estimates by the two methods closely resemble the actual Gaussian blur shown in Figure 7.4d. The identified blur obtained using the SDR approach is given in Figure 7.4c. Comparing Figures 7.4a and b with Figure 7.4c, it is clear that SDR fails to identify the blur adequately. In particular, the coefficients in the upper half of the blur receive exceeding weights, while the central peak coefficient is mislocated. The SDR approach emphasizes on the piecewise smoothness in both the image and blur domains, leading to insufficient texture recoveries in the restored image and consequently poor blur identification.

7.4.2 Identification of 2-D Gaussian Blur from Degraded Image with Additive Noise

In this subsection, we illustrate the effectiveness of the CRL and RSD to deconvolve degraded image with additive noise. The original image in Figure 7.2a was degraded by the exact 5×5 Gaussian mask, coupled with 30 dB additive noise to form Figure 7.5a. We applied the algorithms on the degraded image to obtain the restored images in Figures 7.5b and c. It is observed that the restored images feature fine details at the textured feather regions while suppressing ringing and noise amplification in the smooth backgrounds. This reflects the fact that the random additive noise has been amplified during the restoration process. The restored image using the SDR approach is given in Figure 7.5d. The comparison between Figures 7.5b and c with Figure 7.5d shows the superior restoration results obtained using the two methods. The SDR restored image does not preserve the fine details near the textured regions. Again, ringing starts to appear particularly in the shoulder and the nearby square background. The observations are consistent with the objective measure as our approaches offer ΔSNR improvements of 3.4 and 2.24 dB, respectively, compared to the 1.28 dB obtained by the SDR method. Comparing the restored image using our approach with and without additive noise in Figures 7.5b and c and Figures 7.3c and d, it is clear that they show comparable image quality. On the other hand, the SDR restored images show noticeable deterioration in the quality of the restored images.

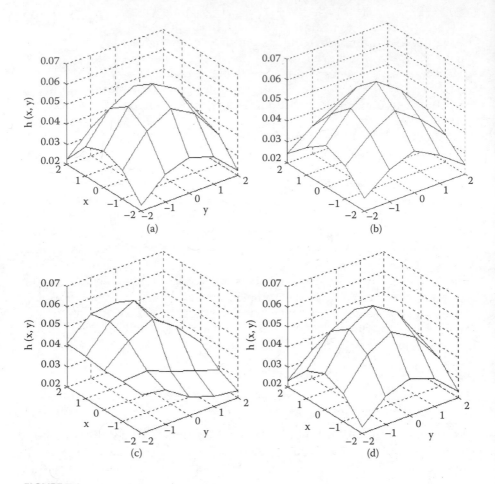

FIGURE 7.4
Blur estimates for blind deconvolution of image degraded by Gaussian blur. (a) Blur estimate by CRL. (b) Blur estimate by RSD. (c) Blur estimate by SDR. (d) Actual 5×5 Gaussian blur with a standard deviation of 2.0.

The corresponding blur estimates by CRL and RSD are presented in Figures 7.6a and b, respectively. We can see clearly see that the blur estimates match the actual blur in Figure 7.4d closely. Compared with the identified blur obtained using the SDR approach in Figure 7.6c, it is obvious that CRL and RSD offer superior blur identifications, and consequently render better image restorations.

7.4.3   Identification of 2-D Uniform Blur by CRL

To illustrate the flexibility of the new approach in dealing with different blur structures, we deconvolved an image degraded by uniform blur with some additive noise using the evolutionary strategy method. The original Lena

(a)　　　　　　　　　　(b)

(c)　　　　　　　　　　(d)

FIGURE 7.5
Blind deconvolution of image degraded by Gaussian blur with additive noise. (a) Image degraded by 5 × 5 Gaussian blur with 30 dB additive noise. (b) Restored image using CRL. (c) Restored image using RSD. (d) Restored image using SDR.

image in Figure 7.2a was degraded by a 5 × 5 uniform blur coupled with 40 dB additive noise to form Figure 7.7a. The restored image after applying the CRL algorithm is given in Figure 7.7b. It is observed that the restored image features detailed textured and edge regions, with no visible ringing in the smooth backgrounds. Compared to the restored image using the SDR approach in Figure 7.7c, it is obvious that our technique achieves better restoration results by rendering details and suppressing ringing in the appropriate image regions. The ΔSNR improvement supports the visual inspection as our approach offers an improvement of 6.4 dB compared to the 5.2 dB obtained by the SDR method.

The evolution of the corresponding blur estimates is given in Figure 7.8. The initial blur estimate, the evolved solution after a few generations, and the final estimate are given in Figures 7.8a, b and c, respectively. The random estimate in Figure 7.8a evolves gradually to the final estimate in Figure 7.8c, assimilating the good properties as it progresses. We notice that the final blur

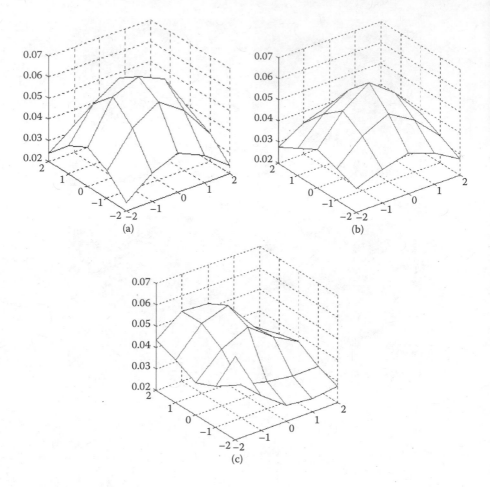

FIGURE 7.6
Blur estimates for blind deconvolution of image degraded by Gaussian blur with Gaussian noise.
(a) Blur estimate by CRL. (b) Blur estimate by RSD. (c) Blur estimate by SDR.

estimate matches the actual uniform blur in Figure 7.8e closely. The identified blur using the SDR approach in Figure 7.8d achieves a reasonably good estimation, albeit at the cost of strong smoothing constraints in the blur and image domains. This leads to insufficient detail recovery in the textured and edge regions shown in Figure 7.7c. In addition, the excessive smoothing constraint prohibits successful identification of spatially varying blurs such as Gaussian mask as shown in Figure 7.6a. The good performances of our approach illustrate that this technique is effective and robust in blind deconvolution of images degraded under different circumstances, namely, various blur structures and noise levels.

(a)

(b)

(c)

FIGURE 7.7

Blind deconvolution of image degraded by uniform blur with additive noise. (a) Image degraded by 5 × 5 uniform blur with 40 dB additive noise. (b) Restored image using CRL. (c) Restored image using SDR.

7.4.4   Identification of Nonstandard Blur by RSD

We illustrate the capability of the soft-decision method to handle nonstandard blur in Figure 7.9. The original image in Figure 7.2a was degraded by the 5 × 5 nonstandard exponential blur given in Figure 7.9f, followed by 30 dB additive noise to form the degraded image in Figure 7.9a. The algorithm was applied to obtain the restored image and identified blur in Figures 7.9b and d, respectively. It is clear that this algorithm is effective in restoring the image by providing clarity in the fine texture regions, and suppressing noise and ringing in the smooth backgrounds. Compared to our approach, the restored image using SDR method in Figure 7.9c does not render enough details near the texture feather regions. Again, ringing starts to appear near the shoulder in the smooth backgrounds. An inspection on our identified blur shows that it captures the overall structure of the actual blur. In contrast, the identified blur using the SDR method in Figure 7.9e shows that SDR is ineffectual in recovering the nonstandard blurs. The satisfactory results of our approach

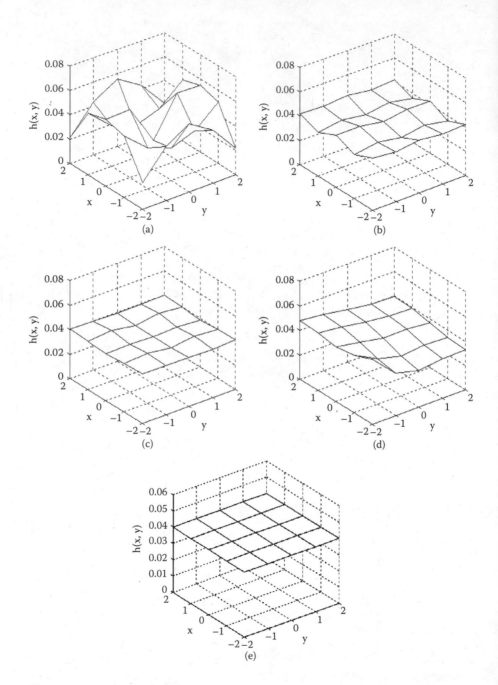

FIGURE 7.8
Blur estimates for blind deconvolution of image degraded by uniform blur with 40 dB additive noise. (a) Initial random blur estimate. (b) Blur estimate after a few generations. (c) Final identified blur using our approach. (d) Identified blur using SDR approach. (e) Actual 5×5 uniform blur.

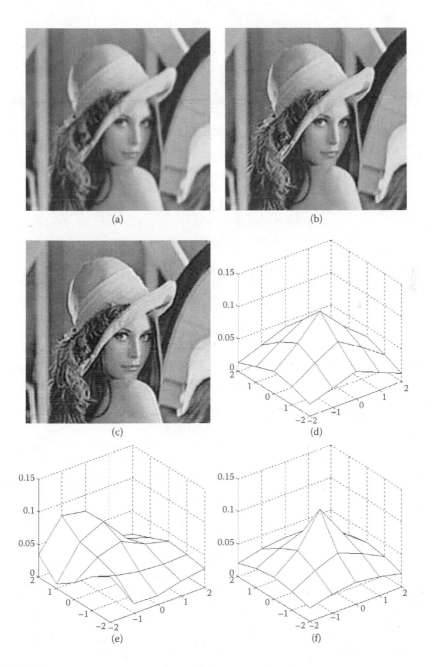

(a)

(b)

(c)

(d)

(e)

(f)

FIGURE 7.9

Blind deconvolution of image degraded by nonstandard blur. (a) Image degraded by 5×5 nonstandard exponential blur with 30 dB additive noise. (b) Restored image using RSD. (c) Restored image using SDR. (d) Identified blur by RSD. (e) Identified blur by SDR. (f) Actual 5×5 exponential blur.

illustrate that CRL and RSD methods are effective and robust in the blind deconvolution of images degraded under different circumstances, namely, various standard blurs, nonstandard blur, and different noise levels.

7.5   Conclusions

This chapter described two methods for blind image deconvolution. The first formulates the blind deconvolution problem into an evolutionary strategy comprising the generation of image and blur populations. A knowledge-based stochastic initialization is developed to initialize the blur population. It utilizes the notion that the more likely search space should be assigned more emphasis. Inter-niche-space recombination is employed to introduce variation across the population spectrum, while intra-niche-space recombination strengthens the existence of solutions within similar spatial residency. A reinforced mutation scheme that combines the classical mutation and reinforced learning is developed. The scheme integrates the *a priori* information of the blur structures as dynamic attractor. This improves the convergence greatly, leading to significant reduction in computational cost.

The second applies a soft-decision blur identification and neural network approach to adaptive blind image restoration. The method incorporates the *a priori* information of well-known blur structures without compromising its flexibility in restoring images degraded by other nonstandard blurs. A multimodal cost function consisting of data fidelity measure, image and blur-domain regularization terms and soft-decision estimation error is presented. A subspace optimization process is adopted where the cost function is projected and optimized with respect to the image and blur domains recursively. The hierarchical cluster model is used to achieve the perception-based restoration by minimizing the image–domain cost function. HCM has sparse synaptic connections, thereby reducing the computational cost of restoration. In addition, its unique cluster configuration is compatible with region-based human visual perception. The technique models the current blur with the best-fit parametric structure, assesses its relevance by evaluating the proximity measure and incorporates the weighted soft estimate into the blur identification criteria. It reconciles the dilemma faced by traditional blind image restoration schemes where a hard decision on the certainty of blur structure has to be made before the algorithm formulation.

Experimental results show that the methods presented in the chapter are robust in blind deconvolution of images degraded under different blur structures and noise levels. In particular, it is effective in identifying the difficult Gaussian blur that troubles most traditional blind restoration methods.

8

Edge Detection Using Model-Based Neural Networks

8.1 Introduction

In this chapter, we adopt user-defined salient image features as training examples for a specially designed model-based neural network to perform feature detection. Specifically, we have investigated an alternative MBNN with hierarchical architecture [48,73,129] which is capable of learning user-specified features through an interactive process. It will be shown that this network is capable of generalizing from the set of specified features to identify similar features in images which have not been included in the training data.

Edge characterization represents an important subproblem of feature extraction, the aim of which is to identify those image pixels with appreciable changes in intensities from their neighbors [2,4]. The process usually consists of two stages: in the first stage, all sites with appreciable intensity changes compared to their neighbors are identified, and in the second stage, the level of intensity change associated with each site is compared against a threshold to decide if the change is "significant" enough such that the current pixel is to be regarded as an edge pixel. In other words, a suitable difference measure is required in the first stage, examples of which include the Roberts operator, the Prewitt operator and the Sobel operator [2,4]. On the other hand, the second stage requires the specification of a threshold which describes the notion of a significant edge magnitude level.

In simple edge detection, the user specifies a global threshold on the edge magnitudes in an interactive way to produce a binary edge map from the edge magnitude map. The result is usually not satisfactory since some of the noisy pixels may be misclassified as edge pixels. More sophisticated approaches like the Canny edge detector [6] and the Shen–Castan edge detector [7] adopt the so-called hysteresis thresholding operation, where a significant edge is defined as a sequence of pixels with the edge magnitude of at least one of its members exceeding an upper threshold, and with the magnitudes of the other pixels exceeding a lower threshold. The localization of an edge within a single pixel width, in addition, requires some form of Laplacian of Gaussian (LoG) filtering [8] to detect the zero crossings, where more associated parameters have to be specified. Adding these to the previous thresholding parameter set gives rise to a large number of possible combinations of parameter values, each of which will result in a very different appearance for the edge map.

In this chapter a new MBNN architecture is adopted to estimate the set of parameters for edge characterization. In anticipation of a still larger parameter set for more sophisticated edge detection operations, a logical choice for its representation would be in the form of the set of connection weights for a neural network. To this end, an MBNN based on the hierarchical architecture proposed by Kung and Taur [48] is developed for the purpose of edge characterization, in which the connection weights play the dual role of encoding the edge-modeling parameters in the initial high-pass filtering stage and the thresholding parameters in the final decision stage.

8.2   MBNN Model for Edge Characterization

In Chapter 5, we have adopted a hierarchical model-based neural network architecture for adaptive regularization. Specifically, the network consists of a set of weight-parameterized model-based neurons as the computational units, and the output of each subnetwork is defined as the linear combination of these local neuron outputs. For the current application, the MBNN architecture will again be adopted, but due to the different requirements of the edge characterization problem, an alternative neuron model, which we refer to as the input-parameterized model-based neuron, is proposed. In addition, a winner-take-all competition process is applied to all the model-based neurons within a single subnetwork to determine the subnetwork output instead of the previous linear combination operation.

8.2.1   Input-Parameterized Model-Based Neuron

Instead of mapping the embedded low-dimensional weight vector $z \in R^M$ to the high-dimensional weight vector $p \in R^N$ as in the case of a weight-parameterized model-based neuron, the *input-parameterized* model-based neuron is designed such that a high-dimensional *input vector* $x \in R^N$ is mapped to a low-dimensional vector $x^P \in R^M$. This is under the assumption that for the problem at hand the reduced vector x^P can fully represent the essential characteristics of its higher-dimensional counterpart x. If such a mapping exists for the set of input vectors x, we can directly represent the weight vector in its low-dimensional form z in the network instead of its embedded form in a high-dimensional vector p.

More formally, we assume the existence of a mapping $\mathcal{P} : R^N \longrightarrow R^M$, such that $x^P = \mathcal{P}(x) \in R^M$, where $M < N$. The operation of this model-based

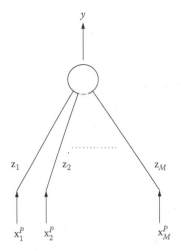

FIGURE 8.1
The input-parameterized model-based neuron.

neuron is then defined as follows:

$$y_s = f_s(\mathbf{x}, \mathbf{p})$$
$$= f_s(\mathbf{x}^P, \mathbf{z}) \tag{8.1}$$

The input-parameterized model-based neuron is illustrated in Figure 8.1.

The two different types of model-based neurons both allow the search for the optimal weights to proceed more efficiently in a low-dimensional weight space. The decision to use either one of the above neurons depends on the nature of the problem, which in turn determines the availability of either the mapping $\mathcal{M}$ for the weights, or the mapping $\mathcal{P}$ for the input: if the parameter space of a certain problem is restricted to a low-dimensional surface in a high-dimensional space, then it is natural to adopt the weight-parameterized model-based neuron. This is the case for the adaptive regularization problem discussed previously, where the valid set of weight vectors for the network forms a path, or alternatively a one-dimensional surface, governed by the single regularization parameter λ.

On the other hand, for cases where such representation for the weight vector is not readily available, while there is evidence, possibly through the application of principal component analysis (PCA) [162–164] or prior knowledge regarding the input vector distribution, of the existence of an information-preserving operator (with respect to the current problem) which maps the input vectors to a low-dimensional subspace, then it is more natural to adopt the input-parameterized model-based neuron. This is the case for the current edge characterization model-based neural network where the original input vector comprising a window of edge pixel values is mapped to a two-dimensional vector representing the two dominant gray levels around the edge.

8.2.2 Determination of Subnetwork Output

For the current application, instead of using a linear combination approach as in the previous adaptive regularization problem, a winner-take-all competition process is applied to all the neurons within a certain subnetwork to determine the *local winner* as follows:

$$\mathbf{p}_{s_r^*} = \arg\min_{s_r} \psi_r(\mathbf{x}, \mathbf{p}_{s_r}) \tag{8.2}$$

where $\mathbf{p}_{s_r^*}$ is the index of the winning neuron.

The subnetwork output $\phi(\mathbf{x}, \mathbf{p}_r)$ is then substituted with the corresponding local neuron output of the winner

$$\phi(\mathbf{x}, \mathbf{p}_r) = \psi_r(\mathbf{x}, \mathbf{p}_{s_r^*}) \tag{8.3}$$

In accordance with this competition process, it is natural to adopt the Euclidean distance for evaluating the local neuron output

$$\psi_r(\mathbf{x}, \mathbf{p}_{s_r}) = \|\mathbf{x} - \mathbf{p}_{s_r}\| \tag{8.4}$$

This class of networks is especially suitable for unsupervised pattern classification, where each pattern class is composed of several disjoint subsets of slightly different characteristics. We can then assign each primary pattern class to a single subnetwork, and each secondary class under the current primary class to a neuron within the subnetwork. The resulting network architecture and the structure of its subnetwork is shown in Figure 8.2.

For the input-parameterized model-based neuron, instead of embedding the model-based vector $\mathbf{z}_{s_r}$ in a higher-dimensional space, we map the input vector $\mathbf{x} \in \mathbf{R}^N$ on to a low-dimensional submanifold $\mathbf{R}^M$ through the operator $\mathcal{P}$ as follows:

$$\psi_r(\mathbf{x}, \mathbf{p}_{s_r}) \equiv \psi_r(\mathbf{x}^P, \mathbf{z}_{s_r}) \tag{8.5}$$
$$= \psi_r(\mathcal{P}(\mathbf{x}), \mathbf{z}_{s_r}) \tag{8.6}$$

where $\mathbf{x}^P = \mathcal{P}(\mathbf{x})$ is the low-dimensional input vector corresponding to $\mathbf{x}$ in $\mathbf{R}^N$.

8.2.3 Edge Characterization and Detection

The adoption of the current network architecture is motivated by our observation of the different preferences of human beings in regarding a certain magnitude of gray-level discontinuity as constituting a significant edge feature under different illuminations. To incorporate this criterion into our edge detection process, it is natural to adopt a hierarchical network architecture where we designate each subnetwork to represent a different illumination level, and each neuron in the subnetwork to represent different prototypes of edge-like features under the corresponding illumination level. In this work, we have defined an *edge prototype* as a two-dimensional vector $\mathbf{w} \in \mathbf{R}^2$ which represents the two dominant gray-level values on both sides of the edge.

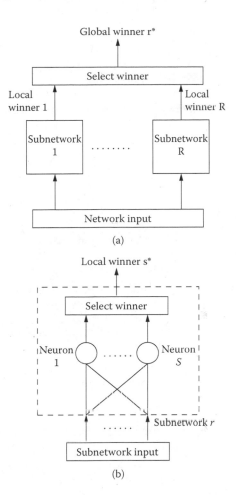

FIGURE 8.2
Hierarchical network architecture for edge characterization: (a) global network architecture, (b) subnetwork architecture.

Due to this necessity of inferring prototypes from the human-supplied training examples, it is natural to adopt unsupervised competitive learning where each prototype is represented as the weight vector of a neuron. The winner-take-all nature of the competition process also favors the use of the subcluster hierarchical architecture, such that only the local winner within each subnetwork is allowed to update its weight vector. In addition, for the rth subnetwork, the local neuron output is evaluated in terms of the Euclidean distance between the edge prototype and the current edge example

$$\psi_r(\mathbf{x}^P, \mathbf{z}_{s_r}) = \|\mathbf{x}^P - \mathbf{z}_{s_r}\| \tag{8.7}$$

where $\mathbf{x}^P \in \mathbf{R}^2$ is the current edge example, and $\mathbf{z}_{s_r}$ is the s_rth edge prototype of the rth subnetwork.

In view of the fact that the current edge example is usually specified in terms of a window of gray-level values $\mathbf{x} \in \mathbf{R}^N$, where $N \gg 2$, it is necessary to summarize this high-dimensional vector in terms of its two dominant gray-level values. In other words, we should derive a mapping $\mathcal{P} : \mathbf{R}^N \longrightarrow \mathbf{R}^2$ such that

$$\mathbf{x}^P = \mathcal{P}(\mathbf{x}) \tag{8.8}$$

which corresponds to an input-parameterized model-based neuron.

8.3   Network Architecture

The proposed MBNN architecture consists of a number of subnetworks, with each neuron of a subnetwork encoding a specified subset of the training samples. In the initial training stage, the edge examples in the training set are adaptively partitioned into different subsets through an unsupervised competitive process between the subnetworks. In the subsequent recognition stage, edge pixels are identified by comparing the current pixel configuration with the encoded templates associated with the various subnetworks.

The adoption of the current HMBNN architecture is due to our observation that, in edge detection, it would be more effective to adopt multiple sets of thresholding decision parameters corresponding to different local contexts, instead of a single parameter set across the whole image as in previous approaches.

We consider the following encoding scheme where each subnetwork is associated with an edge template corresponding to a different background illumination level, and each neuron in the subnetwork encodes possible variations of edge prototypes under the corresponding illumination level. The architecture of the feature detection network is shown in Figure 8.3 and the

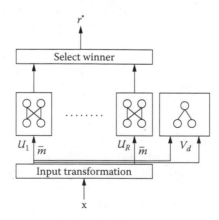

FIGURE 8.3
The architecture of the HMBNN edge detector.

hierarchical encoding scheme for the edge prototypes is described in the following sections.

8.3.1   Characterization of Edge Information

The edge detection is performed on the $N \times N$ neighborhood of the current pixel. Concatenating the corresponding gray-level values into a vector $\mathbf{x} = [x_1 \ldots x_{N^2}]^T \in \mathbf{R}^N$, the mean of the gray is evaluated as follows:

$$\bar{x} = \frac{1}{N^2} \sum_{n=1}^{N^2} x_n \tag{8.9}$$

Given this mean value, the gray-level information in this local window can be further characterized by a vector $\mathbf{m} = [m_1 \ m_2]^T \in \mathbf{R}^2$, the components of which correspond to the two dominant gray-level values within the window and is defined as follows:

$$m_1 = \frac{\sum_{i=1}^{N^2} I(x_i < \bar{x}) x_i}{\sum_{i=1}^{N^2} I(x_i < \bar{x})} \tag{8.10}$$

$$m_2 = \frac{\sum_{i=1}^{N^2} I(x_i \geq \bar{x}) x_i}{\sum_{i=1}^{N^2} I(x_i \geq \bar{x})} \tag{8.11}$$

$$m = \frac{m_1 + m_2}{2} \tag{8.12}$$

where the function $I(\cdot)$ equals 1 if the argument is true, and equals 0 otherwise.

8.3.2   Subnetwork U_r

As described previously, each subnetwork $U_r, r = 1, \ldots, R$ is associated with a prototype background illumination gray level value p_r. Those local $N \times N$ windows in the image with their mean values closest to p_r are then assigned to the subnetwork U_r for further encoding. Specifically, a particular pixel window $\mathcal{W}$ in the image, with its associated mean values $\bar{m}, m_1, m_2$ as defined in Equations (8.10) to (8.12), is assigned to the subnetwork U_{r^*} if the following conditions are satisfied:

$$p_{r^*} \in [m_1, m_2] \tag{8.13}$$

$$|\bar{m} - p_{r^*}| < |\bar{m} - p_r| \quad r = 1, \ldots, R, r \neq r^* \tag{8.14}$$

where $[m_1, m_2]$ is the closed interval with m_1, m_2 as its endpoints. The set of all $N \times N$ pixel windows are thus partitioned into subsets with the members of each subset exhibiting similar levels of background illumination. For convenience, we will refer to the two conditions in Equations (8.13) and (8.14) collectively as $\mathbf{x} \longrightarrow U_{r^*}$. The architecture of the subnetwork is shown in Figure 8.4.

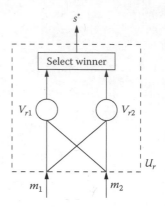

FIGURE 8.4
The architecture of a single subnetwork U_r.

8.3.3 Neuron V_{rs} in Subnetwork U_r

Each subnetwork U_r consists of S neurons V_{rs}, $s = 1, \ldots, S$, which encode the local edge templates representing the possible variations under the general illumination level p_r. Each neuron is associated with a weight vector $\mathbf{w}_{rs} = [w_{rs,1}\ w_{rs,2}]^T \in \mathbf{R}^2$ which summarizes the two dominant gray-level values in each $N \times N$ window $\mathcal{W}$ in the form of a local prototype vector $\mathbf{m}$. A window with an associated vector $\mathbf{m}$ is assigned V_{r*s*} if the following condition holds

$$\|\mathbf{m} - \mathbf{w}_{r*s*}\| < \|\mathbf{m} - \mathbf{w}_{rs}\| \quad s = 1, \ldots, S, s \neq s^* \tag{8.15}$$

We have chosen $S = 2$, that is, each subnetwork consists of two neurons, in order that one of the neurons encodes the local prototype for weakly visible edges and the other encodes the prototype for strongly visible edges. Correspondingly, one of the weight vectors $\mathbf{w}_{r*s}$, $s = 1, 2$ is referred to as the weak edge prototype $\mathbf{w}_{r*}^l$ and the other one as the strong edge prototype $\mathbf{w}_{r*}^u$. The determination of whether a neuron corresponds to a strong or weak edge prototype is made according to the following criteria:

- $\mathbf{w}_{r*}^l = \mathbf{w}_{rs'}$, where $s' = \arg\min_s (w_{r*s,2} - w_{r*s,1})$
- $\mathbf{w}_{r*}^u = \mathbf{w}_{rs''}$, where $s'' = \arg\max_s (w_{r*s,2} - w_{r*s,1})$

Given the weak edge prototype $\mathbf{w}_{r*}^l = [w_{r*,1}^l\ w_{r*,2}^l]^T$, the measure $(w_{r*,2}^l - w_{r*,1}^l)$ plays a similar role as the threshold parameter in conventional edge detection algorithms in specifying the lower limit of visibility for edges, and is useful for identifying potential starting points in the image for edge tracing. The structure of the neuron is shown in Figure 8.5.

8.3.4 Dynamic Tracking Neuron V_d

In addition to the subnetworks U_r and the local neurons V_{rs}, a *dynamic tracking neuron* V_d is associated with the network itself. In other words, this neuron is global in scope and does not belong to any of the subnetworks U_r. The dynamic neuron is a hybrid between a subnetwork and a local neuron in

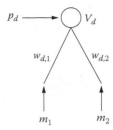

FIGURE 8.5
The structure of the dynamic edge tracking neuron.

that it consists of *both* a *dynamic weight vector* $\mathbf{w}_d = [w_{d,1} \; w_{d,2}]^T \in \mathbf{R}^2$, which corresponds to the weight vector of the local neuron, and a scalar parameter p_d which is analogous to the illumination level indicator p_r of each subnetwork. The structure of the dynamic tracking neuron is shown in Figure 8.5.

The purpose of this neuron is to track the varying background gray levels during the edge tracking operation. The neuron is inactive in the training stage. In the recognition stage and upon the detection of a prominent edge point, the dynamic weight vector $\mathbf{w}_d$ and illumination level indicator p_d are continuously modified to track those less prominent edge points connected to the initial edge point.

8.3.5 Binary Edge Configuration

We suppose that the vector $\mathbf{m}$ for the current window $\mathcal{W}$ is assigned to neuron V_{r*s*} with weight vector $\mathbf{w}_{r*s*}$. For edge detection, the real-valued vector $\mathbf{x} \in \mathbf{R}^{N^2}$ representing the gray-level values of the current window is mapped to a binary vector $\mathbf{b} \in \mathbf{B}^{N^2}$, where $\mathbf{B} = \{0,1\}$. To achieve this purpose, we define the mapping $\mathcal{Q}: \mathbf{R}^{N^2} \times \mathbf{R}^2 \longrightarrow \mathbf{B}^{N^2}$ as follows:

$$\mathbf{b} = \mathcal{Q}(\mathbf{x}, \mathbf{w}_{r*s*}) = [q(x_1, \mathbf{w}_{r*s*}) \ldots q(x_{N^2}, \mathbf{w}_{r*s*})]^T \in \mathbf{B}^{N^2} \qquad (8.16)$$

where the component mappings $q : \mathbf{R} \times \mathbf{R}^2 \longrightarrow \mathbf{B}$ are specified as

$$q(x_n, \mathbf{w}_{r*s*}) = \begin{cases} 0 & \text{if } |x_n - w_{r*s*,1}| < |x_n - w_{r*s*,2}| \\ 1 & \text{if } |x_n - w_{r*s*,1}| \geq |x_n - w_{r*s*,2}| \end{cases} \qquad (8.17)$$

For valid edge configurations, the binary vectors $\mathbf{b}$ assume special forms which are illustrated in Figure 8.6 for $N = 3$.

0	0	0
0	0	0
1	1	1

0	0	1
0	0	1
0	0	1

0	1	1
0	0	1
0	0	0

FIGURE 8.6
Examples of valid edge configurations.

FIGURE 8.7
Illustrating the operation $R_{\frac{\pi}{4}}$.

During the initial training stage, the binarized edge configuration associated with each training edge pattern is stored in an *edge configuration set* C which forms part of the overall network parameter set. The set C is then expanded in order that, when $N = 3$, it is closed under an operation $R_{\frac{\pi}{4}}$. Specifically, this operation permutes the entries of a vector $\mathbf{b} \in \mathbf{B}^9$ in such a way that, when interpreted as the entries in a 3×3 window, $R_{\frac{\pi}{4}}(\mathbf{b})$ is the 45° clockwise rotated version of $\mathbf{b}$. This is illustrated in Figure 8.7. In this way, detection of those rotated edge configurations not present in the training set is facilitated.

8.3.6   Correspondence with the General HMBNN Architecture

The current edge characterization network has a direct correspondence with the general MBNN architecture as follows: the subnetworks $U_r, r = 1, \ldots, R$ can be directly associated with the subnetworks in the general model, and the neurons $V_{rs_r}, s_r = 1, \ldots, S_r$ are associated with those within each subnetwork in the general model, with $S_r = 2$ for all r in the current application.

Instead of embedding a low-dimensional weight vector in a high-dimensional space as with the weight-parameterized model-based neuron, we have instead mapped the relatively high-dimensional input vector $\mathbf{x}$ into a low-dimensional input $\mathbf{m}$ which characterizes the two dominant gray-level values within a local window of pixels. In other words, we adopt the input-parameterized model-based neuron described previously, with the mapping $\mathcal{P} : \mathbf{R}^{N^2} \longrightarrow \mathbf{R}^2$, defined such that

$$\mathbf{m} = \mathcal{P}(\mathbf{x}) \tag{8.18}$$

Despite the above correspondences, there are slight differences between the original formulation of the subcluster hierarchical structure in [48] and the current edge characterization network structure. In the original network, the subnetwork output $\phi(\mathbf{x}, \mathbf{w}_r)$ is directly substituted by the neuron output of the local winner as follows:

$$\phi(\mathbf{x}, \mathbf{w}_r) \equiv \psi_r(\mathbf{x}, \mathbf{w}_{s^*}) \tag{8.19}$$

where s^* is the index of the local winner, and $\psi_r(\mathbf{x}, \mathbf{w}_{s^*})$ is the local neuron output of the winning neuron.

In the current network, on the other hand, the competition process at the subnetwork level is independent of the corresponding process at the neuronal level, and is of a very different nature: the competition between the subnetworks is specified in terms of the conformance of the current local illumination gray-level value with one of the prototype background illumination levels, that is, a comparison between scalar values. The competition at the neuronal level, however, involves the comparison of vectors in the form of two-dimensional edge prototypes existing under a particular illumination gray-level value. As a result, the subnetwork outputs and the local neuron outputs are independently specified at the two hierarchical levels. At the subnetwork level, the following subnetwork output is defined for the background illumination gray-level values:

$$\phi(\overline{m}, p_r) = |\overline{m} - p_r| \tag{8.20}$$

At the neuron level, the following local neuron output is defined for the edge prototype vectors:

$$\psi_r(\mathbf{m}, \mathbf{w}_{rs}) = \|\mathbf{m} - \mathbf{w}_{rs}\| \tag{8.21}$$

8.4 Training Stage

The training stage consists of three substages: in the first substage, the prototype background gray level $p_r, r = 1, \ldots, R$ for each subnetwork U_r is determined by competitive learning [34,44]. In the second substage, each window W is assigned to a subnetwork and a local neuron under this subnetwork based on its associated parameters p_r and $\mathbf{m}$. The weight vector $\mathbf{w}_{rs}$ of the winning local neuron is then updated using competitive learning. In the third stage, the corresponding binary edge configuration pattern $\mathbf{b}$ is extracted as a function of the winning weight vector $\mathbf{w}_{rs}$ based on the mapping Q. Adopting a window size of $N = 3$, we apply the operation $R_{\frac{\pi}{4}}$ successively to $\mathbf{b}$ to obtain the eight rotated versions of this pattern, and insert these patterns into the edge configuration memory C.

8.4.1 Determination of p_{r^*} for Subnetwork U_{r^*}

Assume that the current window with associated mean p_r is assigned to U_{r^*} based on conditions given in Equations (8.13) and (8.14). The value of p_{r^*} is then updated using competitive learning as follows:

$$p_{r^*}(t+1) = p_{r^*}(t) + \eta(t)(\overline{m} - p_{r^*}(t)) \tag{8.22}$$

The learning stepsize $\eta(t)$ is successively decreased according to the following schedule:

$$\eta(t+1) = \eta(0)\left(1 - \frac{t}{t_f}\right) \tag{8.23}$$

where t_f is the total number of iterations.

8.4.2   Determination of $w_{r^*s^*}$ for Neuron $V_{r^*s^*}$

Assuming that the current window with associated feature vector $\mathbf{m}$ is assigned to the local neuron $V_{r^*s^*}$ under subnetwork U_{r^*}, the associated weight vector $\mathbf{w}_{r^*s^*}$ is again updated through competitive learning as follows:

$$\mathbf{w}_{r^*s^*}(t+1) = \mathbf{w}_{r^*s^*}(t) + \eta(t)(\mathbf{m} - \mathbf{w}_{r^*s^*}(t)) \tag{8.24}$$

where the stepsize $\eta(t)$ is successively decreased according to Equation (8.23).

8.4.3   Acquisition of Valid Edge Configurations

After going through the previous stages, the background illumination level indicators p_r for the subnetworks and the weight vectors $\mathbf{w}_{rs}$ for the neurons have all been determined. As a result, all the $N \times N$ windows can be assigned to their corresponding subnetworks and the correct neurons within the subnetworks according to their parameters $\overline{m}$ and $\mathbf{m}$. If the current window is assigned to neuron $V_{r^*s^*}$ under subnetwork U_{r^*}, the gray-level vector $\mathbf{x}$ of the current window W can be transformed into a binary edge configuration vector $\mathbf{b}$ as a function of $\mathbf{w}_{r^*s^*}$ according to Equation (8.17):

$$\mathbf{b} = \mathcal{Q}(\mathbf{x}, \mathbf{w}_{r^*s^*}) \tag{8.25}$$

The binary vector $\mathbf{b}$ is then inserted into the valid edge configuration set C. Using a window size of $N = 3$, the requirement that the set C be closed under the operation $R_{\frac{\pi}{4}}$ can be satisfied by generating the following eight edge configurations $\mathbf{b}_j$, $j = 0, \ldots, 7$ using $R_{\frac{\pi}{4}}$ as follows

$$\mathbf{b}_0 = \mathbf{b} \tag{8.26}$$
$$\mathbf{b}_{j+1} = R_{\frac{\pi}{4}}(\mathbf{b}_j) \quad j = 0, \ldots, 6 \tag{8.27}$$

and inserting all of them into the configuration set C.

8.5 Recognition Stage

In this stage, all pixels in a test image are examined in order that those pixels with edge-like features are identified. This recognition stage consists of two substages. In the first substage, all pixels in the test image with high degree of similarity to the learned edge prototypes are declared as primary edge points. In the second substage, these primary edge points are used as starting points for an edge tracing operation such that the less prominent edge points, which we refer to as secondary edge points, are recursively identified.

8.5.1 Identification of Primary Edge Points

In this substage, all $N \times N$ windows in the test image are examined. All those windows with associated parameters $\overline{m}, m_1, m_2$ satisfying the following conditions are declared as primary edge points.

(A1). $\mathbf{x} \longrightarrow U_{r*}$ (i.e., satisfaction of conditions in Equations (8.13) and (8.14)) for some r^*.

(A2). $m_2 - m_1 \geq w^l_{r*,2} - w^l_{r*,1}$, where $\mathbf{w}^l_{r*}$ is the weak edge prototype vector of U_{r*}.

(A3). $\mathbf{b} = \mathcal{Q}(\mathbf{x}, \mathbf{w}_{r*s*}) \in C$, where $\mathbf{w}_{r*s*}$ is the weight vector associated with the selected neuron V_{r*s*}.

Condition (A1) specifies that the mean gray-level value of the current window should be close to one of the designated levels of the network. Condition (A2) ensures that the edge magnitude, as represented by the difference $m_2 - m_1$, is greater than the difference between the components of the corresponding weak edge prototype. Condition (A3) ensures that the binary edge configuration corresponding to the current window is one of the valid templates in the configuration set C.

8.5.2 Identification of Secondary Edge Points

In this second substage, the dynamic tracking neuron V_d is activated to trace the secondary edge points connected to the current primary edge point. The gray-level indicator p_d and weight vector $\mathbf{w}_d$ of the neuron are initialized using the parameters of the detected primary edge point $\overline{m}_p$ and $\mathbf{m}^p = [m_1^p \ m_2^p]^T$ as follows:

$$p_d(0) = \overline{m}^p \tag{8.28}$$

$$\mathbf{w}_d(0) = \mathbf{m}^p \tag{8.29}$$

After initializing the parameters of the dynamic neuron, a recursive edge tracing algorithm is applied to identify the less prominent edge pixels (the secondary edge points) connected to the primary edge points by applying the following set of conditions at each 8-neighbor of the current primary edge point:

(B1). $\mathbf{b} = \mathcal{Q}(\mathbf{x}, \mathbf{w}_d) \in C$.

(B2). $p_d \in [m_1, m_2]$.

Condition (B1) is similar to condition (A3) for the primary edge point detection, while condition (B2) is a modified version of the condition in Equation (8.13) which forms part of the requirements for $\mathbf{x} \longrightarrow U_r$, to ensure that the mean gray-level value of a potential secondary edge point, as represented by p_d, should be similar to those of the previously traversed edge points. In addition, no specific conditions are imposed on the edge magnitude of the current pixel to allow the possibility of including weak edge points in the final edge segment, as long as it is connected to a prominent edge point.

For each secondary edge point (with corresponding parameters $\overline{m}^s$ and $\mathbf{m}^s$) which satisfies the above conditions, the local illumination level indicator p_d of the dynamic neuron is updated as follows:

$$p_d(t+1) = p_d(t) + \eta(t)(\overline{m}^s - p_d(t)) \tag{8.30}$$

In addition, if this edge point satisfies condition (A2) for primary edge point detection, indicating that its edge magnitude is comparable to that of a primary edge point, the weight vector $\mathbf{w}_d$ of the dynamic neuron is also updated to incorporate the characteristics of the current point:

$$\mathbf{w}_d(t+1) = \mathbf{w}_d(t) + \eta(t)(\mathbf{m}^s - \mathbf{w}_d(t)) \tag{8.31}$$

8.6 Experimental Results

In the training stage, users are requested to select different edge examples from two images depicting a flower and an eagle, respectively. These are then used as training examples for the network. To ensure that the users' preferences are adequately represented, different edge traces have been prepared for each of the two images. The eagle image and three different sets of edge examples are shown in Figure 8.8, while those for the flower image are shown in Figure 8.9.

After the training stage, the MBNN is first applied to images in the training set in order to test the capability of the network to identify all relevant edge-like features from the previous sparse tracings. The result is shown in Figure 8.10a.

By comparing this figure with the tracings in Figures 8.8b to d, we can conclude that the network is capable of generalizing from the initial sparse

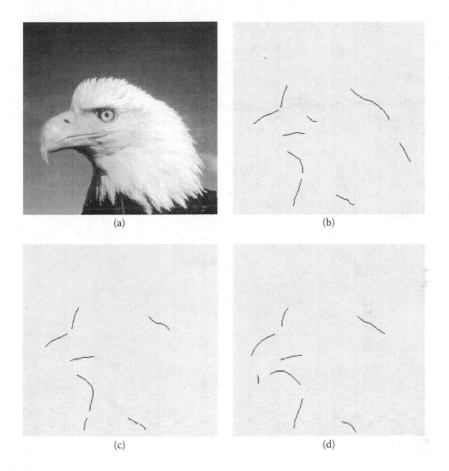

FIGURE 8.8
(a) Eagle image. (b)–(d) User-specified edge examples.

tracings to identify the important edges. In addition, we compare this result with that of a standard edge detector. Since the performance of standard edge detectors depends on the values of a set of tunable parameters, our result can be validated if the performance of our approach is comparable to the standard results under near-optimal parameter settings.

The Shen–Castan edge detector [7] is chosen as the comparison standard in Figure 8.10. This edge detector employs an exponential filter in the smoothing operation prior to the differencing operation, which provides a better degree of edge localization than the Gaussian filter used in the Canny edge detector [6]. For this edge detector, the hysteresis thresholds t_1, t_2 are adjusted in each edge detection operation to provide the best visual result.

FIGURE 8.9
(a) Flower image. (b)–(d) User-specified edge examples.

In Figure 8.10, the performance of the MBNN edge detector is compared with that of the Shen–Castan edge detector under different hysteresis threshold settings. The MBNN result is shown in Figure 8.10a and the Shen–Castan edge detection results are shown in Figures 8.10b to d. The lower hysteresis threshold ranges from $t_1 = 10$ to $t_1 = 40$, and the upper threshold is set to $t_2 = t_1 + 5$.

We can observe from Figures 8.10b to d that the set of detected edges is sensitive to the choice of t_1 and t_2: lower values of t_1 and t_2 reveal more details but result in more false positive detections as seen in Figure 8.10b, while higher threshold values lead to missed features as in Figure 8.10d. In our opinion, Figure 8.10c, with $t_1 = 20$ and $t_2 = 25$, constitutes an acceptable

FIGURE 8.10
Edge detection results for the eagle image: (a) Detected edges using NN, (b)–(d) Detected edges using the Shen–Castan edge detector with different hysteresis thresholds t_1, t_2: (b) $t_1 = 10$, $t_2 = 15$; (c) $t_1 = 20$, $t_2 = 25$; (d) $t_1 = 30$, $t_2 = 35$.

representation of the underlying edge features. These can be compared with the MBNN result in Figure 8.10a, where we can see that the edge map is similar to that of Figure 8.10c which we have previously chosen as the one corresponding to the preferred threshold parameter settings for the Shen–Castan edge detector.

The results for the flower image are shown in Figure 8.11. Figures 8.11b to d show the effect of varying the thresholds t_1, t_2 and the MBNN result is shown in Figure 8.11a. For the Shen–Castan edge detection results, we may consider Figure 8.11c (with thresholds $t_1 = 20$ and $t_2 = 25$) as a near optimal

(a)

(b)

(c)

(d)

FIGURE 8.11

Edge detection results for the flower image. (a) Detected edges using NN. (b)–(d) Detected edges using Shen–Castan edge detector with different hysteresis thresholds t_1, t_2: (b) $t_1 = 10$, $t_2 = 15$; (c) $t_1 = 20$, $t_2 = 25$; (d) $t_1 = 30$, $t_2 = 35$.

representation and Figure 8.11d as an adequate representation, although there are some missing features in this case. In Figure 8.11a, we notice that the MBNN detection result lies in between these two and thus can be considered a close approximation to the optimal result. On the other hand, we can see that the result in Figure 8.11b is over-cluttered with nonessential details due to the low threshold values.

The generalization performance of the NN edge detector is further evaluated by applying the network trained on the eagle and flower images to previous unseen images. In Figure 8.12a, an image of a building is shown, and Figure 8.12c shows an image of a plane. The corresponding detection results are shown in Figures 8.12b and d, respectively. The results are very

FIGURE 8.12
(a) Image of building. (b) Detected edges using NN. (c) Image of plane. (d) Detected edges using NN.

satisfactory, considering that the network is trained only on the eagle and flower images.

To test the robustness of the current approach, we apply the network to noise-contaminated images. Figure 8.13a shows the result of adding zero-mean Gaussian noise with a standard deviation of $\sigma_n = 10$ to the eagle image. The previous MBNN (trained using the noiseless images) is applied to this noisy image without any retraining and alteration of architecture. The result is shown in Figure 8.13b, showing that although some false alarms occurred, the overall effect is not serious and the result is reasonably similar to the noiseless case. On the other hand, for the Shen–Castan edge detector, if we choose the previous optimal threshold of $t_1 = 20$ and $t_2 = 25$ (Figure 8.13c), the effect of

FIGURE 8.13
(a) Eagle image with additive Gaussian noise ($\sigma_n = 10$). (b) Detected edges using NN. (c)–(d) Detected edges using Shen–Castan edge detector with different hysteresis thresholds t_1, t_2: (c) $t_1 = 20$, $t_2 = 25$; (d) $t_1 = 25$, $t_2 = 30$.

noise is clearly noticeable, and the thresholds have to be readjusted to $t_1 = 25$ and $t_2 = 30$ to remove the noise specks (Figure 8.13d).

For the other three images, the results for the noisy case are shown in Figures 8.14b, d and f. We can again notice that the effect of noise contamination is not significant.

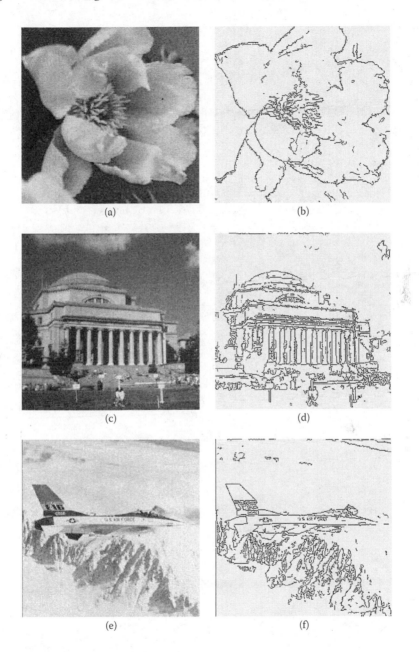

FIGURE 8.14
(a) Flower image with additive Gaussian noise ($\sigma_n = 10$). (b) Detected edges using NN. (c) Building image with additive Gaussian noise ($\sigma_n = 10$). (d) Detected edges using NN. (e) Plane image with additive Gaussian noise ($\sigma_n = 10$). (f) Detected edges using NN.

8.7 Summary

We have developed a model-based feature detection neural network with hierarchical architecture (HMBNN) which directly learns the essential characteristics of user-specified features through a training process. The specific architecture of the network allows the division of the training edge examples into subclasses such that the different preferences of users in regarding intensity discontinuities as edges under different illumination conditions are taken into account. To achieve this, the MBNN implicitly represents these required parameters for edge detection in the form of network weights which are updated during the training process. The current approach also takes into account the local variations in intensity distributions along edge segments, and the rules which capture this nonstationarity are learned by the proposed architecture. This HMBNN edge detector has been successfully applied to both the set of training images and to previously unseen images with promising results. In addition, no retraining of the network and no alteration of architecture are required for applying the network to noisy images.

9

Image Analysis and Retrieval via Self-Organization

9.1 Introduction

This chapter aims at introducing a family of unsupervised algorithms that have a basis in self-organization, yet are somewhat free from many of the constraints typical of other well known self-organizing architectures. Within this family, the basic processing unit is known as the self-organizing tree map (SOTM). The chapter will provide an in-depth coverage of this architecture and its derivations. It will then move through a series of pertinent real world applications with regards to the processing of image and video data—from its role in more generic image processing techniques such as the automated modeling and removal of impulse noise in digital images, to problems in digital asset management including the modeling of image and video content, indexing, and intelligent retrieval.

9.2 Self-Organizing Map (SOM)

The self-organizing map (SOM) is an unsupervised neural network model that implements a characteristic nonlinear projection from the high dimensional space of input signal onto a low dimensional array of neurons. It was proposed by Kohonen [165]. The principle of the network is based on a similar model developed earlier by Willshaw and Malsburg [166], explaining the formation of direct topographic projections between two laminar structures known as retinotectal mappings.

Kohonen discovered an astounding new phenomenon: there exist adaptive systems which can automatically form one or two dimensional maps of features that are present in the input signals. The input signals can be presented in a random order. If they are metrically similar in some structured way, then the same structure, in a topologically correct form, will be reflected in the spatial relations of the output responses. This phenomenon follows closely

FIGURE 9.1
The K-shape distribution of the input vectors.
(*Adapted from* [193], *with permission of publisher SPIE—The International Society for Optical Engineering.*)

to the observations in biological systems in which different sensory inputs are known to be mapped onto neighboring areas of the brain in an orderly fashion.

However, it can be demonstrated that if the input pattern distribution is uniform, the neuron set neatly adapts to the input data. when the input pattern distribution has a prominent shape, as shown in Figure 9.1, the result of the best-match computations tends to be concentrated on a fraction of neurons in the map. In the learning process, it may easily happen that the neurons lying in zero-density areas are affected by input patterns from all

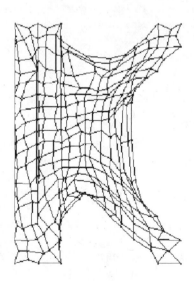

FIGURE 9.2
The final representation of the SOM for K-shape distribution of the input vectors. (*Adapted from* [193], *with permission of publisher SPIE—The International Society for Optical Engineering.*)

the surrounding parts of the nonzero distribution [167]. Although the SOM's topology exhibits the distribution of the structured input vectors, it also introduces false representations outside the distribution of the input space as shown in Figure 9.2 from our experiments. In Figure 9.2, there are grid-points in the regions with zero probability of information and it is difficult to find a better solution without clustering a lot of points near each inside corner. The outlying neurons are pulled towards the wining neuron. As this happens repeatedly, the window size shrinks and the fluctuation ceases. As a result, the outlying points remain outliers since they have no stable places to go.

9.3   Self-Organizing Tree Map (SOTM)

In view of the above observation of the SOM, a new mechanism named the self-organizing tree map (SOTM) is proposed. The motivation for the new method is twofold: (a) to keep the SOM's property of topology preserving while strengthening the flexibility of adapting to changes in the input distribution and maximally reflecting the distribution of the input patterns; (b) to add the ability of the adaptive resonance theory (ART) [168] to create new output neurons dynamically while overcoming the global threshold setting problem. In the SOTM, the relationships between the output neurons can be dynamically defined during learning. It allows the learning of not only the weights of the connections, but also the structure of the network including the number of neurons, the number of levels, and the interconnections among the neurons.

9.3.1   SOTM Model: Architecture

The SOTM is basically a competitive neural network in which the neurons are organized as a multilayer tree structure as shown in Figure 9.3.

Following the statistics of the input patterns, the neurons grow from scratch to form the self-organizing tree, capturing the local context of the pattern space

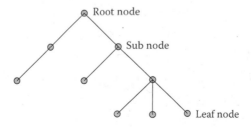

FIGURE 9.3
The structure of the tree hierarchy. (*Adapted from* [193], *with permission of publisher SPIE—The International Society for Optical Engineering.*)

and mapping it onto the structure of the tree. This principle can be expressed from a different viewpoint: the clustering algorithm starts from an isolated neuron and coalesces the nearest patterns to form the tree from the root node to the leaf nodes according to a hierarchy control function. There are two different levels of adaptation in the SOTM:

- Weights adaptation: If the input vector is close to one of the neurons in the network and the distance between them is within the hierarchy control function range, the winner's weight vector can be adjusted to the incoming vector.

- Structure adaptation: Adapt the structure of the network by changing the number of neurons and the structural relationship between neurons in the network. If the distance between the incoming vector and the winner neuron is outside the hierarchy control function range, the input vector becomes the winner's son, forming a subnode in the tree.

9.3.2 Competitive Learning Algorithm

The basic principle underlying competitive learning is vector quantization. Competitive learning neural networks adaptively quantize the pattern space R^N. Neurons compete for the activation induced by randomly sampled pattern vectors $x \in R^N$. The corresponding random weight vectors $\mathbf{w}_j$ represent the Voronoi regions[1] about $\mathbf{w}_j$. Each weight vector $\mathbf{w}_j$ behaves as a quantization vector or reference vector. Competitive learning distributes the reference vectors to approximate the unknown probability density function $p(x)$. The network learns as weight vectors $\mathbf{w}_j$ change in response to random training data.

In the SOTM, each input pattern vector $\mathbf{x} = \{x_1, \ldots x_n\} \in R^N$ is projected onto a tree node. With every node j, a weight vector $\mathbf{w}_j = [w_{1j} \ldots w_{Nj}] \in R^N$ is associated. When an input vector is presented to the network, it is compared with each of the $\mathbf{w}_j$, and the best matching node is defined as the winner. The input vector is thus mapped onto this location. The best matching node is defined using the smallest of the Euclidean distances $\| \mathbf{x} - \mathbf{w}_j \|$. We modify the closest weight vector or the "winning" weight vector $\mathbf{w}_{j*}$ with simple difference learning law which is similar to Kohonen's SOM algorithm. A scaling factor of $\mathbf{x}(t) - \mathbf{w}_{j*}(t)$ is added to $\mathbf{w}_{j*}(t)$ to form $\mathbf{w}_j(t+1)$. The "losers" are not modified: $\mathbf{w}_j(t+1) = \mathbf{w}_j(t)$.

$$\mathbf{w}_j(t+1) = \mathbf{w}_{j*}(t) + \alpha(t)[\mathbf{x}(t) - \mathbf{w}_{j*}(t)],$$
$$\mathbf{w}_j(t+1) = \mathbf{w}_j(t) \qquad \text{if } j \neq j^*, \tag{9.1}$$

---

[1] Voronoi tesselation partitions the pattern space into regions around reference vectors; such regions are called Voronoi regions.

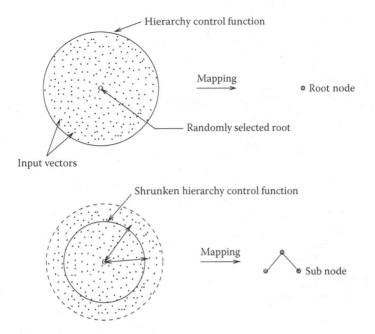

FIGURE 9.4
The formation of the tree map.

where $\alpha(t)$ denotes a monotonically decreasing sequence of learning coefficients. We can also update the nearest-neighbors of the winning node. However, in this chapter we modify only one weight vector at a time.

A hierarchy control function is proposed to control the growth of the tree. At the initial stage, there is only one neuron whose weight vector is randomly chosen from the input pattern space to become the root of the tree. The stimulus which activates the neurons has a wide dynamic range; that is, the hierarchy control function is set large enough to enable all the neurons to be activated by the input stimulus. Then the hierarchy control function shrinks monotonically with time. The decreasing speed can be linear or exponential depending on the applications. During each decreasing stage, the input vector is checked to see if it is within the range of the hierarchy control value. If the input vector falls into the range, the weight vector is adjusted to the input vector. Otherwise, a new node (subnode) is formed in the tree. The formation of the tree map is shown in Figure 9.4.

Within a given period of the hierarchy control function, the learning tends to be a stochastic process, the weight vectors will eventually converge in the mean-square sense to the probabilistic mean of their corresponding input vectors as the learning rate $\alpha(t)$ decreases. The hierarchical classification organizes the samples such that each node represents a subset of samples which share some similar features. Such hierarchical classification can be quickly searched to find a matching pattern for a new input.

The SOTM algorithm is summarized as follows:

- **Step 1.** Initialize the weight vector with a random value (randomly take a training vector as the representative of the root node).
- **Step 2.** Get a new input vector, and compute the distances d_j between the input vector and all the nodes using

$$d_j = \sqrt{\sum_{i=1}^{N}(x_i(t) - w_{ij}(t))^2} \qquad (j = 1, \ldots, J) \qquad (9.2)$$

where J is the number of nodes.

- **Step 3.** Select the wining node j^* with minimum d_i.

$$d_{j^*}(\mathbf{x}, \ \mathbf{w}_j) = \min_i d_j(\mathbf{x}, \ \mathbf{w}_j) \qquad (9.3)$$

- **Step 4. If** $d_{j^*}(\mathbf{x}, \ \mathbf{w}_j) <= H(t)$
  where $H(t)$ is the hierarchy control function which decreases with time. $H(t)$ controls the level of the tree.
  Then assign x to the jth cluster, and update the weight vector $\mathbf{w}_j$ according to the following learning rule:

$$\mathbf{w}_j(t + 1) = \ \mathbf{w}_j(t) + \alpha(t)[\mathbf{x}(t) - \ \mathbf{w}_j(t)]; \qquad (9.4)$$

  where $\alpha(t)$ is the learning rate which decreases with time, $0 < \alpha(t) < 1$.
  Else form a new subnode starting with x.
- **Step 5.** Check terminate condition:
  Exit if one of the following three conditions is fulfilled.
  - The specified number of iterations is reached.
  - The specified number of clusters is reached.
  - No significant change to the tree map has occurred.
  Otherwise, repeat by going back to Step 2.

Learning in the SOTM takes place in two phases: the locating phase and the convergence phase. The adaptation parameter $\alpha(t)$ controls the learning rate which decreases with time as weight vectors approach the cluster centers. It is given by either a linear function $\alpha(t) = (1 - t/T_1)$ or an exponential function $\alpha(t) = e^{(-t/T_2)}$. T_1 and T_2 are constants which determine the decreasing rate. During the locating phase, the global topological adjustment of the weight vectors $\mathbf{w}_j$ takes place. $\alpha(t)$ stays relatively large during this phase. Initially, $\alpha(t)$ can be set as 0.8 and it decreases with time. After the locating phase, a small $\alpha(t)$ for the convergence phase is needed for the fine tuning of the map.

The hierarchy control function $H(t)$ controls the levels of the tree. It begins with a large value and decreases with time. It adaptively partitions the input vector space into smaller subspaces. In our experiments, $H(t)$ is defined as

$H(t) = (1 - t/T_3)$ or $H(t) = e^{-t/T_4}$. T_3 and T_4 are also constants that control the decreasing rate.

With the decreasing of the hierarchy control function $H(t)$, a subnode comes out to form a new branch. The evolution process progresses recursively until it reaches the leaf node. The entire tree structure preserves topological relations from the root node to the leaf nodes.

Learning in SOTM follows the stochastic competitive learning law which is expressed as a stochastic differential equation [169–172].

$$\dot{\mathbf{w}}_j = \mathbf{I}_{D_j}[\mathbf{x}_j - \mathbf{w}_j] \tag{9.5}$$

where $\mathbf{w}$ is the weight vector and $\mathbf{x}$ is the input vector. I_{D_j} denotes the zero-one indicator function of decision class D_j.

$$I_{D_j} = \begin{cases} 1 & \text{if } \mathbf{x} \in D_j \\ 0 & \text{if } \mathbf{x} \notin D_j \end{cases} \tag{9.6}$$

I_{D_j} indicates whether pattern $\mathbf{x}$ belongs to decision class D_j.

The stochastic differential equation describes how weight random processes change as a function of input random processes. According to the centroid theorem [173], the competitive learning converges to the centroid of the sampled decision class. The probability of $\mathbf{w}_j = \overline{\mathbf{x}}_j$ at equilibrium is

$$Prob(\mathbf{w}_j = \overline{\mathbf{x}}_j) = 1. \tag{9.7}$$

where the centroid $\overline{\mathbf{x}}_j$ of decision class D_j equals its probabilistic center of mass, given by

$$\overline{\mathbf{x}}_j = \frac{\int_{D_j} \mathbf{x} p(\mathbf{x}) \, d\mathbf{x}}{\int_{D_j} p(\mathbf{x}) \, d\mathbf{x}}$$
$$= E[\mathbf{x} | \mathbf{x} \in D_j] \tag{9.8}$$

The centroid theorem concludes that the average weight vector $E[\mathbf{w}_j]$ equals the jth centroid $\overline{\mathbf{x}}_j$ at equilibrium

$$E[\mathbf{w}_j] = \overline{\mathbf{x}}_j. \tag{9.9}$$

The weight vector $\mathbf{w}_j$ vibrates in a Brownian motion about the constant centroid $\overline{\mathbf{x}}_j$. Our simulated competitive weight vectors have exhibited such Brownian wandering about centroids.

9.3.3 Dynamic Topology and Classification Capability of the SOTM

The evolution processes of the SOTM and the convergence of the stochastic competitive learning are illustrated in this section by some snapshots. Dynamic topology and classification capability are two prominent characteristics of the SOTM which will be demonstrated by an application.

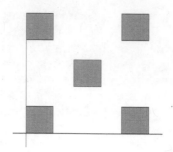

FIGURE 9.5
Uniformly distributed input vectors in five squares. (*Adapted from* [193], *with permission of publisher SPIE—The International Society for Optical Engineering.*)

Dynamic Topology of the SOTM

The dynamic SOTM topology is demonstrated in the following examples. In Figure 9.5, the learning of the tree map is driven by sample vectors uniformly distributed in the five squares. The tree mapping starts from the root node and gradually generates its subnodes as $H(t)$ decreases. The waveform of $H(t)$ function and learning rate $\alpha(t)$ is shown in Figure 9.6. Each time $H(t)$ decreases, $\alpha(t)$ starts from the initial state again. For every $H(t)$, α decreases with time. By properly controlling the decreasing speed of $\alpha(t)$, the SOTM finds the cluster center. For example, in the beginning of learning, the initial $H(t)$ is set large, it covers the whole space of the input patterns. As $\alpha(t)$ decreases with time, the SOTM converges to the root node shown in Figure 9.7a. When $H(t)$ shrinks with time, the subnodes are formed at each stage shown in Figure 9.7b to 9.7h. In the node organizing process, the optimal number of the output nodes can be obtained from visualizing the tree map evolution.

Another merit of the SOTM topology is shown in Figure 9.8. The input space is the same as Figure 9.1. From Figure 9.8, it can be seen that all the tree nodes are situated within the area of the distribution. The entire tree truthfully reflects the distribution of the input space. Using these nodes as code vectors, the distortion of vector quantization can be kept to a minimum.

9.3.4   Summary

In this section, a self-organizing tree map is presented. This model not only enhances the SOM's topology preserving capability, but also overcomes its weaknesses. Based on the competitive learning algorithm the SOTM adaptively estimates the probability density function $p(x)$ from sample realizations in the most faithful fashion and tries to preserve the structure of $p(x)$. Since weight vectors during learning tend to be stochastic processes, the centroid theorem is used to demonstrate its convergence. In conclusion, the weight vectors converge in the mean-square sense to the probabilistic centers of input

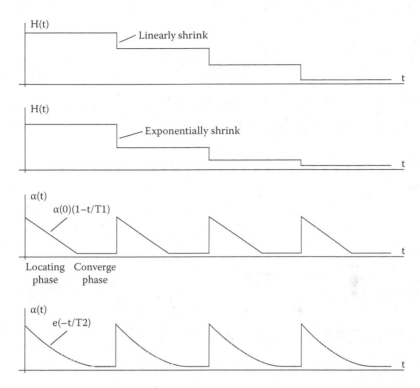

FIGURE 9.6
Waveform of the hierarchy control function $H(t)$ and the learning rate $\alpha(t)$. (*Adapted from* [193], *with permission of publisher SPIE—The International Society for Optical Engineering.*)

subsets to form a quantization mapping with a minimum mean squared distortion. A number of examples were used to illustrated the dynamic topology of the SOTM and its classification ability in the following sections.

9.4 SOTM in Impulse Noise Removal

9.4.1 Introduction

Removal of noise from images while preserving as much fine detail as possible is an essential issue in image processing. A variety of linear and nonlinear filtering techniques have been proposed for improving the quality of images degraded by noise [174–177] In the early stages of signal and image processing, linear filters were the primary tools for noise cleaning. However, linear filtering reduces noise at the expense of degrading signals. It tends to blur the edges and does not remove impulse noise effectively. The drawbacks of linear filters spurred the development of nonlinear filtering techniques for signal and image processing [178].

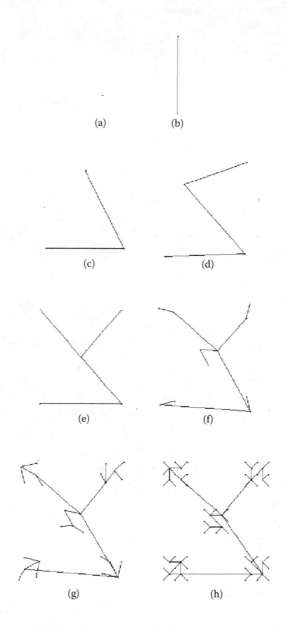

FIGURE 9.7
Different quantizers produced by the SOTM with different $H(t)$ for Figure 9.5: (a) the root node which represents the mean of all the input vectors; (b) to (g) the evolution processes of the self-organizing tree; (h) the final representation of the SOTM. (*Adapted from* [193], *with permission of publisher SPIE—The International Society for Optical Engineering.*)

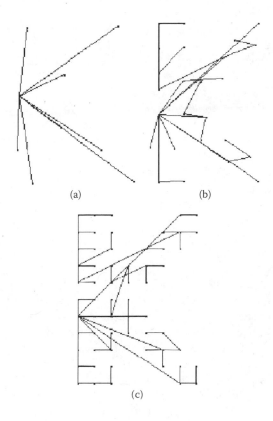

(a) (b)

(c)

FIGURE 9.8
The representation of SOTM for K-shape distribution of the input vectors. (a) to (b) the evolution of the tree map; (c) the final representation. (*Adapted from* [193], *with permission of publisher SPIE— The International Society for Optical Engineering.*)

In 1971, the best known and most widely used nonlinear filter, the median filter, was introduced by Tukey as a time series analysis tool for robust noise suppression [179]. Later, this filter came into use in image processing [174]. Median filtering has the advantages of preserving sharp changes in signals and of being effective for removing impulse noise when the error rate is relatively low. This is performed by moving a window over the image and replacing the pixel at the center of the window with the median value inside of the window. Due to the computational simplicity and easy hardware implementation, it is recognized as an effective alternative to the linear smoother for the removal of impulse noise.

One problem with the median filter is that it removes very fine details in the images and changes signal structures. In many applications, the median filter not only smooths the noise in homogeneous image regions, but also tends to produce regions of constant or nearly constant intensity. These regions are usually linear patches (streaks) or amorphous blotches. These side effects of the median filter are highly undesirable since they are perceived as either lines or contours that do not exist in the original image. To improve the performance of the median filter, many generalized median filters and alternatives to the median filter have been proposed. These include the weighted median (WM) filter [180], center weighted median (CWM) filter [181],weighted order statistic (WOS) filter [182], max/median filter [183], multistage median filter [19,184], nonlinear mean filters [185,186], etc. The generalized median filters tend to have better detail preserving characteristics than the median filter, but they preserve more details at the expense of poor noise suppression. Their performance depends strongly on the noise model and the error rate. Nonlinear mean filters exhibit better impulse suppression ability than the regular median filter when only positive or negative impulse noise exists. The results of nonlinear mean filters are unsatisfactory for mixed impulse noise [187,188].

In this chapter, we propose a novel approach for suppressing impulse noise in digital images while effectively preserving more details than previously proposed methods. The method presented is based on impulse noise detection and noise exclusive restoration. The motivation is that if impulses can be detected and their positions can be located in the image, then it is possible to replace the impulses with the best estimates by using only the uncorrupted neighbors. The noise removing procedure consists of two steps: the detection of the noise and the reconstruction of the image. As the SOTM network possesses the capability of classifying pixels in an image, it is employed to detect the impulses. A noise-exclusive median (NEM) filtering algorithm and a noise-exclusive arithmetic mean (NEAM) filtering algorithm are proposed to restore the image. By noise-exclusive it is meant that all the impulses in the window do not participate in the operation of order sorting or do not contribute to the operation of mean calculation. The new filtering scheme is different from the traditional median-type filtering because for the median-type filters, all the noise pixels inside the window are involved in the operation of ordering. This is the fundamental difference between the new filtering techniques and the traditional ones. According to the distribution of the remaining pixels in the window the point estimation can be adaptively chosen either from the NEM filter or the NEAM filter. As the method is able to detect noise locations accurately, the best possible restoration of images corrupted by impulse noise is achieved. This filtering scheme also has the characteristic that it can be generalized to incorporate any median type filtering techniques into the second step after noise detection.

9.4.2 Models of Impulse Noise

When an image is coded and transmitted over a noisy channel, or degraded by electrical sensor noise, as in a vidicon TV camera, degradation appears as salt-and-pepper noise (i.e. positive and negative impulses) [189,190]. Two models have been proposed for the description of such impulse noise [191]. The first model assumes fixed values for all the impulses. In the literature, corrupted pixels are often replaced with values that are equal to either the minimum or the maximum of the allowable dynamic range. For 8-bit images, this corresponds typically to fixed values equal to 0 or 255. In this chapter, a more general model in which a noisy pixel can take on arbitrary values in the dynamic range is proposed.

Impulse Noise: Model 1.

$$x_{ij} = \begin{cases} d & \text{with probability } p \\ s_{ij} & \text{with probability } 1 - p \end{cases} \tag{9.10}$$

where s_{ij} denotes the pixel values of the original image, d denotes the erroneous point value, x_{ij} denotes the pixel values of the degraded image.

A modified model assumes that both positive and negative impulses are present in the image.

$$x_{ij} = \begin{cases} d_p & \text{with probability } p_p \\ d_n & \text{with probability } p_n \\ s_{ij} & \text{with probability } 1\text{-}(p_p+p_n) \end{cases} \tag{9.11}$$

where d_p and d_n denote positive and negative impulse noise.

The second model allows the impulses to follow a random distribution.
Impulse Noise: Model 2.

$$x_{ij} = \begin{cases} d_p + n_{gu} & \text{with probability } p_p \\ d_n + n_{gu} & \text{with probability } p_n \\ s_{ij} & \text{with probability } 1\text{-}(p_p+p_n) \end{cases} \tag{9.12}$$

where n_{gu} follows a random distribution with zero mean.

The impulse noise models with fixed values, a Gaussian distribution and a uniform distribution are shown in Figure 9.9.

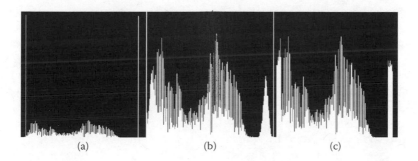

(a)   (b)   (c)

FIGURE 9.9

Histograms of impulse noise with different distributions: impulse noise (a) with fixed value distribution, (b) with Gaussian distribution, and (c) with uniform distribution. (*Adapted from [193], with permission of publisher SPIE—The International Society for Optical Engineering.*)

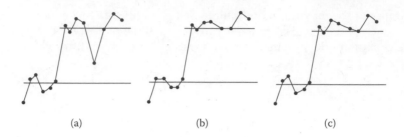

(a)　　　　　　　　(b)　　　　　　　　(c)

FIGURE 9.10
a) A signal sequence. b) The signal sequence proposed by the CWM filter with window size equal to 5, thin lines are smeared. c) The signal sequence processed by noise-exclusive filters, spike is removed while thin lines of the signal structure are kept very well.

As stated previously, the median filter has been widely used in image processing due to its properties of removing impulse noise while preserving sharp edges.

Because the design of the median filter is based on the theory of order statistics, it possesses the characteristic of robust estimation. However, it also has the ordering process that removes the fine details and changes the signal structure. Thus the generalized median-type filters have been introduced to retain the advantage of the robustness of the median filter and to take into account the details as well.

Typical median filters include the basic median filter and the generalized median-type filters such as the CWM filter, the max/median filter and the multistage/median filter.

9.4.3   Noise-Exclusive Adaptive Filtering

The problems with the median-type filters are the following: a) all pixels in the image are processed irrespective of whether they are impulses or normal pixels; b) impulse noise pixels are included in the ordering and filtering. The net effect is the degradation of the originally correct pixels and biased estimation due to the inclusion of the impulses in the filtering. In order to avoid unnecessary processing on pixels which are not corrupted by impulse noise and to improve the correct reconstruction rate, the impulse noise suppression is divided into two steps in our proposed approach. In the first step, we focus on noise detection and aim at locating noise pixels in the image. In the second step, to make full use of the robust estimation of order statistics and to take into account the simplicity of computing the sample mean, two noise-exclusive filtering algorithms are introduced. By performing these operations, the fine details can be preserved and proper estimations for the corrupted pixels can be derived. The principle of the methods is schematically illustrated in Figure 9.10c. Experimental results will be given in the following section.

Feature Selection and Impulse Detection

People usually attempt to use the local statistical characteristics to remove the impulse noise. The local statistics can help characterize the impulse noise to a certain degree, but the impulse noise cannot be completely determined by the local information, as sometimes the local information does not represent the characteristics of the impulse noise very well. For instance, in Equations (9.13) and (9.14), both pixels in the center of the window represent positive impulses. It is hard to say that the central pixel in Equation (9.14) is due to impulse noise.

$$
\begin{matrix} 69 & 99 & 120 \\ 241 & 238 & 121 \\ 244 & 56 & 241 \end{matrix} \qquad md = pixelvalue - median = 238 - 121 = 117 \qquad (9.13)
$$

$$
\begin{matrix} 235 & 19 & 21 \\ 236 & 241 & 237 \\ 249 & 17 & 237 \end{matrix} \qquad md = pixelvalue - median = 241 - 236 = 5 \qquad (9.14)
$$

In order to effectively detect the impulse noise, the local statistics are used as feature vectors in this chapter. Impulse noise detection is, in fact, a special application of the vector quantization. From a statistical point of view, the impulse noise should be characterized by its distribution of the probability density function. Our goal is to find the reference vectors that represent the impulse noise instead of using local statistics to compare each pixel with a threshold [185]. To extract features from local statistics, a 3×3 square window is used to pass through the entire noise-degraded image. The size and shape of the window is not restricted to 3×3 and square. It may vary, for example, 5×5 or 7×7, and in any shape such as a cross or circle. The consideration for choosing a 3×3 square window is twofold:

- The square window can best reflect the correlations between the pixel at the center and the neighbors. As the central pixel has neighbors in each direction, it contains sufficient information for the statistical analysis.
- Working within a smaller window is computationally efficient. From the pixel values within a sliding window, many useful local features, such as mean, median, variance, range, extreme range can be gathered. Using these features, the concept of noise detection is expressed as feature extraction and pattern classification. Suppose that N features are to be measured from each input pattern. Each set of the N features can be considered as a vector $\mathbf{x}$ in the N-dimensional feature space.

To fulfil pattern classification, the aforementioned SOTM neural network is employed. Of the local features, two are chosen as an input pattern to be fed into the neural network in our experiment. One feature is the pixel

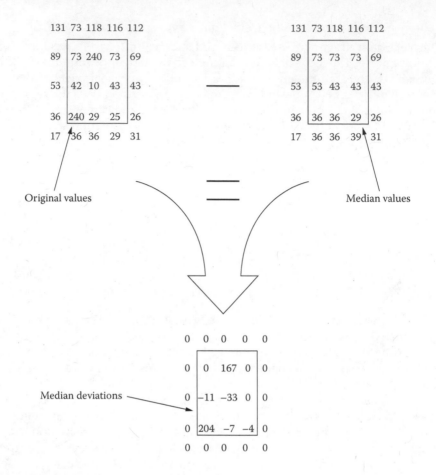

FIGURE 9.11
The median deviation is obtained from the difference of the original value and the median value.

value, and the other is termed the median deviation which is obtained from the difference between the pixel value and the median in the window as shown in Figure 9.11. These two features are used to distinguish the impulse noise from the signals, as they can effectively reflect the characteristics of the impulse noise. The training process of the weight vectors illustrated in Figure 9.12 helps understand the impulse noise detection. Intuitively, the image normally has homogeneous regions and the median deviations of the homogeneous regions should have small values. The median deviation has a large value if and only if there is impulse noise or a thin line existing within the window. The first feature helps the neural network to train the weight vectors to represent impulses of both directions (positive impulse and negative impulse) according to the modified impulse noise model 1. Since the median value is a robust estimate of the original pixel value based on the theory of

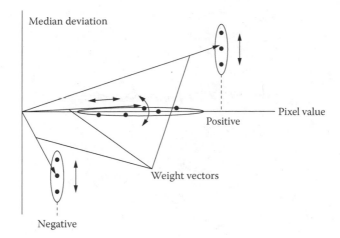

FIGURE 9.12
The training process of the weight vectors.

order statistics, the second feature provides accurate information about the likelihood of whether the current pixel is corrupted.

The approach of using SOTM in impulse detection is obviously superior to that of the local detection approach [185]. The advantage of the SOTM is that it not only considers the local features, but also takes into account the global information in the image. In addition, it detects not only fixed value impulse noise but also impulse noise with a random intensity distribution, as the SOTM can classify input vectors into clusters and find the center of each cluster. For the image with impulse noise model 2, the distribution of the input vectors is shown in Figure 9.13 based on the 2-D probability density function. The vertical axis denotes the pixel value and the horizontal axis denotes the distribution of the median deviation. Negative impulse noise coalesces at the top left of the figure. Signals are shown as a bright vertical bar in the center, and positive impulse noise is at the right bottom. With the classified clusters obtained by the SOTM, the cluster centers that have the largest and smallest values of median deviation are chosen to represent positive and negative impulse noise, respectively. Going back to the 1-D distribution in Figures 9.9b and 9.9c, it is clear that the cluster centers with a proper variance cover all of the impulse noise.

Noise Removal Filters

Since the cluster centers which represent the means of the impulse noise have been detected, recovering of image becomes the process of matching pixels with the cluster centers. If the pixel value lies in the interval of one of the impulse noise cluster centers with the variance, and the pixels median deviation is large enough, then the pixel is determined to be impulse noise. The NEM and the NEAM are two noise-exclusive filters [192,193] that are introduced to

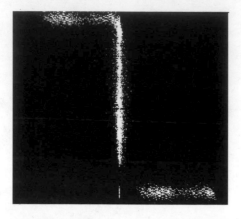

FIGURE 9.13
2-D distribution of the input vectors. (*Adapted from* [193], *with permission of publisher SPIE—The International Society for Optical Engineering.*)

restore the impulse corrupted images. The window size of these filters may vary as the traditional median-type filters. The noise exclusive scheme is incorporated in filtering by applying the filters to the impulses, but without affecting the uncorrupted pixels. The NEM filter only enables uncorrupted pixels inside the window to participate in ordering, while the NEAM filter calculates arithmetic means from these uncorrupted pixels. Since both filters use true information in the window to estimate the corrupted pixels, the quality of the restored image is ensured. For example, in Equations (9.13) and (9.14), the results obtained by the NEM filter and NEAM filter are

$$y_i = med(56, 69, 99, 120, 121) = 99, \tag{9.15}$$

$$y_i = mean(69, 99, 120, 121, 56) = 92, \tag{9.16}$$

and

$$y_i = med(17, 19, 21) = 19, \tag{9.17}$$

$$y_i = mean(19, 21, 17) = 19, \tag{9.18}$$

respectively, while the original pixel values in the center of Equations (9.13) and (9.14) are 104 and 15.

It can be seen that the NEM filter is a natural and logical extension of the median filter. The median filter takes the median value within a window that may contain some contaminated pixels. Its aim is to reduce the probability of taking the extreme values, whereas the NEM filter first eliminates the extreme values in the window, then takes the median. Since the estimation of the NEM filter is based on the reduced original sample space, its estimation accuracy is better than that of the median filter. Therefore, the NEM filter possesses all the good properties that the median filter has, such as edge preserving, robust estimation, etc., and also preserves the integrity of the image without changing the structure of the signals. The statistical analysis of the NEM filter is given in [194]. Combining the NEM and the NEAM filters can even further improve their performance. Among the remainder of the pixels, a comparison of a predefined value V with the range value is performed. If the value is bigger than the V, the NEM filter is chosen, otherwise the NEAM filter is chosen. For instance, for Equations (9.15) and (9.16) the range value is equal to 65 which is rather large, so we use the NEM filter. The adaptive procedure further ensures the accuracy of the estimation. If the NEAM filter is applied to a degraded image, the operation can be simplified and computing time can be reduced, because no sorting operation is involved. The α-trimmed mean filter [195] is, in fact, a special case of NEAM filter when both negative and positive impulses exist.

Proof We give the general expression of the NEAM filter as follows,

$$y = \frac{1}{n[1 - (p_p + p_n)]} \sum_{i=1}^{n[1-(p_p+p_n)]} x_i$$

if we do the sorting before eliminating the noise pixels in the window, then calculating the exclusive mean, the above formula becomes

$$y = \frac{1}{n[1 - (p_p + p_n)]} \sum_{i=[np_n]+1}^{n-[np_p]} x_i$$

Let $p = p_n = p_p$, then

$$y = \frac{1}{[n(1 - 2p)]} \sum_{i=[np]+1}^{n-[np]} x_i$$

This is the case of the α-trimmed mean [195].

$$X_\alpha = \frac{1}{n - 2 * [\alpha n]} \sum_{i=[n\alpha]+1}^{n-[\alpha n]} x_i$$

where $0 \leq 0.5$ and $[\cdot]$ is the nearest integer to $\cdot$.
Q.E.D. ∎

The NEAM filter has two merits superior to the α-trimmed mean filter. The first is that it does not require the sorting operation. It simply calculates the mean from the uncorrupted pixels. The second is that it can handle unevenly distributed positive and negative impulses.

9.4.4 Experimental Results

The noise-exclusive filtering schemes described in Section 9.4.3 have been extensively tested and compared with several popular median-type filtering techniques, such as the median, the CWM, the max/median and the multistage median filters. A number of experiments were performed in the presence of negative, positive, or both negative and positive impulse noise with low and high probabilities. The comparisons of the performance are based on the normalized mean square error (NMSE) and the subjective visual criteria. Quantitative error results are presented and their filtered images are shown for subjective evaluation. The normalized mean square error is given by

$$NMSE = \frac{\sum_{i=0}^{N-1} \sum_{j=0}^{N-1} [f_{ij} - y_{ij}]^2}{\sum_{i=0}^{N-1} \sum_{j=0}^{N-1} [f_{ij}]^2} \qquad (9.19)$$

where f_{ij} is the original image, y_{ij} is the filtered image and N is the width and height of the image.

The experiments performed include the following:

- an assessment of the proposed filters with respect to the percentage of impulse noise corruption;
- a demonstration of the robustness of the proposed filtering scheme with respect to the model of impulse noise;
- an evaluation of the NEM and the NEAM with respect to the edge preservation, fine details preservation and signal distortion;
- an overall comparison of the features of both the NEM and the NEAM with the median-type filters;
- a demonstration of the perceptual gains achieved by the filtering scheme.

The restoration performances of different methods for fixed impulse noise with even distribution are compared in Table 9.1. Subjective evaluations for image "Flower" with 20% impulse noise and image "Lena" with 40% impulse noise are given in Figures 9.14 and 9.15, respectively. All the filters use 3×3 windows. The CWM filter has its central weighted coefficient $k + 1 = 3$. To generate a set of test image, four pictures (256×256 pixels, 8 bits/pixel) with different types including a lake, a tower, a flower, and Lena are used in the experiment.

TABLE 9.1

Performance Comparison of Filters in Terms of NMSE with Window Size 3×3 and Condition of Even Impulse Noise Distribution

	Lake	Tower	Flower	Lena
Image size	256×256	256×256	256×256	256×256
Noise rate	0.13	0.18	0.20	0.40
Median	0.004885	0.006075	0.003854	0.0036475
CWM $k + 1 = 3$	0.007615	0.00698	0.008749	0.070176
Max	0.088714	0.069088	0.126696	0.395211
Multistage	0.010214	0.007905	0.01685	0.081947
NEM	0.000558	0.001064	0.000161	0.003061
NEAM	0.000566	0.000998	0.000232	0.002928

FIGURE 9.14

All filtered images with window size 3×3. (a) The original image; (b) 20% mixed impulse corrupted image; (c) the median filtered image; (d) the CWM filtered image; (e) the max/median filtered image; (f) the multistage median filtered image; (g) the NEM filtered image; (h) the NEAM filtered image. (*Adapted from* [193], *with permission of publisher SPIE—The International Society for Optical Engineering.*)

FIGURE 9.15
All filtered images with window size 3 × 3. (a) The original image; (b) 40% mixed impulse corrupted image; (c) the median filtered image; (d) the CWM filtered image; (e) the max/median filtered image; (f) the multistage median filtered image; (g) the NEM filtered image; (h) the NEAM filtered image.

The original images are shown in Figures 9.14a and 9.15a, respectively. Images corrupted by even positive and negative impulse noise with different probabilities are shown in Figures 9.14b and 9.15b.

From the visual result of the filtered images, one can easily tell that the median filter has the property of robust estimate against impulse noise and ability of preserving sharp edges. However, it removes fine details and thin lines and causes a slight blur effect as shown in Figures 9.14c and 9.15c. These images also show that as the impulse noise ratio gradually increases, the equalities of the filtered images decreases. When the impulse noise ratio increases to 40%, the filtered image becomes unacceptable as shown in Figure 9.15c. In general, the median filter can only suppress the impulse noise effectively when the noise rate is lower than 20% and this must be in the mixed impulse noise case because of the positive and negative impulse noise offset effect.

With the CWM filter, as expected, the ability of the signal preservation increases because more emphasis is placed on the central weight. Unfortunately, the noise suppression decreases correspondingly. Compared with the median filter, the CWM filter preserves fine details better, such as the leaves in the "lake," and the detail appearance in the "tower" images. But it produces visually unpleasing effects as shown in Figures 9.14d and 9.15d.

The max/median filter, as mentioned in the previous section, can be effective in the negative impulse noise case, but it fails in removal of mixed impulse noise, because it tends to enhance positive spikes in the image as shown in Figures 9.14e and 9.15e.

From the figures, it is not difficult to discover that the multistage median filtered images visually appear very similar to those of the CWM. The multistage median filter has good edge and detail preserving properties. In fact, during filtering, it not only preserves edges, but also enhances the contrast of the edge effect, which can be seen in Figures 9.14f and 9.15f. However, its noise suppression characteristic is worse than that of the CWM. All the filters mentioned above have the same problem in that their noise suppression ability decreases as error rate increases. They also tend to produce blur effects, as these approaches are implemented uniformly across the image. The pixels that are undisturbed by impulse noise are most likely modified. The blur is more obvious especially when a large window is applied.

In contrast, the results obtained by the NEM and the NEAM filters are superior to other filters. The results are shown in Figures 9.14g, 9.15g, 9.14h and 9.15h. The two filters possess good properties of noise suppression, edge and fine details preservation, and very little signal distortion. The edge and fine detail characteristics of the filters are compared with those of the above filters. The results for the flower image are shown in Figure 9.16. The max-median filter is not suitable for mixed impulse noise, while the median, the CWM and the multistage median filters are not good in the single type impulse case. However, the proposed filters are not affected by these limitations. In order to clearly see the reconstructed images and compare them with the original image, the 3-D plots of the original image and filtered images are shown in Figure 9.17. From the results, it is obvious that the NEM and NEAM filters

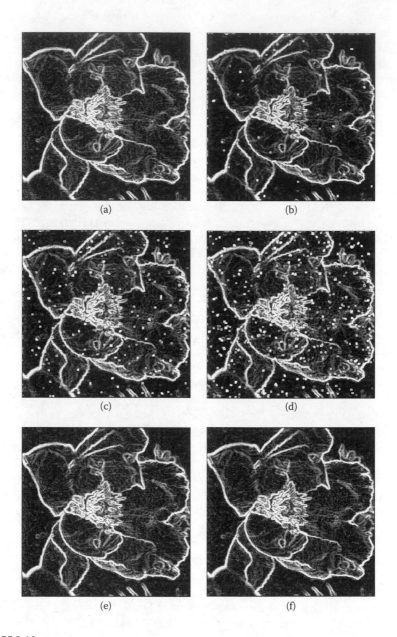

FIGURE 9.16
Edge extraction of the flower image shown in Figure 9.14. (a) The original image; (b) the median filtered image; (c) the CWM filtered image; (d) the multistage median filtered image; (e) the NEM filtered image; (f) the NEAM filtered image.

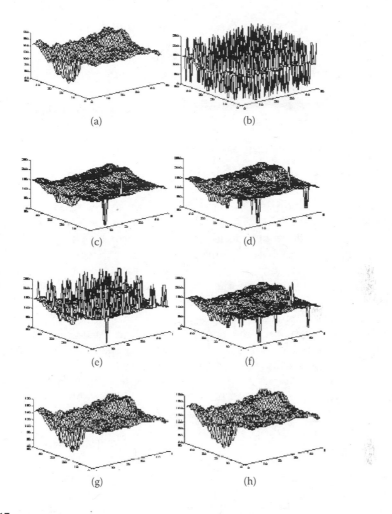

FIGURE 9.17

3-D representation of the flower image. (a) the original image; (b) The 20% impulse noise corrupted image; (c) the median filtered image; (d) the CWM filtered image; (e) the max/median filtered image; (f) the multistage median filtered image; (g) the NEM filtered image; (h) the NEAM filtered image.

are superior to the traditional median-type filters. The achieved restorations result in total impulse cleaning, fine detail preservation and minimum signal distortion. One of the most significant properties of the two filters is their robustness. The qualities of the reconstructed images by the two filters do not deteriorate when the noise ratio is getting higher.

Table 9.1 summarizes the NMSEs for all filters. From the table, it is clear that the proposed two filters are hardly affected by the error rate. Although the NMSEs of the two filters increase slightly as the error rate increases, visually, the filtered images still have good qualities. The increased error rate is because of too few true pixels left in the estimation window after the noise excluding operation, which causes certain distortion. Although the experiments for other images with varying percentage of impulse noise are not shown, the proposed filtering scheme produces superior results.

9.5   SOTM in Content-Based Retrieval

Relevence feedback (RF) is a popular and effective way to improve the performance of content-based retrieval (CBR). However, RF needs a high level of human participation which often leads to excessive subjective errors. Also, RF cannot be efficiently used to search in distributed digital libraries. This section presents automatic interactive content-based retrieval (AI-CBR) methods using the SOTM which minimizes user participation, provides a more user-friendly environment and avoids errors caused by excessive human involvement. In the conventional CBR systems, users are required to bring the systems up to the desired level of performance. These systems depend on a large number of query submissions (in the form of user relevance feedback samples) to improve retrieval capability and impose a great responsibility on the user. Although existing interactive learning methods can effectively learn with a small set of samples, it is still necessary to reduce the user workload. Moreover, the reduced feedback also decreases the number of transmitted images or videos from the service provider, reducing the required transmission bandwidths, an important requirement for retrieval in distributed digital libraries.

The main purpose in this section is to integrate the CBR system with compressed domain processing in support of remote access to distributed digital libraries. In this context, it is assumed that users are most interested in accessing many sources of information distributed and shared in the digital databases across the Internet. Although images and videos stored in those databases may be indexed and organized in various ways, data is usually stored in a compressed format, employing a number of existing compression techniques. Given this, it is therefore appropriate to introduce compressed domain indexing and retrieval. This enables the realization of direct access to compressed data without employing any preorganized database indexing.

Unlike the conventional CBR systems, where the user's direct input is required in the execution of the RF algorithms, the SOTM estimate is now adopted to *guide* the adaptation of the RF parameters. As a result, instead of imposing a greater responsibility on the user, independent learning can be integrated to improve retrieval accuracy. This makes it possible to obtain either a fully automatic or a semiautomatic RF system suitable for practical applications. Consequently, the interactive CBR system of Figure 9.18a is generalized to include a self-learning component, as shown in Figure 9.18b. The interaction and relevance feedback modules are implemented in the form of specialized neural networks. In these fully automatic models, the learning

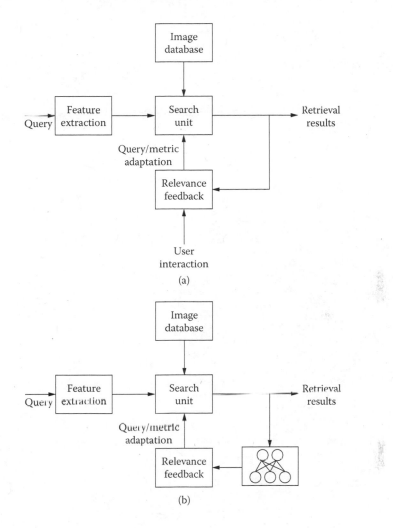

FIGURE 9.18
(a) The conventional CBR system. (b) The AI-CBR system.

capability associated with the networks and their ability to perform general function approximations offers improved flexibility in modeling the user's preferences according to the submitted query.

9.5.1 Architecture of the AI-CBR System with Compressed Domain Processing

The system architecture, shown in Figure 9.19, is composed of a search unit using features extracted in the compressed domain and a self-organizing adaptation network. In the initial stage, the search unit utilizes compressed-domain visual descriptors to access the compressed databases, and the similarity is

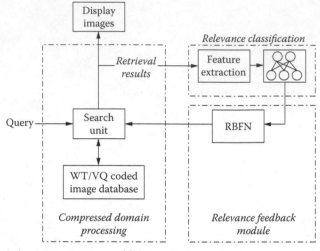

(a) Automatic interaction.

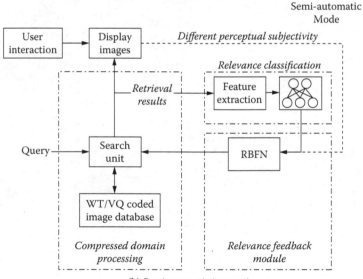

(b) Semiautomatic interaction.

FIGURE 9.19
AI-CBR architecture with compressed domain processing.

simultaneously computed. In the following adaptation stage, the self-organizing network is applied to the retrieved samples to make a decision for advising on the relevance feedback module and for improving retrieval results. The system can either run in fully automatic mode (Figure 9.19a) or can be incorporated into user interactions to learn further different user subjectivities (Figure 9.19b). During automatic learning, there is no transmission of the sample files from the server side to a client, as only the improved retrieval set will be delivered to the user. Thus, the required transmission bandwidth can be reduced or eliminated during the retrieval process.

9.5.2 Automatic Interaction by the SOTM

The user's judgment of image relevancy is closely related to the classification problems associated with the supervised learning of artificial neural networks. To automate the RF process, unsupervised learning has to be considered. In general any unsupervised learning method can be adopted for this purpose. We select the SOTM due to its ability to work well with sparsely distributed data shown in the previous sections and also demonstrated in image segmentation in [196]. The source of such sparse data in the current application is the high dimension of the data space (e.g., image feature vectors). There is also a limit to the size of training data to which we can afford to label individual items of the retrieved data, for subsequent relevance feedback learning.

Feature Space for Relevance Identification

Having selected the learning architecture, we focus on a feature space which can effectively distinguish between images of relevance and irrelevance. The general guideline for selecting the feature space in relevance identification is as follows: in automatic retrieval, two feature spaces are required: $\mathcal{F}_C$ and $\mathcal{F}_S$. $\mathcal{F}_C$ is for retrieval, and $\mathcal{F}_S$ is for relevance identification. This reflects the fact that we have different requirements for $\mathcal{F}_C$ and $\mathcal{F}_S$. Since retrieval is performed on huge volumes of data, the primary requirement for $\mathcal{F}_C$ is high speed and reasonable accuracy. Normally, compressed domain features can satisfy this requirement. On the other hand, relevance identification requires high accuracy, since we want the process to simulate human performance. Hence the features in $\mathcal{F}_S$ should be of high quality. While use of the feature space $\mathcal{F}_S$ may substantially increase the computational complexity required per image, the added cost is minor, since only a small number of retrieved images undergo such computation.

During relevance feedback, the set of retrieved images is employed as a training data set, $\mathcal{T}_{(t)} = \{\mathbf{v}_1, \mathbf{v}_2, \ldots, \mathbf{v}_{N_{RT}}\}$, which will be used for relevance identification by the SOTM algorithm. The algorithms used to obtain the descriptors in the spaces $\mathcal{F}_C$ and $\mathcal{F}_S$ are explained in Section 9.5.3. As we learned in the previous sections, the SOTM is a competitive neural network which uses a multilayer tree structure to organize the neurons, as shown in Figure 9.3. Following the statistics of the input pattern, different neurons are constructed to capture the local context of the pattern space and mapped into the tree structure. At the same time, the models become ordered on the tree map so that similar models are close to each other and dissimilar models far from each other. Consequently, the different descriptions of relevant and nonrelevant images can generate their own centers on the tree map, with the application of SOTM to the relevance identification.

In order to construct a suitable map, the SOTM offers two levels of adaptation: weight and structure. The weight adaptation is the process of adjusting the weight vector of the "winning" neurons. Structure adaptation is the process of adjusting the structure of the network by changing the number of neurons and the structure relationships between them. Given a training data set $\mathcal{T}_{(t)} = \{\mathbf{v}_1, \mathbf{v}_2, \ldots, \mathbf{v}_{N_{RT}}\}$, the convergence of the adaptation map is reached using the SOTM algorithm presented in Section 9.3.

Subnode Selection

After convergence, the SOTM algorithm produces new node vectors $\breve{\mathbf{w}}_j$, $j = 1, 2, \ldots, L$. In CBR, let $\mathbf{v}_{query} \in \mathcal{F}_S$ be the feature vector associated with a given query image in the current retrieval session. By going along the tree structure, we can find nodes close to the given query. We calculate a set of distances, d_j, $j = 1, 2, \ldots, L$, between the query vector and the node vectors:

$$d_j = \left((\breve{\mathbf{w}}_j - \mathbf{v}_{query})^T (\breve{\mathbf{w}}_j - \mathbf{v}_{query})\right)^{\frac{1}{2}}. \tag{9.20}$$

Based on the calculation result, the nodes:

$$\{\breve{\mathbf{w}}_{j^*} | \; j^* = 1, \ldots, L', \quad L' < L\} \tag{9.21}$$

that are closest to the query vector are then defined to be relevant, so that any input vector partitioned in these clusters is assessed as relevant. Ideally, the number of nodes L' chosen will optimize the classification performance.

Corresponding to these relevant nodes, we define regions $R_{j^*} \subset \mathcal{F}_S$ to associate with the relevant nodes $\breve{\mathbf{w}}_{j^*}$. These regions are then used for classification of the retrieved images. A class label, $y \in \{1, 0\}$, assigned to a given retrieved image, $\mathcal{I}_i, i \in \{1, \ldots, N_{RT}\}$ is then obtained:

$$y_i = \begin{cases} 1 & \mathbf{v}_i \in R_{j^*}, \forall j^* \\ 0 & \text{otherwise} \end{cases} \tag{9.22}$$

The network output is formally defined as $(\mathcal{I}_i, y_i)$, $i = 1, 2, \ldots, N_{RT}$, where y_i is the class label obtained for the retrieved image $\mathcal{I}_i$. This result will guide the adaptation of relevance feedback modules in the interactive retrieval sessions.

Automatic and Semiautomatic Interaction Procedures

SOTM has been used to guide the adaptation of RF-based learning methods, in both automatic and semiautomatic modes. The learning methods include: (1) single-RBF method [197], (2) Adaptive-RBF Network (ARBFN) [197], and (3) relevance feedback method (RFM) [198,199]. The retrieval procedures using machine interactions are summarized as follows:

> **Step 1.** *Self-organizing adaptation*: For a given query image, the most similar images, $\mathcal{I}_i$, $i = 1, \ldots, N_{RT}$, are obtained by using a Euclidean distance measure based on the predefined vectors in the

feature space $\mathcal{F}_C$. The retrieved images then serve as the input to the SOTM algorithm to identify relevance, employing a new feature space $\mathcal{F}_S$. This constructs the training set $(\mathcal{I}_i, y_i)$, $y_i \in \{0, 1\}$, $i = 1, 2, \ldots, N_{RT}$.

Step 2. *RF adaptation*: A new search operation is performed on the database using Single-RBF, ARBFN, or RFM, employing the training data, $(\mathcal{I}_i, y_i)$, $i = 1, 2, \ldots, N_{RT}$, in the feature space $\mathcal{F}_C$. The process may either be repeated by inputting the new retrieved image set to the SOTM in Step 1, or stopped. This process may also be switched to run in semiautomatic mode by incorporating user interactions.

Apart from the automatic adaptation, an important advantage to using a self-organizing network is the flexibility in choosing the number of training samples, N_{RT}. It allows for learning with a large number of training samples, which is usually not possible in the conventional CBR. As a result, interactive learning methods, (e.g., single-RBF and ARBFN), increase their learning capabilities with a large sample number.

9.5.3 Features Extraction for Retrieval

Although the gap in human perception and visual descriptors could be more significant in the transform domain than in the spatial domain, compressed-domain descriptors are widely used in image retrieval [200–205]. The main reason for this is that feature extraction algorithms can be applied to image and video databases without full decompression. This provides fast and efficient tools that are appropriate for real-time applications.

Discrete Wavelet Transform (DWT) Features for Retrieval

The emergence of DWT in image and video compression such as MPEG-4 Visual Texture Coding (VTC) [206], JPEG 2000 [207], and SPIHT has caused much of the latest work on indexing and retrieval to focus on algorithms that are compatible with the new standards [201,203,208].

Since different coders use different coding schemes, in order to make the feature extraction algorithms compatible with all of the coders, it is best to extract the descriptors directly after the decomposition process, before encoding the DWT coefficients. Two types of features are considered.

The first type of features is termed the fundamental statistic feature (FSF). This feature is obtained by computing the mean and the standard deviation of wavelet coefficients in each subband [209]. Let the wavelet coefficients at the ith subband be described by $x_{i,1}, x_{i,2}, \ldots, x_{i,P}$ for three level decompositions, that is, $i = 10$. The two parameters are then computed for each subband, $m^{(i)}$, $\alpha^{(i)}$ where $m^{(i)} = \bar{x}_i$ and $\alpha^{(i)} = var\{x_{i,1}, x_{i,2}, \ldots, x_{i,P}\}$. These descriptors are referred to as *fundamental statistic descriptors* (FSD), which can be applied to all types of wavelet-based coding standards.

The second is designed for extracting features from hybrid WT/VQ coders. Such coders have been proven to achieve high compression while maintaining good visual quality [210,211]. The decomposition produces three subbands at resolution level 1, and four subbands at resolution level 2. To capture the orientational characteristics provided by the wavelet-based subband decomposition, a codebook is designed for each subband. This results in a multiresolution codebook that consists of subcodebooks for each resolution level and preferential direction.

The outcome of the coding process, referred to as coding labels, is used to constitute a feature vector via the computation of the labels histograms. Each subband is characterized by one histogram and represents the original image at a different resolution; thus, the resulting histograms are known as multiresolution histogram indexing (MHI) [212]. MHI features make use of the fact that the usage of codewords in the subcodebook reflects the content of the encoded input subimage. To address the invariant issues of the illumination level, only five subbands containing the wavelet "detail" coefficients are concatenated to obtain the MHI features:

$$MHI = \left[h_1^{(1)} h_2^{(1)} \cdots h_{256}^{(1)} h_1^{(2)} h_2^{(2)} \cdots h_{256}^{(5)}\right]^T \qquad (9.23)$$

where $h_i^{(j)}$ denotes the number of times codeword i in subcodebook j is used for encoding an input image.

9.5.4 Features for Relevance Classification

For relevance classification, it is required that a visual descriptor should be of high quality, in the sense that it can approximate a relevance evaluation as performed by human users. A sophisticated visual descriptor extracted by spatial domain techniques, objects, and region-based segmentations would be appropriate for this requirement. Although, in general, these feature extraction techniques are associated with high computational complexity, they are sufficiently fast for a small set of images presented in the interactive session, as compared to the whole database.

This work has chosen a Gabor wavelet transform technique [213,214] as a visual descriptor for relevance classification. This technique is considered a powerful low-level descriptor for image search and retrieval applications. It has been used by the MPEG-7 to provide a quantitative characterization of homogeneous texture regions for similarity retrieval [215]. The Gabor wavelet method has also been shown to be very effective in such applications as texture classification and segmentation tasks [216,217].

9.5.5 Retrieval of Texture Images in Compressed Domain

The following experiments are designed to compare the performances of four methods: noninteractive CBIR, user interaction retrieval, automatic retrieval, and semiautomatic retrieval. The experimental results are obtained using a texture image database, the Brodatz database (DB1), which contains 1856 texture images. This is a typical, medium-sized database that is potentially

TABLE 9.2

Average Retrieval Rate (AVR) of 116 Query Images on DB1, Obtained by Automatic Interactive Learning. The Initial AVR Results (i.e., 0 Iter.) Were Obtained by Euclidean Metric for ARBFN and Single-RBF, and by Cosine Measure for RFM

Method	0 Iter.	1 Iter.	2 Iter.	3 Iter.	4 Iter.	Parameters
			Based on FSD Descriptor			
ARBFN	58.78	69.02	72.85	76.24	77.21	-
Single-RBF	58.78	66.32	68.80	70.04	71.87	$\alpha_N = 0.1$
RFM	53.88	57.11	59.00	60.45	60.78	$\alpha = 1, \gamma = 1, \varepsilon = 0.5$
			Based on MHI Descriptor			
ARBFN	63.42	71.66	75.22	75.86	76.51	-
Single-RBF	63.42	70.31	72.74	73.11	73.06	$\alpha_N = 0.5$
RFM	60.35	67.89	71.07	72.63	72.79	$\alpha = 1, \gamma = 2.5, \varepsilon = 0.8$

accessible remotely through Internet environments, without preorganization of the stored images [200].

The feature extraction algorithms described in Section 9.5.3 were applied to the compressed images. For the DB1 test set, the visual descriptor used is texture. Both FSD and MHI feature representations were tested, for the characterization of the wavelet-compressed images. The FSD descriptor was obtained before the invert-DWT process of the wavelet-baseline coder, whereas the MHI descriptor was obtained before VQ-decoding of the WT/VQ coder.[2]

In the simulation study, 116 images from all texture classes were used as the query images for testing DB1. The performance was measured by the average retrieval rate, AVR (%), obtained from the top 16 retrievals. The relevance judgments were conducted using two criteria: (1) the ground truth; and (2) the subjectivity of the individual user. For the first criterion, the retrieved images were judged to be relevant if they were in the same class as the query. For the second criterion, the retrieved images were judged as relevant to the perception of the individual user.

Noninteractive Retrieval versus Automatic Interactive Retrieval

Table 9.2 provides the numerical results illustrating the performances of the automatic interaction under different learning conditions. The results were obtained from the DB1 database employing FSD and MHI descriptors. In all cases, relevance judgment was based on the ground truth. The ARBFN method, the single-RBF method, and the RFM were tested. For each learning method, $N_{RT} = 20$ from the top ranked retrievals were utilized as training data. These samples were then input into the SOTM algorithm for the identification of relevance. The output of the unsupervised network was in turn used

---

[2] Note that a feature normalization algorithm was applied to the feature database.

as the supervisor for a learning method to update learning parameters and to obtain a new set of retrievals. The learning procedure was allowed to continue for four iterations. The first column of AVR results corresponds to the initial results obtained by noninteractive retrieval methods. The remaining columns are obtained with the automatic interaction methods.

Evidently, the use of automatic learning techniques resulted in a significant improvement in retrieval performance over that of the simple CBIR technique. For the automatic ARBFN, 18.4% AVR improvement was achieved through four interactions and 13% from automatic single-RBF. These retrievals used the FSD descriptor. The results for each learning method, with the MHI descriptor, show the same general trend. Maximum improvement was 13% with automatic ARBFN and 12% with automatic RFM.

The results in Table 9.2 show that ARBFN gave the best retrieval performance, compared to the other leaning methods, regardless of the descriptor types. In contrast, for single-RBF and RFM, the values for α_N and $(\alpha, \gamma, \varepsilon)$ significantly affected the results.

The learning performances were also affected by the training samples, N_{RT}. With a fixed number of training samples, N_{RT} for every iteration, allowing many iterations, meant that performance deteriorated gracefully. For example, it was observed that the AVR results, obtained by ARBFN after 9 iterations, were reduced by 0.7% with FSD descriptor. This is because at later iterations most of the relevant images had already been found. Thus, if all relevant images are input into the SOTM algorithm as a training set, they will be split into two classes [according to Equation (9.22)]; that is, misclassification will have occurred.

Figure 9.20 provides an example of a retrieval session performed by the automatic ARBFN learning method, using FSD descriptor. Figure 9.20a shows retrieval results without learning, and Figure 9.20b shows the result after automatic learning. The improvement provided by the automatic retrieval method is apparent.

User Interaction versus Semiautomatic Retrieval

In order to verify the performance of the automatic interaction learning of the AI-CBR system, its performance was compared with that of the conventional CBR methods. The learning systems are allowed to interact with the user to perform the retrieval task, and the AVR results obtained are provided in Table 9.3. It was observed that user interaction gave better performance: 3.34% to 6.79% improvement after one iteration, and 3.66% to 4.74% after four iterations. However, it should be taken into account that the users had to provide feedback on each of the images returned by a query in order to obtain these results.

The retrieval performance can be progressively improved by repeated relevance feedback from the user. The semiautomatic approach reported here greatly reduced the number of iterations required for user interaction. This significantly improved the overall efficiency of the system. In this case, the retrieval system first performed an automatic retrieval for each query to

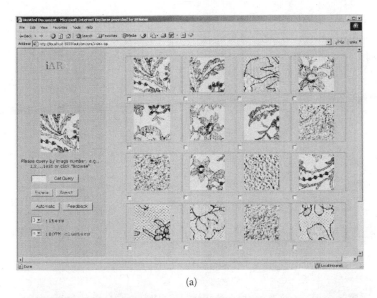

(a)

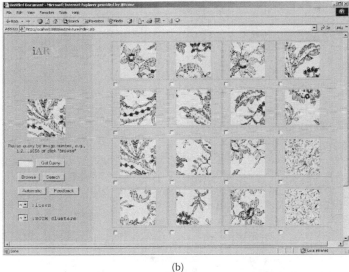

(b)

FIGURE 9.20
(a) Retrieval results without interactive learning; (b) retrieval results after application of automatic ARBFN.

adaptively improve its performance. After four iterations, the retrieval system was then assisted by the users. Table 9.4 provides the summary of the retrieval results, based on one round of user interaction. It was observed that the semiautomatic method is superior to the automatic method and the user interaction method. The best performance was given by semiautomatic ARBFN at 83.41% using FSD descriptors, and 81.14% using MHI descriptors.

TABLE 9.3

A Comparison of AVR(%) Between the AI-CBR Method and the Conventional CBR Method, Using DB1, and MHI Descriptors, Where Δ Denotes AVR Differences Between the Two Methods

Algorithm	Interaction Method	AVR (%)			No. of User RF (Iter.)
		0 Iter.	1 Iter.	4 Iter.	
	a: MCI	63.42	71.66	76.51	—
ARBFN	b: HCI	63.42	77.64	80.17	4
	$\Delta = b - a$	—	+5.98	+3.66	
	a: MCI	63.42	70.31	73.06	—
Single-RBF	b: HCI	63.42	73.65	77.43	4
	$\Delta = b - a$	—	+3.34	+4.37	
	a: MCI	60.35	67.89	72.79	—
RFM	b: HCI	60.35	74.68	77.53	4
	$\Delta = b - a$:	—	+6.79	+4.74	

In Figure 9.21a–c, the results are shown when each method reached convergence. The improvement resulting from the adoption of the semiautomatic approach is indicated by a correspondingly small amount of user feedback for convergence. In particular, the semiautomatic RFM, single-RBF, and ARBFN can reach or surpass the best performance of the conventional CBR within only one to two interactions of user feedback.

It is noticed that the superiority of the semiautomatic technique comes not only from the high initial result, but also from its use of the self-organizing method to identify other relevant samples. These are usually not easily presented by the user interaction method (i.e., it was possible to use a larger number of training samples, N_{RT} when working with the self-organizing method). This can be observed from the figures, where the semiautomatic methods converged with the higher AVR results, when these are compared to the user-controlled interactions.

TABLE 9.4

A Comparison of AVR (%) Between Semiautomatic and the Conventional CBR Methods, Using DB1, Based on One Round of User Interaction

Method	Initial Result	The Conventional CBR	Semiautomatic CBR
Based on FSD Descriptor			
ARBFN	58.78	77.53	83.41
Single-RBF	58.78	75.59	78.34
RFM	53.88	61.75	63.25
Based on MHI Descriptor			
ARBFN	63.42	77.64	81.14
Single-RBF	63.42	73.65	77.05
RFM	60.35	74.68	76.39

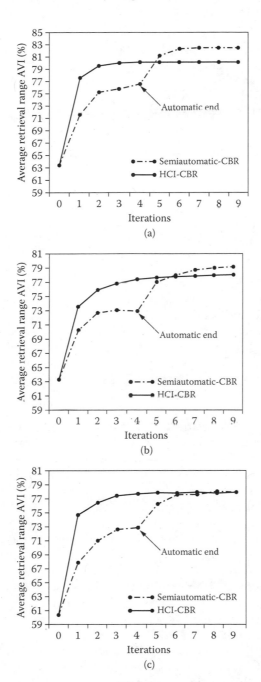

FIGURE 9.21

A comparison of retrieval performance at convergence, between the semiautomatic and the conventional CBR methods, where the similarity learning methods used are (a) ARBFN; (b) single-RBF; and (c) RFM. The semiautomatic method can attain the convergence within one to two iterations of *user feedback*. These results are based on the MHI descriptor.

TABLE 9.5

A Comparison of AVR Where the Relevance Judgment Is Based on Ground Truth and User Subjectivity, Based on ARBFN Method, FSD Descriptor, and DB1

	Before Learning	After Automatic Learning
User1	66.60	83.14
User2	64.82	82.71
User3	64.39	82.60
User4	61.42	79.20
User5	63.41	83.46
User6	60.13	77.75
Average	*63.46*	*81.48*
Ground truth	58.78	77.21

User Subjectivity Tests

In a retrieval process, the term "similarity" usually suggests different things to different users. In this experiment, the automatic retrieval system was examined for its ability to deal with user subjectivity. Six users were invited to test the retrieval system. Each user was asked to judge the relevance of the 16 top-ranked images according to his or her own understanding and information needs. The judgments were made at two points: after the first round of retrieval, in order to evaluate the performance of the noninteractive CBR technique; and after the fourth round of retrieval, in order to evaluate the performances of automatic retrieval methods. All users were requested to provide similarity judgements, but not relevance feedback. No user had any *a priori* information about which retrieved images were in the same class as the given query. Thus relevance judgments were made only according to user subjectivity.

This system was implemented based on the ARBFN learning method for retrieval with the FSD descriptor. The test was conducted for all 116 query images, as used in the previous experiments. The AVR results examined by each user are summarized in Table 9.5. It was observed that the AVR average over all the users was 81.5% after four iterations, an improvement of 18% from the initial value of 63.5%. This result indicates that all users rated the automatic RF approach as the one which performed better than the one-shot retrieval approach (in terms of capturing their perception subjectivity and information needs). *It also shows how the self-organizing machine understood the information requested based on a submitted query image.*

It was observed that AVRs fluctuated between 77.8% and 83.5% according to the users. The level of fluctuation is even more dramatic when we compared the AVRs from each user to the one based on the ground truth criterion. This was to be expected, since no fixed similarity measure can cope with different relevant sets across users.

10

Genetic Optimization of Feature Representation for Compressed-Domain Image Categorization

10.1 Introduction

In recent years, with the growing popularity of digital photography and multimedia systems, the sizes of many image databases have increased dramatically. To adapt to such changes, a number of international compression standards have been developed for effective multimedia information storage. At present, most digital images are stored in compressed formats corresponding to a number of different standards. Among those standards, JPEG, which is based on the discrete cosine transform (DCT) is one of the most popular, and many images on the web are compressed and stored in this format [218–220].

Previously, techniques developed for content-based image classification and retrieval [221–224] are usually performed in the spatial domain, in which features are extracted by inspecting an image pixel by pixel. Therefore, to perform feature extraction for compressed images, decompression is required in order to access the underlying image contents. In addition, performing content-based image classification and retrieval in the pixel domain is expensive in terms of computational and storage requirements. To obviate the need for full decompression, research work focusing on image processing in compressed domain [200,225–229] has been carried out. Examples include feature extraction in DCT domain [200,229] and the wavelet domain [226]. These techniques typically perform image indexing and retrieval by processing the coefficients directly in the compressed domain. For our work, due to the universal popularity of DCT-based JPEG compression standard, we primarily focus on the classification of JPEG-compressed images. In other words, our research focuses on image processing in the DCT domain. One of the benefits of using DCT domain features is that the computationally expensive inverse DCT [220,230] operation can be avoided.

Research has shown that a selected set of DCT coefficients can be indicative of the underlying content of the original image to some extent and thus can be used for image classification and retrieval [200,227]. For example, the DC coefficient in the upper left corner of a DCT block represents the average intensity of all 64 pixels in the block, and thus can be used as color features. Some selected subsets of DCT coefficients are also useful for edge and texture descriptions [200]. Based on these observations, several compressed-domain processing approaches based on DCT coefficients have been proposed. In [231], the author uses the compressed-domain coefficients to develop a quad tree structure for image retrieval. Each node in the tree stores a set of specific coefficients, and similarities between the query image and database images are performed by comparison of the corresponding quad trees using a suitable distance measure. The proposed approach is tolerant of translations between images due to the adopted quad tree structure that focuses on regional properties. In [232], subsampled images directly extracted from the compressed domain are used for image browsing and retrieval. However, the main motivation of adopting compressed domain features in these approaches is due more to the possibility of efficiency improvement than any anticipated enhancement in content characterization capability.

Compared with spatial domain processing approaches where the extracted features are directly related to the perceived luminance and chrominance values, the compressed-domain features may not be able to characterize the image equally well. As a result, accurate characterization of visual contents becomes a critical issue for successful content-based image classification and retrieval in the compressed domain. To address both the accuracy and efficiency issues for compressed image classification, we apply evolutionary computation techniques to search for suitable transformations of the compressed domain features, such that the classification results based on these transformed features are optimally improved while the original simplicity of the compressed domain approaches can be retained.

The rest of this chapter is organized as follows. In Section 10.2, we outline the main approaches for feature comparison in the spatial and compressed domain. In Section 10.3, we model the compressed-domain feature values as a random variable, and formulate our approach in terms of the search for an optimal transformation on this variable. We then describe how this transformation can be indirectly realized by the merging and removal of histogram bin counts, and how we can encode these transformations as individual strings in a population for evolutionary optimization. In Section 10.4, we extend this approach by adopting individually optimized transformations for each of the image classes, in the form of a set of separate classification modules associated with each of these transformations. The experimental results based on our proposed approach are presented in Section 10.5, and concluding remarks are provided in Section 10.6.

10.2 Compressed-Domain Representation

Generally, image classification in the spatial domain [233] is performed by extracting low-level visual features and translating these features into numerical attributes for convenient comparison between images. Specifically, each image in the database is associated with a feature vector $\mathbf{f} = [f_1, f_2, \ldots, f_N]$, and the similarity between images is characterized by a suitable metric between the feature vectors. In content-based image classification and retrieval, different types of histograms have been used to summarize the statistics of the extracted features of an image [222–235]. These histograms are normally used for image indexing and regarded as feature vectors. The color histogram [236] is one of the most popular among those because of invariance of these types of representations to translations and rotations. Comparisons between histograms [237] can also be easily performed using simple distance measures that are especially important for a large database where computational efficiency is important.

Typically, in the spatial domain, a color histogram is constructed based on the occurrence frequencies of particular color intensities by scanning the image pixel by pixel. Similarly, in the compressed domain, we can construct a DCT coefficient histogram [200] by directly accessing the compressed-domain coefficients. In order to obtain the coefficient values from the DCT blocks, we first have to perform the Huffman decoding, de-zigzagging and dequantization [218,220] on the encoded data of the compressed images. These steps are fast and computationally inexpensive compared with the IDCT step. Given a DCT-based JPEG compressed image with n 8×8 DCT coefficient blocks, the histogram of a specific coefficient can be constructed easily by counting the occurrence frequency of a particular coefficient value in these n DCT coefficient blocks. In other words, we are essentially modeling the chosen DCT coefficient as a random variable, and consider the histogram as an approximation of the variable's associated probability mass function (pmf).

While previous approaches then proceed to perform classification by using this histogram directly as classifier input, we propose to further search for and apply an optimized transformation on the underlying random variable, with the resulting reshaped histogram serving as a better representation of the image contents. In addition, instead of considering the complete set of bin counts corresponding to the entire random variable domain, we select a domain subset, in the form of a set of nonoverlapping intervals, to construct a conditional probability mass function (cpmf). The main idea of this approach is to select intervals that can facilitate classification and at the same time avoid intervals which may affect the classification result in a negative way. In our work, we have mainly adopted the luminance and chrominance features, in the form of their associated histograms, for classification purpose. In the DCT blocks, these are represented by the DC coefficients in the Y luminance block and the Cb and Cr chrominance blocks [218,220].

10.3   Problem Formulation

Grouping images into meaningful classes for a large image database is essential for content-based image-retrieval applications, where the system can quickly focus on the relevant subset to reduce the retrieval time. For a database, suppose there are K classes labeled as $C_1, C_2, \ldots, C_K$. Each image $x_i, i = 1, \ldots, I$ in the database is then assigned a class membership label $k \in \{1, \ldots, K\}$ that indicates the class it belongs to. With the knowledge of the correct image class structure, we try to approximate this class structure in the compressed domain feature space. As expected, it is found that there is a discrepancy between classifications based on compressed domain features and the original image class structure. Due to this discrepancy, we propose to transform the compressed domain features such that this discrepancy is minimized.

Specifically, given a discrete random variable u that takes values in the set $U = \{u_1, \ldots, u_n, \ldots, u_N\}$ and with a probability mass function $P(u_n) = \text{Prob}(u = u_n)$, we can define a new random variable v, which takes values in the set $V = \{v_1, \ldots, v_r, \ldots, v_R\}$, applying a nonlinear transformation $T(\cdot)$ to u, that is, $v = T(u)$. The resulting pmf of the transformed variable v is given by

$$P(v_r) = \text{Prob}(v = v_r) = \sum_{n:v_r=T(u_n)} P(u_n) \tag{10.1}$$

From this equation, it is seen that, if we consider the histogram as an approximation of the original pmf, the transformation T can be indirectly realized by selective addition of the bin counts. As a result, the simplicity and efficiency of the original compressed domain approach can be retained. If the histogram is used as the classifier input, then it is apparent that different transformations will result in different sets of classification results. It is also expected that there should exist a set of suitably chosen transformations, with the corresponding reshaped histograms attaining better classification results than the original ones.

In addition, for our classification task, it is useful to focus on a particular set of coefficient values corresponding to specific color ranges that best distinguish the classes. In other words, instead of including the bin counts associated with the entire random-variable domain, we should select a suitable subset of the variable domain, and adopt the resulting conditional pmf as the classifier input.

As a result, our objective is to select multiple intervals $A_1, \ldots, A_m, \ldots, A_M$ from the original variable domain such that $A_{m_1} \cap A_{m_2} = \phi$, where $m_1 \neq m_2$. Given these selected intervals, the conditional pmf (cpmf) is defined as $P(v_r|A)$ with support on which satisfies $\sum_r P(v_r|A) = 1$. In other words, our objective is to select the optimal transformation and the associated interval subsets to construct a cpmf as classifier input, such that the classification accuracy is maximally improved.

Since our classification is performed by comparing the histograms of specific DCT coefficients, which can be considered as empirical probability mass functions of the respective random variables, the proposed approach is equivalent to the following steps:

1. Perform the random variable transformation by selective merging of histogram bin counts.

2. Select multiple intervals from the original support of the histogram.

3. Construct a new histogram from the bin counts of the selected intervals.

4. Renormalize the histogram to form an approximation of the conditional pmf.

The reshaped histograms associated with the candidate transformations will be used to classify the images in the database, and their performances are evaluated by a fitness function. The one with a better content characterization capability that can improve the classification accuracy will be selected during the evolutionary process, while the less accurate ones will be displaced from the population.

Genetic algorithm (GA) [40,61,141,142] is one of the most commonly used optimization techniques among the various evolutionary computational algorithms. The main idea of the technique is to generate a population of potential solutions to an optimization problem. This population-based approach ensures that the search space is adequately sampled and that local minima can be avoided. Each candidate is usually encoded as a string, and the optimality of each string is evaluated by a so-called fitness function. The candidates associated with high fitness scores are selected, and new candidates are generated from them by the reproductive operations of crossover and mutation during the evolutionary process [40,61,142]. On the other hand, the less optimal candidates are displaced from the population. In other words, the algorithm mimics the process of natural evolution with the definition of fitness depending on the external imposed optimization criteria. Since we have no prior knowledge of the optimal way to perform the bin count merging and interval removal to construct a suitable cpmf, for classification accuracy improvement, we adopt GA to search for the suitable transformation. As mentioned earlier, each potential solution will be encoded as a string. In our case, a candidate transformation is represented as a string $b^s, s = 1, \ldots, S$, where S is the size of the population. The length of the string is equal to the number of bins N in the original histogram, and the nth position of the string stands for the nth bin of the original histogram. Each character in the string takes values in the alphabet $\{0, 1, 2\}$. The exact form of the transformation is determined by observing the pattern of values in the string. The bins in an interval corresponding to a consecutive run of 0s are removed. For the remaining bins, those corresponding to 1s will be retained and left unchanged, while those corresponding to a consecutive run of 2 will have their bin counts merged together. An example is shown in Figure 10.1 to illustrate this encoding scheme.

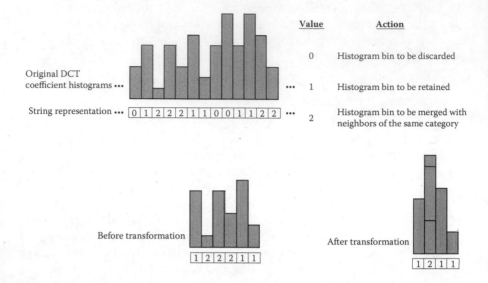

FIGURE 10.1
The encoding scheme of our proposed genetic optimization approach.

For each selected transformation T^s and the associated set of intervals $A^s = \bigcup_m A^s_m$ corresponding to the string b^s, we can construct the cpmf $P_i(v_r | A^s)$ of the DCT coefficient values, where $v_r = T^s(u_n)$ for each image x_i. For each class C_k, we can define the following prototype conditional pmf:

$$\bar{P}^k(v_r | A^s) = \frac{1}{|C_k|} \sum_{x_i \in C_k} P_i(v_r | A^s) \tag{10.2}$$

where $|C_k|$ is the cardinality of class C_k.

Given these prototype cpmfs, a new class structure $C^s_1, C^s_2, \ldots, C^s_K$ can be imposed on the images as follows:

$$\begin{aligned} x_i \in C^s_k \quad &\text{if } d(P_i(v_r | A^s), \bar{P}^k(v_r | A^s)) \leq d(P_i(v_r | A^s), \\ &\bar{P}^l(v_r | A^s)), l = 1, \ldots, K, l \neq k \end{aligned} \tag{10.3}$$

where $d(\cdot, \cdot)$ is the Euclidean metric.

To choose an optimal classifier from a large number of potential candidates, we need to specify a suitable optimization criterion in the form of a fitness function, such that the resulting selected transformation based on this function can optimally improve the classification accuracy. In other words, the fitness function should ensure that the discrepancies between the imposed class structure through A^s and the original class structure should be minimized. Given this requirement, we propose the following function $\Phi^s = \Phi(b^s)$ for fitness evaluation:

$$\Phi^s = \sum_k \left| C_k \cap C_k^s \right| \tag{10.4}$$

The value of Φ^s will be maximized when the two class structures exactly coincide, and will decrease as the discrepancies between the two structures increase.

In summary, the main idea of the proposed technique is to apply genetic optimization once to a small subset of image samples as a training set, such that the color characteristics of this subset can be captured through the optimal histogram transformation. This optimized transformation can then be directly applied to the complete database of images with similar characteristics, *without the need for further optimization due to its generalization capability.* Specifically,

- A small subset of image samples is used as training set to determine the optimal histogram transformation through GA as a *one-off process.*
- The optimized transformation is then applied to the rest of the image database *without requiring further optimization.*
- Since the transformation requires *only the addition and removal of bin counts* in the original histogram, no extensive computation is required for the compressed domain indexing of the remaining entries in the image database.

10.4 Multiple-Classifier Approach

The previous section describes the classification with the conditional pmf by searching for a global set of transformations and interval sets applicable to all classes. In this section, we extend this approach by producing a multiple classifier system, in which a specific subclassifier is designed for each image class. This is due to the possibility that a global optimal transformation and interval subset for all classes might not exist. In other words, the pmf based on the transformed variables may improve the classification accuracy for some classes but not the others.

To remedy this problem, a multiple classifier system is developed. The idea is to develop K subclassifiers for each individual class using a GA training process. During classification, their outputs are aggregated to produce a single classification output. Unlike the previous case where a single transformation and interval subset is adopted for all the classes, we can now search for the optimal transformation and interval subset for each of the classifiers. As a result, the conditional pmf $P(v_r^{s,k} | A^{s,k})$ is used as input for the kth classifier, where $v_r^{s,k} = T^{s,k}(u_n)$ and $A^{s,k}$ are, respectively, the transformation and interval subset associated with the string b^s.

Corresponding to this modified objective, we need to redefine the fitness function accordingly. Since each classifier is now associated with a particular class, the kth classifier should be able to accurately classify images of class k, and at the same time able to avoid assigning other classes' images to this class.

In view of this, the new fitness function is defined as follows:

$$\Phi_k^s = \left(|C_k \cap C_k^s| + |C_k' \cap C_k^{s'}| \right) \tag{10.5}$$

where C_k' and $C_k^{s'}$ denote the complements of C_k and C_k^s, respectively . The first term thus counts the number of correct positive classifications, that is, the number of patterns in class k actually classified as belonging to class k. On the other hand, the second term counts the number of correct negative classifications, that is, the number of patterns not belonging to class k that are correctly classified as not belonging to the class. The multiple classifier structure is shown in Figure 10.2.

With this multiple classifier setting, the issue on how to combine these outputs to produce a final classification decision becomes important. Different classifier fusion methods have been proposed in various research areas [238,239]; examples include majority vote [240–242], decision template [241] and belief networks [240,241]. However, it is expected that the classifier fusion operation should be highly application dependent since different applications use different types of feature sets, and the individual classifiers are also trained differently.

Each individual classifier will provide correct information and at the same time contribute error to the system; therefore, the fusion scheme should be able to improve the final classification accuracy as a whole. For an image x_i, each classifier will classify the image according to Equation (10.3) by assigning a class label $k \in \{1, \dots, K\}$ to x_i.

Given this information, a possible approach of classifier fusion is to select the class with the largest number of votes, based on the rule that the result of each classifier stands for one vote. We refer to this as the class membership-based fusion approach.

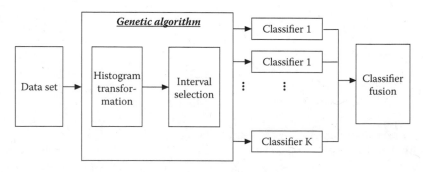

FIGURE 10.2
Proposed genetic optimization approach based on multiple classifiers.

Instead of just using this class membership information, we also make use of additional information provided by each individual classifier. Specifically, the kth classifier can produce an ascending list of distance measure values $d_{i,l}^{s,k}, l = 1, \ldots, K$ of the histogram to each class prototype according to the following equation:

$$d_{i,l}^{s,k} = d\left(P_i\left(v_r^{s,k}|A^{s,k}\right), \bar{P}^l\left(v_r^{s,k}|A^{s,k}\right)\right) \tag{10.6}$$

With the ordered list, we can define a classifier score for each classifier

$$\alpha_i^{s,k} = \frac{\sum_{l \neq k} d_{i,l}^{s,k}}{d_{i,k}^{s,k}}.$$

A high score will be obtained if the image histogram is close to the center associated with the kth classifier and far away from other centers. The final class membership $\gamma(x_i)$ of the image x_i will be determined as follows:

$$\gamma(x_i) = \arg\max_k \alpha_i^{s,k} \tag{10.7}$$

10.5 Experimental Results

We have applied our proposed approach to an image database that contains 1500 JPEG-compressed images. The database was divided into six classes based on their visual contents. Half of the images were randomly selected from each class and used as training set, and the remaining half were used as the test set. Figure 10.3 shows the representative images of the different classes. We can observe that the images encompass a variety of visual themes including sunset, boats, fireworks, and different foliage and flower types. This class structure will then be used as the reference for the GA process to design the feature transformation in such a way that the resulting induced class structure by the optimal transformation will closely approximate this original class structure.

For the GA, we have adopted the parameter settings shown in Table 10.1. The crossover probability corresponds to the probability of the occurrence of crossover between two selected chromosomes and the mutation probability corresponds to the probability of a gene being flipped to another value. The crossover probability is set to a high value, 0.9, while the mutation probability is set very low, 0.005, in accordance with the minor role played by mutation in GA compared with crossover.

The algorithm is terminated if there is no further improvement of the best fitness score over 100 generations, and the current best solution will then be returned. The best solutions found for each class are then used to realize the transformation.

The overall result using different classification approaches on the database is shown in Table 10.2. We can observe that, compared with the case where no

Class	Representative image
1	
2	
3	
4	
5	
6	

FIGURE 10.3
Representative examples of each image class.

TABLE 10.1

Parameter Setting for GA

Parameter	Setting
Population size	50
Chromosome length	384
Gene allele	0, 1, 2
Crossover type	One-point
Crossover probability	0.9
Mutation probability	0.005
Selection mechanism	Roulette wheel
Scaling scheme	Linear
Replacement mechanism	Elitist
Termination	Fitness value does not change over 100 generations

TABLE 10.2

Classification Results Based on Different Approaches

Classification Approach	Training Set	Test Set
Original pmf	77.60%	75.73%
Single-classifier approach	87.69%	80.22%
Multiple-classifier approach (class membership-based voting)	86.13%	80.92%
Multiple-classifier approach (distance-based voting)	89.05%	81.85%

optimization has been performed, our proposed approach, even for the single classifier case, has resulted in a significant improvement in classification accuracy for both training and test sets. The adoption of multiple classifiers with the class distance-based voting mechanism further increases the classification rate as seen in the last row of the table. This is due to the capability of the multiple classifier system to capture diverse aspects of compressed domain color features that are essential for the delineation of the classes, in the form of individually optimized transformation for each of the classifiers.

At the same time, it is observed that the choice of a suitable classifier fusion approach is also essential for optimal classification improvement, as illustrated in the case of the multiple-classifier approach adopting the class membership-based voting mechanism (third row of table). It is seen that the improvement in classification accuracy is comparatively less significant. A possible explanation of this is that, in the multiple-classifier approach, each classifier is optimized for the identification of the members of a specific class, but no further attempts are made to improve the delineation of the other classes in the complement of this specific class, as these will be individually handled by the other classifiers. As a result, if the minimum distance criterion is applied and the current classifier produces a negative response with respect to its designated class, all we can conclude is that, according to this classifier, the current pmf does not belong to this class, but there is no guarantee that further applying the minimum distance criterion within the complement of this class will result in the identification of its correct membership within this set.

To illustrate the performance of our proposed approach in more details, we show the classification result of each class on the training set and test set in Tables 10.3 and 10.4, respectively, based on the multiple-classifier approach. It can be observed that our proposed approach is able to improve the classification accuracies of most classes to a significant extent, especially classes 3, 5, and 6, even though there are variations between the levels of improvement across the classes due to differences in the intrinsic distinctiveness of the classes in terms of their color compositions.

To further demonstrate the effect of our proposed approach on the histogram, we show the original histogram and the transformed histogram of some chosen image samples in Figure 10.4. It can be seen that the merging and removal of histogram bin counts can accentuate the salient features of the original histograms to provide a more succinct characterization of the underlying

TABLE 10.3

Classification Result (Training Set)

Class	Original pmf Accuracy	Transformed pmf Accuracy
1	71.72%	81.31%
2	87.67%	95.48%
3	85.71%	96.14%
4	80.51%	88.56%
5	77.56%	89.87%
6	56.32%	80.69%

TABLE 10.4

Classification Result (Test Set)

Class	Original pmf Accuracy	Transformed pmf Accuracy
1	71.72%	77.58%
2	90.41%	91.10%
3	87.86%	92.93%
4	73.85%	79.95%
5	73.72%	80.06%
6	56.32%	68.62%

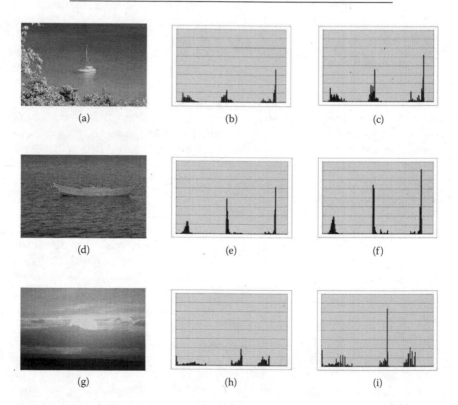

(a)   (b)   (c)

(d)   (e)   (f)

(g)   (h)   (i)

FIGURE 10.4

Original and reshaped histograms of images from different classes. (a) Image boat 1 from class 2; (b) Original histogram; (c) Reshaped histogram; (d) Image boat 2 from class 2; (e) Original histogram; (f) Reshaped histogram; (g) Image sunset from class 5; (h) Original histogram; (i) Reshaped histogram.

TABLE 10.5

Comparison of Distance Measure Values Prior
to and After Transformation

	Original Histogram	Transformed Histogram
Boat1 & Boat2	1.198	1.178
Boat1 & Sunset	1.513	1.525

image contents. In addition, for these examples, it is seen that the distance measures between images of the same class have decreased while images from different classes have increased after performing histogram transformation, as shown in Table 10.5.

As described previously, one of the main objectives of our approach is to select intervals that can help facilitate classification. This can be further verified in Figure 10.5 where we remove the image blocks with associated chrominance values belonging to the discarded color intervals in the histogram. It is observed that the image blocks corresponding to the selected color intervals characterize each image class in a meaningful way. For example, the algorithm identifies the orange color of the sunset as more important in class 5. Similarly, the blue color of the sea in class 3 is considered essential for identifying images from the corresponding class, which is intuitively reasonable from the user's viewpoint.

In Table 10.6, we have listed the percentage of histogram bin reduction resulting from the optimization process for the different classifiers. From the

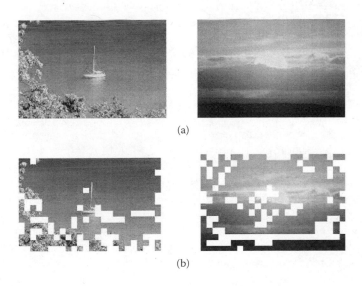

(a)

(b)

FIGURE 10.5

Selection of color intervals in histograms. (a) Representative images form each class; (b) Image blocks corresponding to the selected color intervals in histograms.

TABLE 10.6

Amount of Bin Reduction of Each Classifier

Classifier	Bin Reduction (%)
1	33.70%
2	33.88%
3	33.13%
4	33.39%
5	34.27%
6	33.83%

table, it is seen that a significant number of bins have been removed from the original histogram, which contributes to the improvement of efficiency for the histogram comparison process during classification.

10.6   Conclusion

The adoption of compressed-domain features in image classification facilitates the efficient processing and indexing of image database entries without the need for performing the inverse DCT transform. However, in many cases, the compressed-domain features are only approximately indicative of the actual image contents. Given this shortfall, we propose to perform a transformation for the compressed-domain features such that the classification results based on the transformed features can be maximally improved. More importantly, it is shown that, if the compressed-domain feature values are modeled as realizations of a random variable, then the corresponding transformation on this random variable can be indirectly performed by selectively merging and removing the feature histogram bin counts. In this way, the original simplicity and efficiency of the compressed-domain approach can be retained. Since there is a large number of possible transformations that can be applied to the compressed-domain features, we adopt genetic algorithm to search for the optimal one by generating a population of candidate transformations, and then apply the operations of crossover and mutation to generate improved candidates. Each candidate classifier is encoded as a string and its fitness is defined as a function of the discrepancies between the compressed-domain classification results and the original image class structure. This approach is then further extended by searching for individually optimized transformations for the different classes, in the form of multiple classification modules, in order to capture the distinct visual characteristics associated with each class. As shown by the experiments, our proposed approach is capable of improving the classification accuracy to a significant extent even in the case when only a single global transformation for all classes is used, and the results are further enhanced by the adoption of multiple local transformations for each individual class, thus indicating the effectiveness of the optimization process.

11

Content-Based Image Retrieval Using Computational Intelligence Techniques

11.1 Introduction

In recent years, the world has witnessed a significant growth in the volume of image contents due to increasing popularity of imaging devices such as mobile phones and digital cameras, the ease of these devices to capture, store, and transfer images, and the proliferation of image contents over the Internet and wireless communications. A Reuters report showed that worldwide mobile phone subscriptions reached 3.25 billion in 2007, equivalent to around half of the world's population [243]. Among these, camera phones are expected to account for 87% of all mobile phone handsets shipped by 2010 [244]. This suggests that there are potentially millions or even billions of image content producers and consumers (in short prosumers) worldwide. In addition, a recent study showed that the total number of images captured using camera phones will reach 228 billion by 2010 [244]. Flickr.com, a popular image-sharing portal, has 42 million visitors worldwide each month, and more than 2 billion photos stored in 2008 [245]. The convergence of image content creation and consumption, and community-based media sharing, supported by better networking infrastructure and bandwidth has resulted in a strong demand for efficient techniques to search, browse, filter, index, and retrieve image contents.

Current information-retrieval systems such as the Internet-based Google and Yahoo search engines are very successful in performing text-based Web document search and retrieval. However, the performance of keyword-based image retrieval is currently still falling short of expectation. Conventional text-based image retrieval requires a large number of dedicated annotators to annotate the images with keywords before they can be searched. This process, however, is time consuming and manually expensive. In addition, it also experiences subjectivity of the annotators in interpreting the images. As opposed to manual annotation, the Google and Yahoo image search engines rely on web crawling where the images embedded in the Web documents are analyzed and annotated with selected keywords after careful study of the surrounding texts and metadata such as filename, timestamp, etc. However, it is challenging to obtain a high precision rate as the automatic process to extract keywords in order to annotate the embedded images proves to be elusive. Hence, it is clear that text-based analysis and retrieval of the image contents alone is inadequate. In view of this, content-based image retrieval (CBIR) has been proposed as an alternative to address the problem. CBIR aims to retrieve a set of desired images from the database based on analysis of visual contents in the images. The visual features may include color, texture, shape, and spatial relationships that are present in the images. In CBIR, the process of feature extraction and image indexing can be fully automated, hence alleviating the difficulty of human annotation in conventional text-based retrieval systems. More importantly, CBIR can also be integrated with the text-based techniques to enhance the performance of image retrieval.

Various CBIR systems have been proposed and developed in recent years, including QBIC [246], MARS [198], Virage [223], Photobook [247], VisualSEEk [248], PicToSeek [249], and PicHunter [250]. Among the CBIR schemes, relevance feedback is a popular technique that has been introduced to bridge the semantic gap between the high-level human perception and low-level visual features [198,249–266]. Relevance feedback is an interactive mechanism where the users provide feedback on the relevance of the retrieved images, and the system learns the user information needs based on this feedback. Many relevance feedback algorithms have been developed in CBIR and demonstrated considerable performance improvement [198,249–266]. These include query refinement [198], feature reweighting [251,252], statistical learning [250,254,255], neural networks [256–260], and support vector machine (SVM) [261–265].

Query refinement and feature reweighting are two commonly used techniques in relevance feedback. Query refinement aims to estimate the optimal query point by moving it toward positive/relevant images and away from the negative/irrelevant ones. The technique is adopted in the multimedia analysis and retrieval system (MARS) [198]. On the other hand, reweighting technique updates the weights of the feature vectors in such a way that it strengthens the feature's components that help to retrieve relevant images. Another class of CBIR algorithms is centered on statistical modeling of the probability distribution of image features in the database [250,254,255]. For instance, feedback image samples have been used to train Bayesian classifiers in [255]. Positive samples are used to estimate the Gaussian distribution of the desired images for a given query, while the negative samples are used to modify the ranking of the retrieved images. Neural networks have also been employed in interactive image retrieval due to their learning capability and generalization power [256–260]. SVM has also emerged as a popular relevance feedback technique for CBIR [261–265]. The objective of SVM classifiers is to find an optimal hyperplane that maximizes the separation margin between the positive and negative classes in a kernel-induced feature space.

Even though relevance feedback is instrumental in improving the performance of CBIR systems, there are still a number of issues that need to be addressed. Among these are (1) the uncertainty of user perception on the relevance of the feedback images, and (2) the small sample problem of feedback images. In a classical CBIR search operation, a user will choose a query image, and the system will return a list of feedback images that are visually similar to the query image. The user will then make a binary decision for each feedback image to determine whether it is "fully relevant" or "totally irrelevant." This hard-decision approach, however, is inconsistent with human visual perception, which is uncertain at times. For instance, in a search operation, a user is looking for red cars. If the feedback images contain a yellow car, he or she may be unsure whether to classify it as fully relevant or totally irrelevant as the image satisfies, up to a certain extent, his or her information needs. Therefore, it is important for a relevance feedback scheme to be able to evaluate and handle these uncertainties. The second issue in CBIR centers on the small sample problem encountered in relevance feedback. During the feedback process, users are usually unwilling to label too many images as it is a repetitive task. This gives rise to the small sample problem where there is insufficient number of training data to train the classifier adequately. This in turn will diminish the performance of the classifier. In this chapter, we will demonstrate how computational intelligence techniques can be used to address these two challenging problems in CBIR.

Computational intelligence, which offers instrumental mathematical modeling of artificial intelligence, is employed in this work due to its effectiveness. The requirements for systematic signal identification, intelligent information integration, and robust optimization point to computational intelligence techniques. Several techniques including neural networks, clustering, fuzzy reasoning, and SVM are adopted and developed in this chapter.

11.2 Problem Description and Formulation

The aim of CBIR is to retrieve a set of images that are of interest to the user based on a query. The query can be in the form of an example image, a sketch, a color description, etc. This chapter will focus on the popular approach of query-by-example (QBE). A typical CBIR search session is shown in Figure 11.1. A user chooses an image as the query. The system then performs feature extraction and analysis to retrieve a set of best-matched images and displays them to the user for feedback. The user then labels those images that match his or her information needs as relevant, and the others as irrelevant. The system will take the feedbacks and use them to train a classifier. The trained classifier will then be used to retrieve the best-matched images in the next round that will be displayed to the user for further feedback. The process will continue until the user is satisfied with the retrieved results.

As highlighted in the previous section, existing relevance feedback schemes in CBIR face two issues: (1) the uncertainty in user perception of feedback images, and (2) the small sample problem. The first issue is centered with user's

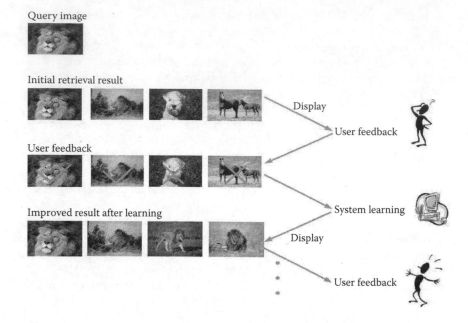

FIGURE 11.1
Illustration of relevance feedback in interactive CBIR system.

fuzzy perception on the relevance of feedback images. Classical relevance feedback techniques require the user to label a feedback image as either "fully relevant" or "totally relevant." This binary hard-decision approach, however, is inconsistent with human visual perception, which is at times ambiguous and imprecise. In a search operation, a user may often have multiple levels of information needs, which can be viewed as an information hierarchy. For example, if a user is looking for red cars during a CBIR session, the primary focus will be centered on the object, namely the car, whereas the secondary focus will be on the attribute, namely the red color in this case. Therefore, given the feedback image of a yellow car, under the existing binary labeling scheme, the user would encounter a dilemma as whether to label it as "fully relevant" or "totally irrelevant" since the yellow car satisfies the primary but not the secondary need. It is clear that the crisp binary labeling scheme cannot resolve this dilemma adequately. In view of that, this chapter will propose a fuzzy labeling scheme that allows the user to choose an image as either "relevant," "irrelevant," or "fuzzy." Under the new scheme, the user can now label an image as fuzzy if it satisfies only partial information needs of the user. An issue arising out of this framework is how to determine the relevance of the fuzzy data. In this chapter, an a posteriori probability estimator will be developed to determine the soft relevance/significance score of each fuzzy image. These fuzzy images are combined with the relevant and irrelevant samples to train a classifier. A network called recursive fuzzy radial basis function network (RFRBFN) is then constructed to learn the user information needs. A gradient descent-based algorithm will be developed to optimize a carefully chosen cost function.

The second issue to be addressed in this chapter is centered on the small sample problem. During relevance feedback, users are usually unwilling to label too many images. This gives rise to the small sample problem where there may be insufficient number of training data to train the classifier adequately. In view of this, this chapter will develop a predictive-label fuzzy support vector machine (PLFSVM) framework to address this problem. The main idea of the proposed framework is to deduce some predictive-labeled images from the unlabeled images by studying the correlation between the labeled and unlabeled images. The unlabeled images are chosen carefully and assigned predictive labels of either "relevant" or "irrelevant." As these deduced images are not labeled explicitly by the users, there is an inherent imprecision embedded in their class labels. Therefore, a fuzzy membership function is employed to estimate the relevance/reliability of these predictive-labeled images. The new training set consists of both the explicit-labeled samples as well as the predictive-labeled samples. A fuzzy support vector machine (FSVM) that is capable of handling imprecision in class labels is then developed to take the enlarged training set to train the classifier.

11.3 Soft Relevance Feedback in CBIR

This section will discuss how computational intelligence can handle fuzzy user perception in relevance feedback. The structure and rationale of the proposed recursive fuzzy radial basis function network (RFRBFN) will be discussed. It will also explain how the relevance/significance of the uncertain samples can be estimated through fuzzy reasoning. The positive, negative, and fuzzy samples will then used to train the RFRBFN. Experimental results will be given at the end of this section to demonstrate the effectiveness of the proposed framework.

11.3.1 Overview and Structure of RFRBFN

To design an efficient CBIR system, the following factors need to be taken into consideration: (1) the CBIR system should be real-time, (2) the feedback process is performed in sessions, hence the classifier should be updated based on batches of new training data over time, (3) the positive data tend to form clusters, hence the classifier should have a cluster-based structure. An ideal choice that satisfies the above-mentioned criteria is the radial basis function (RBF) networks. RBF networks enjoy the following characteristics that make them a good choice in the context of CBIR: fast learning speed, simple network structure, multiple local modeling feature, and global generalization power [34].

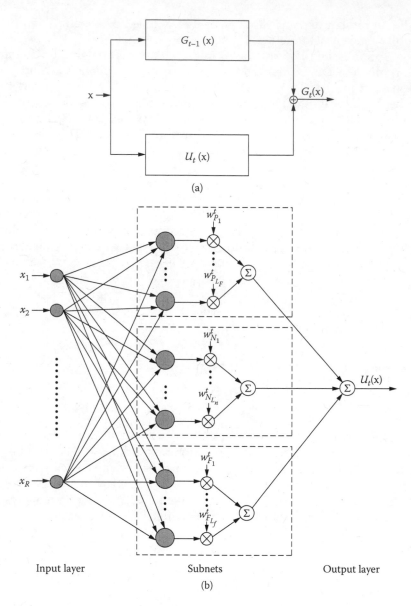

FIGURE 11.2
Schematic diagram of RFRBFN: (a) recursive structure of RFRBFN; (b) update network.

The proposed RFRBFN has a structure that resembles an RBF network. The schematic diagram of the RFRBFN is given in Figure 11.2. From the figure, it is observed that the tth iteration RFRBFN is constructed recursively based on two modules: (1) the $(t-1)$th iteration RFRBFN, and (2) the tth iteration update network. The output of the network at the tth iteration $G_t(\mathbf{x})$ is given by

$$G_t(\mathbf{x}) = \begin{cases} G_{t-1}(\mathbf{x}) + U_t(\mathbf{x}) & t = 1, 2, \ldots, T \\ 0 & t = 0 \end{cases} \tag{11.1}$$

where $\mathbf{x} = [x_1, x_2, \ldots, x_R]^T$ is an R-dimensional input feature vector, $G_{t-1}(\mathbf{x})$ is the $(t-1)$th iteration RFRBFN output, $U_t(\mathbf{x})$ is the update network output, and T is the total number of feedback iterations. The update network consists of an input layer, three subnets, and an output layer. The input layer has a set of R units, which accepts the feature elements of R-dimensional input feature vectors. The input units are connected to the three subnets that contain a number of RBF units. The subnets are constructed from the positive, negative, and fuzzy samples, respectively. The output layer has a single output neuron. The output value $U_t(\mathbf{x})$ is computed based on the weighted combination of the subnet responses. $\{w_{Pi}^t\}_{i=1}^{L_p}$, $\{w_{Ni}^t\}_{i=1}^{L_n}$, $\{w_{Fi}^t\}_{i=1}^{L_f}$ are output connection weights associated with the positive, negative, and fuzzy samples at the tth iteration. L_p, L_n, and L_f are the number of RBF units for the positive, negative, and fuzzy samples, respectively.

There are several considerations when developing the RFRBFN: (1) to reduce the training time of the network, (2) to take advantage of the unique batch-based user feedback, and (3) to integrate the potential fuzzy feedback from the users. In view of these considerations, the proposed RFRBFN is constructed recursively based on the previously trained network and an update network. In other words, the tth iteration RFRBFN is constructed based on a combination of the $(t-1)$th iteration RFRBFN and the tth iteration update network. The purpose of the recursive structure is to minimize unnecessary retraining of the whole network for each feedback iteration as the previous $(t-1)$th iteration RFRBFN has been trained to learn the user information needs. Therefore, an update network is introduced instead to track the current user information needs. As only a small update network is trained now, the training time in each feedback iteration will be greatly reduced. An a posteriori estimator is used to evaluate the relevance of the fuzzy images. These are then combined with the relevant and irrelevant feedbacks in network training. The details of the training process will be explained in Section 11.3.2.

The RFRBFN is constructed recursively based on the update network. Hence, to understand the structure of the RFRBFN, we need to focus on the construction of the update network. The update network consists of three layers: the input layer, the hidden subnet layer, and the output layer. The three subnets correspond to the positive, negative, and fuzzy samples. The first step in the formation of the update network is to determine the number and centers of the RBF nodes for each subnet. This will be achieved using clustering. In this work, the clustering process will be conducted in two stages, namely, the subtractive clustering [267] followed by the fuzzy C-means (FCM) clustering [56]. The rationale for using this two-stage clustering technique to estimate the number and centers of the RBF units is given as follows. FCM clustering is a popular technique used to perform clustering and grouping in many applications, including the estimation of the RBF unit centers. However, FCM clustering requires the number of clusters to be known a priori. This requirement, clearly, is not appropriate in the context of relevance feedback for CBIR as it is difficult to predetermine the number of clusters that would be required to model the data distribution. In view of this, subtractive clustering is adopted as the first step since it is fast, efficient, and does not require the number of clusters to be known in advance. Further, its clustering results can be used as a good initial estimate for the subsequent FCM clustering.

Subtractive clustering assumes each sample as a potential cluster center. It computes a potential field that determines the likelihood of a sample being the cluster center. The potential fields exerted by all the data points are added, and an iterative process is used to determine the samples that have the highest field density. The process continues until a predetermined threshold condition is satisfied. The resulting sample points would then be taken as the estimate of the cluster centers, which are passed to the subsequent FCM clustering. The two-stage clustering for the construction of the three subnets can be summarized as follows.

For each relevant, irrelevant, and fuzzy cluster type,

- Perform subtractive clustering to obtain an estimate of the number of clusters and the cluster centers.
- Perform FCM clustering based on the estimates obtained from the subtractive clustering.
- Construct three subnets by initializing the number and centers of RBF units based on the clustering results of FCM clustering.

The outcomes from the clustering process are three separate sets of clusters: relevant, irrelevant, and fuzzy sets. Let X be the set of all training samples. The representations for the cluster formation of X are given as

$$X = P \cup N \cup F \tag{11.2}$$

$$P = \bigcup_{i=1}^{L_p} P_i \qquad N = \bigcup_{j=1}^{L_n} N_j \qquad F = \bigcup_{k=1}^{L_f} F_k \tag{11.3}$$

where P, N, and F denote the positive, negative, and fuzzy cluster types, respectively. P_i, N_j, and F_k are the ith positive cluster, jth negative cluster, and kth fuzzy cluster.

11.3.2   Network Training

In this work, CBIR is formulated as an online supervised learning process. An important issue in the development of the proposed framework is the estimation of the relevance/significance of the fuzzy samples. We formulate this problem as finding a soft relevance function $h(\mathbf{x}_j) : \mathbf{R}^R \to [0, 1]$ that evaluates and maps each fuzzy sample into a value between 0 and 1. The interpretation of the output range is given as follows. If a user selects a feedback sample as relevant, then the output will be 1. On the other hand, if the user selects the sample as irrelevant, then the output will be 0. Therefore, the fuzzy sample should be mapped to an output value between 0 and 1, depending on its degree of correlation with the positive and negative samples. The desired output values $Y_t(\mathbf{x}_j)$ at the t-iteration for the RFRBFN can be given as follows:

$$
Y_t(\mathbf{x}_j) = \begin{cases} 0 & \mathbf{x}_j \in N \\ 1 & \mathbf{x}_j \in P \\ h(\mathbf{x}_j) & \mathbf{x}_j \in F \end{cases} \tag{11.4}
$$

where P is the positive, N is the negative, and F is the fuzzy feedback images.

We propose an a posteriori estimator to estimate $h(\mathbf{x}_j)$. The problem is formulated into the estimation of the conditional probability $P(\omega_r | \mathbf{x}_j)$ that a fuzzy sample $\mathbf{x}_j$ belonging to the relevant class ω_r. $\mathbf{x}_j$ can be represented as $\mathbf{x}_j = [\mathbf{x}_{j1}^T, \mathbf{x}_{j2}^T, \ldots, \mathbf{x}_{jC}^T]^T$ that consists of C feature subvectors $\mathbf{x}_{jc}$, $c = 1, 2, \ldots, C$. Each $\mathbf{x}_{jc}$ is a d_c-dimensional feature subvector extracted from color histogram, wavelet moments, etc. The following estimation principle is adopted:

$$
h(\mathbf{x}_j) = P(\omega_r | \mathbf{x}_j) = \frac{1}{C} \sum_{c=1}^{C} P(\omega_r | \mathbf{x}_{jc}) \tag{11.5}
$$

$P(\omega_r | \mathbf{x}_{jc})$ is the a posteriori probability of the cth subvector $\mathbf{x}_{jc}$. It is assumed that each of the C features contributes equally to the overall estimation. Using the Bayesian theorem, the a posteriori probability in Equation (11.5) can be rewritten as

$$
h(\mathbf{x}_j) = \frac{1}{C} \sum_{c=1}^{C} \frac{p(\mathbf{x}_{jc} | \omega_r) P(\omega_r)}{p(\mathbf{x}_{jc} | \omega_r) P(\omega_r) + p(\mathbf{x}_{jc} | \omega_i) P(\omega_i)} \tag{11.6}
$$

where ω_r and ω_i represent the relevant and irrelevant classes, respectively. $P(\omega_r)$ and $P(\omega_i)$ are the prior probabilities of the relevant and irrelevant classes. They are calculated as follows:

$$
P(\omega_r) = \frac{l_r}{l_r + l_i} \tag{11.7}
$$

$$
P(\omega_i) = 1 - P(\omega_r) \tag{11.8}
$$

where l_r and l_i are the total number of relevant and irrelevant samples accumulated over previous feedback iterations, $p(\mathbf{x}_{jc} | \omega_r)$ and $p(\mathbf{x}_{jc} | \omega_i)$ are the class conditional probability density functions (pdfs) of $\mathbf{x}_{jc}$ for the relevant and irrelevant classes, respectively. In this work, we assume that each feature vector is Gaussian-distributed. The class pdf can then be expressed as

$$
p(\mathbf{x}_{jc} | \omega_s) = \frac{1}{(2\pi)^{d_c/2} |\Sigma_c^s|^{1/2}} \exp\left[-\frac{1}{2} (\mathbf{x}_{jc} - \mu_c^s)^T {\Sigma_c^s}^{-1} (\mathbf{x}_{jc} - \mu_c^s) \right] \tag{11.9}
$$

where $\omega_s \in \{\omega_r, \omega_i\}$, μ_c^s is the mean vector and $\sum_c^s$ is the covariance matrix for the cth feature vector of class ω_s. They are estimated as follows:

$$\mu_c^s = \frac{1}{N_s} \sum_{j=1}^{N_s} x_{jc}^s \tag{11.10}$$

$$\sum_c^s = \frac{1}{(N_s - 1)} \sum_{j=1}^{N_s} \left(x_{jc}^s - \mu_c^s\right)\left(x_{jc}^s - \mu_c^s\right)^{\mathrm{T}} \tag{11.11}$$

where N_s is the number of training samples for each relevant or irrelevant class ω_s. It is noted that the covariance matrix $\sum_c^s$ may encounter the problem of singularity as the number of training samples is usually small in relevance feedback. To address the issue, we employ an approach called linear shrinkage estimator that incorporates a regularized diagonal matrix [268,269]:

$$\hat{\sum}_c^s = (1 - \lambda)\sum_c^s + \frac{\lambda}{d_c} tr\left[\sum_c^s\right]\mathbf{I} \tag{11.12}$$

where $tr\left[\sum_c^s\right]$ denotes the trace of $\sum_c^s$, d_c is the dimension of the feature space, and $0 < \lambda < 1$ controls the amount of shrinkage toward the identity matrix $\mathbf{I}$.

The tth iteration error function of the RFRBFN, E_t is defined as

$$E_t = \frac{1}{2}\sum_{j=1}^{M} e_{jt}^2 = \frac{1}{2}\sum_{j=1}^{M}(Y_t(\mathbf{x}_j) - G_t(\mathbf{x}_j))^2 \tag{11.13}$$

where e_{jt} is the error for the jth training sample $\mathbf{x}_j$ at the tth iteration, M is the total number of training samples, $G_t(\mathbf{x}_j)$ and $Y_t(\mathbf{x}_j)$ are the actual and desired network output for $\mathbf{x}_j$, respectively. The kernel function of the RBF units in the RFRBFN is given by

$$f\left(\mathbf{x}, \mathbf{v}_{\beta i}^t, \sigma_{\beta i}^t\right) = \exp\left(-\frac{\left(\mathbf{x} - \mathbf{v}_{\beta i}^t\right)^{\mathrm{T}}\Lambda\left(\mathbf{x} - \mathbf{v}_{\beta i}^t\right)}{2\left(\sigma_{\beta i}^t\right)^2}\right),$$

$$t = 1, 2, \ldots, T; i = 1, 2, \ldots, L_\beta \tag{11.14}$$

where $\beta \in \{P, N, F\}$ is the cluster type representing the feedback, i and L_β are the index and number of RBF units for feedback type β, t is the iteration number, and T is the total number of feedback iterations. $\mathbf{v}_{\beta i}^t$ and $\sigma_{\beta i}^t$ are the center and width for the ith RBF unit of the β subnet type at the tth iteration. The values $\mathbf{v}_{\beta i}^t$ are initialized using the results obtained through two-stage clustering. The matrix $\Lambda = diag[1/\sigma_1, 1/\sigma_2, \ldots, 1/\sigma_R]$ is a diagonal matrix used to characterize the relative contributions of different feature components. $\sigma_r, r = 1, 2, \ldots, R$ is the standard deviation of all relevant samples along the rth dimension. The rationale behind this approach is that consistency of a particular feature component among all the relevant samples indicates its importance in determining image similarity. Therefore, larger weighting will be assigned for feature component with consistent values [252].

The update network output $U_t(\mathbf{x})$ is given as

$$U_t(\mathbf{x}) = \sum_{\beta \in \{P,N,F\}} \sum_{i=1}^{L_\beta} w_{\beta i}^t f\left(\mathbf{x}, \mathbf{v}_{\beta i}^t, \sigma_{\beta i}^t\right) \tag{11.15}$$

where $w_{\beta i}^t$ is the output connection weight. Substituting Equation (11.15) into Equation (11.1), the RFRBFN output function $G_t(\mathbf{x})$ is given as

$$G_t(\mathbf{x}) = \begin{cases} G_{t-1}(\mathbf{x}) + \displaystyle\sum_{\beta \in \{P,N,F\}} \sum_{i=1}^{L_\beta} w_{\beta i}^t f\left(\mathbf{x}, \mathbf{v}_{\beta i}^t, \sigma_{\beta i}^t\right) & t = 1, 2, \ldots, T \\ 0 & t = 0 \end{cases} \tag{11.16}$$

The training of the RFRBFN at each feedback iteration is performed by minimizing the cost function in Equation (11.17) with respect to the network parameters of $U_t(\mathbf{x})$, namely, $\theta = \{w_{\beta i}^t, \mathbf{v}_{\beta i}^t, \sigma_{\beta i}^t | \beta \in \{P, N, F\}, i = 1, 2, \ldots, L_\beta\}$. The learning process can be represented as

$$\theta = \operatorname{argmin}_{\theta \in \Theta}(E_t) = \operatorname{argmin}_{\theta \in \Theta}\left(\frac{1}{2} \sum_{j=1}^{M}(Y_t(\mathbf{x}_j) - G_t(\mathbf{x}_j))^2\right) \tag{11.17}$$

where Θ is the solution space of the parametric vector θ.

In this work, a gradient-descent algorithm is adopted to estimate the network parameters using all the feedback samples accumulated over all previous iterations. The update equations for $w_{\beta i}^t$, $\mathbf{v}_{\beta i}^t$, and $\sigma_{\beta i}^t$ are summarized as follows:

(1) Weight estimation at the lth learning iteration

$$\frac{\partial E_t(l)}{\partial w_{\beta i}^t(l)} = -\sum_{j=1}^{M} e_{jt}(l) f\left(\mathbf{x}_j, \mathbf{v}_{\beta i}^t(l), \sigma_{\beta i}^t(l)\right) \tag{11.18}$$

$$w_{\beta i}^t(l+1) = w_{\beta i}^t(l) - \eta_1 \frac{\partial E_t(l)}{\partial w_{\beta i}^t(l)}, \quad \beta \in \{P, N, F\}, \quad i = 1, 2, \ldots, L_\beta \tag{11.19}$$

(2) Center estimation at the lth learning iteration

$$\frac{\partial E_t(l)}{\partial \mathbf{v}_{\beta i}^t(l)} = -w_{\beta i}^t(l) \sum_{j=1}^{M} e_{jt}(l) f\left(\mathbf{x}_j, \mathbf{v}_{\beta i}^t(l), \sigma_{\beta i}^t(l)\right) \frac{\Lambda\left(\mathbf{x}_j - \mathbf{v}_{\beta i}^t(l)\right)}{\left(\sigma_{\beta i}^t(l)\right)^2} \tag{11.20}$$

$$\mathbf{v}_{\beta i}^t(l+1) = \mathbf{v}_{\beta i}^t(l) - \eta_2 \frac{\partial E_t(l)}{\partial \mathbf{v}_{\beta i}^t(l)}, \quad \beta \in \{P, N, F\}, \quad i = 1, 2, \ldots, L_\beta \tag{11.21}$$

(3) Width estimation at the lth learning iteration

$$\frac{\partial E_t(l)}{\partial \sigma_{\beta i}^t(l)} = -w_{\beta i}^t(l) \sum_{j=1}^{M} e_{jt}(l) f\left(\mathbf{x}_j, \mathbf{v}_{\beta i}^t(l), \sigma_{\beta i}^t(l)\right)$$

$$\times \frac{\left(\mathbf{x}_j - \mathbf{v}_{\beta i}^t(l)\right)^{\mathrm{T}} \Lambda \left(\mathbf{x}_j - \mathbf{v}_{\beta i}^t(l)\right)}{\left(\sigma_{\beta i}^t(l)\right)^3} \tag{11.22}$$

$$\sigma_{\beta i}^t(l+1) = \sigma_{\beta i}^t(l) - \eta_3 \frac{\partial E_t(l)}{\partial \sigma_{\beta i}^t(l)}, \quad \beta \in \{P, N, F\}, \quad i = 1, 2, \ldots, L_\beta \tag{11.23}$$

(4) Repeat steps (1–3) until convergence or a maximum number of iterations is reached.

The term $e_{jt}(l)$ is the error signal of the jth training sample $\mathbf{x}_j$ at the lth learning iteration, and η_1, η_2, and η_3 are the learning parameters for $w_{\beta i}^t$, $\mathbf{v}_{\beta i}^t$, and $\sigma_{\beta i}^t$, respectively.

11.3.3 Experimental Results

To evaluate the performance of the framework, we use an image database obtained from the Corel Gallery product. The database consists of 10,000 natural images with 100 different categories. The categories are predefined by the Corel Photo Gallery based on their semantic concepts as shown in Figure 11.3.

The following feature representations are employed in the system: color histogram [270], color moments [271] and color autocorrelogram [272] as the color features representation, Gabor wavelet [213] and wavelet moments [273] as the texture feature representation. They have been shown to be effective in other retrieval systems and are listed in Table 11.1. After all the color and texture features have been extracted, the feature elements from all the individual features are concatenated into an overall feature vector with a dimension of 170. Next, Gaussian normalization is used to provide equal emphasis on each feature component [252].

The RFRBFN method is applied in our system to perform CBIR. The two-stage clustering consisting of subtractive clustering and FCM clustering is used to estimate the initial RBF unit centers. The learning step-sizes of the constructed RFRBFN are $\eta_1 = 0.01$, $\eta_2 = 0.0001$, and $\eta_3 = 0.02$. The values of the parameters are determined using the following rule of thumb. The approximate values of the parameters are first selected based on previous literature and empirical studies. These are followed by experimental fine-tuning.

There are two parts in the experiments, namely, objective performance evaluation and subjective performance evaluation. For objective evaluation, 85 queries from 100 semantic concepts are selected for evaluation. For subjective evaluation, four users are invited to evaluate the retrieval system. A total of 85 queries from 100 semantic concepts are used for evaluation.

FIGURE 11.3
Selected sample images from the database.

Performance Evaluation Based on Objective Measure

When there is no fuzzy feedback, we use an objective measure to evaluate the performance. The measure is based on Corel's predefined ground truth where the retrieved images are considered to be relevant if they come from the same category as the query image. We compared the RFRBFN method with the adaptive radial basis function network (ARBFN) method [258]. The ARBFN method adopts an inherent strategy of local modeling, and associates those relevant images as the models. The irrelevant samples are then used to modify the multiclass models in such a way that the models are moved away from the irrelevant samples.

We adopted the performance metric called precision-versus-recall (PR) in our experiments [199]. The definitions of precision and recall are given

TABLE 11.1

Visual Features Used in the System

Color Descriptors	
Color histogram (dimension = 32)	This descriptor represents the first-order color distribution in an image. The RGB color space of each image is converted into its HSV equivalence. Each H,,S, V component is then uniformly quantized into 8, 2, 2 bins, respectively, to get a 32-dimensional feature vector.
Color moments (dimension = 6)	The first two moments (mean and standard deviation) from the R, G, B color channel are extracted as the color feature.
Color autocorrelogram (dimension = 64)	This descriptor is a two-dimensional spatial extension of the color histogram. The chessboard distance is chosen as the distance metric. The image is quantized into $4 \times 4 \times 4 = 64$ colors in the RGB space.
Texture Descriptors	
Gabor wavelet (dimension = 48)	Gabor wavelet filters spanning four scales: 0.05, 0.1, 0.2, and 0.4 and six orientations: $\theta = 0, \pi/6, 2\pi/6, 3\pi/6, 4\pi/6, 5\pi/6$ are applied to the image. The mean and standard deviation of the Gabor wavelet coefficients are used to form the feature vector.
Wavelet moments (dimension = 20)	Wavelet transform is applied to the image with three-level decomposition. The mean and standard deviation of the transform coefficients are used to form the feature vector.

as follows:

$$\text{Precision} = \frac{\text{number of relevant images retrieved}}{\text{total number of images retrieved}} \qquad (11.24)$$

$$\text{Recall} = \frac{\text{number of relevant images retrieved}}{\text{total number of relevant images in the database}} \qquad (11.25)$$

Eighty-five queries were used for evaluation. The precision and recall rates were averaged over all the queries. The average precision-versus-recall (APR) graph after five feedback iterations is shown in Figure 11.4. It can be observed that the RFRBFN method outperformed the ARBFN method. The RFRBFN method achieved higher precision rate at the same recall level, and higher recall rate at the same precision level. For instance, at 20% recall rate, the RFRBFN method offered a precision rate of 82% as compared with 71% provided by the ARBFN method.

We also employed another performance metric called retrieval accuracy (RA) to evaluate the retrieval performance [255,274]. Its definition is given as follows:

$$\text{Retrieval accuracy} = \frac{\text{relevant images retrieved in top } T \text{ returns}}{T} \qquad (11.26)$$

where T is the number of retrieved images. The retrieval accuracy averaged over the 85 queries with seven iterations of feedback is shown in Figure 11.5. In the experiments, T was set to 25, namely, the number of feedback images

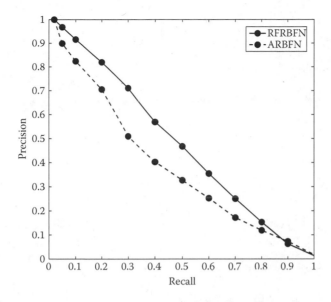

FIGURE 11.4
The APR graph of the RFRBFN and ARBFN methods (after five feedback iterations).

in each iteration is 25. It is observed that the retrieval accuracy of the RFRBFN method increased quickly in the first few iterations. This is desirable as the user can obtain satisfactory results within a few iterations. Further, the RFRBFN method reached steady-state retrieval accuracy of 82% in about five feedback iterations. In comparison, the ARBFN method achieved lower steady-state retrieval accuracy of 76%. This reflects the effectiveness of the proposed RFRBFN method.

Performance Evaluation Based on Subjective Measure

When there is ambiguity in the user perception of image similarity, or in other words, there are fuzzy feedbacks, we employ a subjective performance measure. This is because the fuzzy images selected by the users may span across different image categories. Therefore, we cannot adopt ground truth-based objective measures. We invited four users to evaluate the retrieval system. In total, 85 queries were selected for evaluation. We introduced the following performance measures: total retrieval accuracy (TRA), and relevant retrieval accuracy (RRA) to evaluate the system performance:

$$\text{Relevant retrieval accuracy} = \frac{\text{relevant images retrieved in top } T \text{ returns}}{T}$$

$$(11.27)$$

$$\text{Total retrieval accuracy} = \frac{\text{relevant and fuzzy images retrieved in top } T \text{ returns}}{T}$$

$$(11.28)$$

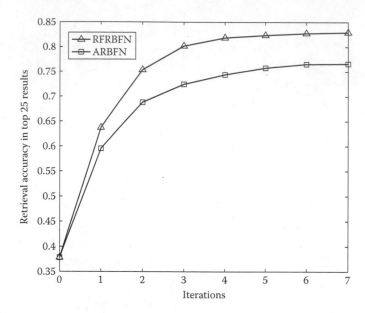

FIGURE 11.5
Retrieval accuracy of the RFRBFN and ARBFN methods in the top 25 results using objective measure.

The RRA is similar to the conventional RA when there are no fuzzy feedbacks. However, the RRA alone cannot adequately evaluate the performance of the RFRBFN method when there are fuzzy feedbacks. This is because the fuzzy images selected by the users also satisfy the user information needs up to a certain extent. Therefore, we introduced the measure of TRA that includes both the relevant and fuzzy images as the desired images. The TRA and RRA can be considered as the upper and lower bounds of effective retrieval accuracy for the RFRBFN method when fuzzy feedbacks are incorporated.

We evaluated the performance of the RFRBFN method and compared it with the ARBFN and MARS methods [252]. The MARS method adopts multilevel labeling in relevance feedback. It involves five discrete levels of (ir)relevance, namely, highly relevant, relevant, neutral, irrelevant, and highly irrelevant. The scores associated with each level are: 3, 1, 0, -1, -3, respectively. It is noted that neutral images are assigned the score of 0, which implies that they do not provide any contribution during the training process. In order to provide a comparative performance measure, we defined the retrieval accuracy of the MARS method as the ratio of the sum of both "highly relevant" and "relevant" images in the top T returned images. The retrieval accuracy averaged over all the test queries and users is given in Figure 11.6. From the figure, it is observed that the RFRBFN method consistently outperformed the ARBFN and MARS methods. The RFRBFN method offered a steady-state RRA of 81% and TRA of 86% in about five feedback iterations. The effective retrieval accuracy lay within the upper bound of TRA and lower bound of

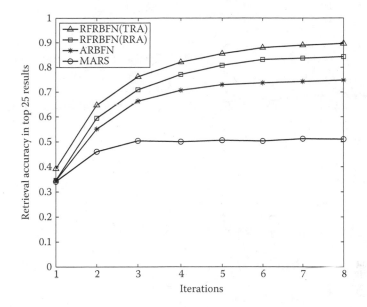

FIGURE 11.6
Retrieval accuracy of the RFRBFN, ARBFN, and MARS methods in the top 25 results based on user subjectivity.

RRA. In comparison, the ARBFN and MARS methods achieved lower steady-state retrieval accuracy of 73% and 51%, respectively. The MARS method uses heuristic updates of intra- and interfeature weights. Further, it requires the users to classify each retrieved image into one of the multiple levels which is tedious. In contrast, the proposed RFRBFN method only requires simple feedback from the users, while maintaining the priority of the user information needs.

11.4 Predictive-Label Fuzzy Support Vector Machine for Small Sample Problem

This section will discuss how computational intelligence can be used to address the small sample problem encountered in the relevance feedback process. It will outline the structure of the proposed predictive-label fuzzy support vector machine (PLFSVM), and explain the rationale behind this classifier. Then, it will explain the PLFSVM training procedure including the selection of the unlabeled data for label prediction, the estimation on the relevance of the predictive samples, and the PLFSVM training. Finally, experimental results will be given to demonstrate the effectiveness of the proposed method.

11.4.1   Overview of PLFSVM

The main modules of PLFSVM comprise the following: (1) label prediction for selected unlabeled samples in the database, (2) evaluation on the relevance/significance of these predictive samples, and (3) integration of these predictive samples with the labeled samples to perform CBIR through fuzzy support vector machine (FSVM) [275]. The overview of the framework is given in Figure 11.7.

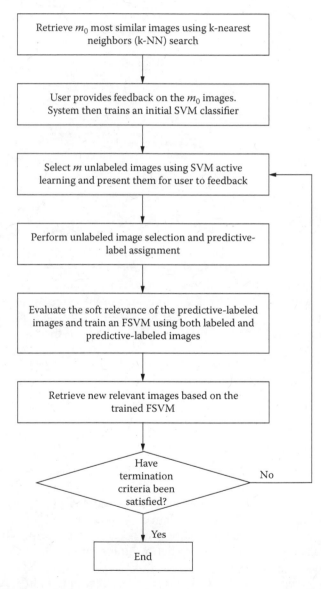

FIGURE 11.7
Flowchart on overview of the PLFSVM.

Given a query image chosen by the user, the m_0 best-matched images are retrieved from the database using the k-NN search. The user then labels these images as either relevant or irrelevant. Based on these labeled images, an initial SVM classifier is trained. The active learning approach is employed whereas m unlabeled images that are closest to the current SVM decision hyperplane are selected for user feedback. After user's labeling, these m-labeled images are added into the training set. Next, a clustering process is performed on the labeled relevant and irrelevant images separately. The formed clusters are then used for unlabeled image selection and predictive-label assignment. Fuzzy reasoning is employed to evaluate the soft relevance of the predictive-labeled images. An FSVM is trained using a combination of the labeled and predictive-labeled images with an appropriate weighting function. The learning process repeats until the user is satisfied with the re-trieved results.

11.4.2 Training of PLFSVM

Predictive Data Labeling

This section will discuss the following two issues in the proposed framework: (1) the selection of unlabeled images for label prediction, and (2) the estima-tion on the relevance of these predictive-labeled images. During the relevance feedback process, the number of feedback/labeled images is usually small. As a classifier requires a reasonable number of training samples for it to be adequately trained, the idea of propagating the labels of the feedback images to the unlabeled images becomes appealing. However, careful selection of the unlabeled images to increase the training population is important as im-properly chosen unlabeled images may not improve the performance. In this work, a new method is presented to select the unlabeled images for predictive-labeling by studying the characteristics of the labeled images. The main idea is to discover those informative samples among the unlabeled images that are highly correlated to the labeled images. The labels and relevance of these images are then estimated. The enlarged data set will now consist of both predictive-labeled and user-labeled samples for training the FSVM classifier.

From previous studies, it is observed that relevant images usually exhibit local characteristics of image similarity. Therefore, it is desirable to adopt a multicluster local modeling strategy. Taking into consideration the multiclus-ter nature of local image similarity, a two-stage clustering technique similar to that in Section 11.3.1 is used here to determine the local clusters. The labeled samples are clustered for the relevant and irrelevant samples, respectively. The two-stage clustering consists of subtractive clustering followed by K-means clustering.

After the clustering process, two sets of relevant and irrelevant clusters will be obtained. An unlabeled image-selection scheme based on the princi-ple of k-NN is adopted, namely, data points that are closer in feature space are more likely to share similar class labels. In this scheme, for each cluster, K unlabeled images are chosen based on the smallest Euclidean distance to the center of the labeled cluster. The label of each cluster is then propagated to these unlabeled neighbors. These images with deduced labels are called predicative-labeled images. As the computational load will increase with larger number of predicative-labeled images, only the nearest neighbor ($K = 1$) is used in this work.

In consideration of the potential fuzziness associated with the predictive-labeled images, we aim to determine a soft relevance membership function $g(\mathbf{x}_p) : \mathbf{R}^R \rightarrow [0, 1]$ that maps each predictive-labeled image $\mathbf{x}_p$ to a proper relevance value between 0 and 1. The estimated relevance of the predictive-labeled images is then used in the FSVM training. In this study, $g(\mathbf{x}_p)$ is determined by two measures, $h_C(\mathbf{x}_p)$ and $h_A(\mathbf{x}_p)$. $h_C(\mathbf{x}_p)$ estimates the relevance of the predictive-labeled images based on the clustering information. On the other hand, $h_A(\mathbf{x}_p)$ determines the reliability of the predictive-labeled images based on the agreement between the predicted labels obtained through the clustering process and the trained FSVM. These two measures are fused together to produce the final soft relevance estimate as follows:

$$g(\mathbf{x}_p) = h_C(\mathbf{x}_p)h_A(\mathbf{x}_p) \tag{11.29}$$

The formulation of $h_C(\mathbf{x}_p)$ is discussed as follows. Intuitively, if a predictive-labeled image is closer to the nearest cluster of the same label type (relevant or irrelevant), then it is more likely the predicted label is reliable. In contrast, if the predictive-labeled image is closer to the nearest cluster of the opposite label type, then the predicted label becomes less reliable. Based on this argument, an exponentially based fuzzy function is selected:

$$h_C(\mathbf{x}_p) = \begin{cases} \exp\left(-r_1 \frac{\min_i(\mathbf{x}_P - \mathbf{c}_{Si})^T(\mathbf{x}_P - \mathbf{c}_{Si})}{\min_j(\mathbf{x}_P - \mathbf{c}_{Oj})^T(\mathbf{x}_P - \mathbf{c}_{Oj})}\right), & \text{if } \frac{\min_i(\mathbf{x}_P - \mathbf{c}_{Si})^T(\mathbf{x}_P - \mathbf{c}_{Si})}{\min_j(\mathbf{x}_P - \mathbf{c}_{Oj})^T(\mathbf{x}_P - \mathbf{c}_{Oj})} < 1 \\ 0, & \text{otherwise} \end{cases}$$
$$\tag{11.30}$$

where $\mathbf{c}_{Si}$ is the centroid of the ith cluster that has the same class label as the predictive-labeled image $\mathbf{x}_P$. $\mathbf{c}_{Oj}$ is the centroid of the jth cluster that has the opposite class label with $\mathbf{x}_P$. $\min_i(\mathbf{x}_P - \mathbf{c}_{Si})^T(\mathbf{x}_P - \mathbf{c}_{Si})$ and $\min_j(\mathbf{x}_P - \mathbf{c}_{Oj})^T(\mathbf{x}_P - \mathbf{c}_{Oj})$ denote the distance between $\mathbf{x}_P$ and the nearest cluster centers with the same and opposite class labels, respectively. $r_1 > 0$ is a scaling factor. There are two scenarios for $h_C(\mathbf{x}_p)$. If the distance ratio given in the top half of Equation (11.30) is less than 1, this will suggest that the predictive-labeled image is closer to the nearest cluster with the same class label. In this case, we will estimate the soft relevance of the predictive-labeled image based on the top term in Equation (11.30). Otherwise, a value of zero is assigned.

The second factor $h_A(\mathbf{x}_p)$ is chosen as a sigmoid function and expressed as follows:

$$h_A(\mathbf{x}_p) = \begin{cases} \dfrac{1}{1 + \exp(-r_2 y)}, & \text{predictive-label is positive} \\ \dfrac{1}{1 + \exp(r_2 y)}, & \text{otherwise} \end{cases} \tag{11.31}$$

where y is the directed distance of the predictive-labeled image $\mathbf{x}_p$ from the SVM decision hyperplane (i.e., the output value of SVM for $\mathbf{x}_p$) and $r_2 > 0$ is a scaling factor. The rationale for using the metric in Equation (11.31) is explained as follows. First consider the case when the predictive-label of the image has been deduced as positive during the predicative-labeling process. In this case, the upper term in Equation (11.31) will be used. If y has a large positive value, this would suggest that there is a strong likelihood that the selected image is indeed a relevant one. It is observed that an agreement exists between the predictive-label obtained through the clustering information and the trained SVM. Therefore, a large fuzzy membership with value close to unity should be assigned to it to indicate high reliability of the predicted label. On the other hand, if y has a large negative value, suggesting a strong disagreement between the predictive-label obtained through the clustering process and the trained SVM, then a small fuzzy membership with value close to zero should be assigned to it. The same arguments apply when the predictive-label of the selected image is negative.

PLFSVM Training

SVM is a powerful machine learning classifier centered on the idea of structural risk minimization (SRM) [276]. It has been successfully adopted and employed in many real-world applications. The aim of SVM classification is to find an optimal hyperplane that maximizes the separation margin between two classes in a kernel-induced feature space. Although SVM has been shown to achieve good performance in solving classification problems, a disadvantage associated with it is that it can only handle a crisp classification problem, namely, each training sample has to be classified as either positive or negative data with equal emphasis. Nevertheless, in real-world applications, there are various cases where the samples do not fall neatly into discrete classes, and there are uncertainties in the labels of the training samples. In order to address this problem, FSVM has been developed to address the uncertainty in the labels of the training samples [275]. FSVM is an extended model of SVM that takes into account different significance/reliability of the training samples. It exhibits the following properties that motivate us to adopt it in our framework: integration of fuzzy data, strong theoretical foundation, and excellent performance.

In this work, a unified PLFSVM framework is developed by integrating the advantages of predictive-labeling and FSVM. It uses predictive-labeled samples to augment the small set of labeled training data. There are several differences between the proposed PLFSVM and the conventional SVM. First, the PLFSVM makes use of predictive-labeled images to resolve the small sample problem, whereas traditional SVM mainly focuses on explicitly labeled images. Second, PLFSVM takes into account the relative importance of the training samples, thus making it more general and flexible. Active learning that aims to achieve maximal information gain or minimize uncertainty in decision making is utilized in the proposed CBIR system. It attempts to select the most informative samples and asks the users to label them. SVM-based active learning has been introduced to conduct relevance feedback in image retrieval [261]. It selects samples that can reduce the version space of SVM whenever possible. In this scheme, the most informative samples are chosen from those that are closest to the current SVM decision boundary and presented to the user for labeling.

Assume that a training set of n samples is given as $S = \{\mathbf{x}_i, y_i\}_{i=1}^n$, where $\mathbf{x}_i \in \mathbf{R}^R$ is an R-dimensional sample in the input space, and $y_i \in \{-1, 1\}$ denotes the class label of $\mathbf{x}_i$. The separating hyperplane of the SVM can be represented as

$$\mathbf{w} \cdot \mathbf{u} + b = 0 \tag{11.32}$$

where $\mathbf{u} = \varphi(\mathbf{x})$ is the mapping function that transforms data in the original input space to a higher dimensional feature space, $\mathbf{w}$ is the normal vector of the hyperplane, and b is the bias that is a scalar. To construct the optimal hyperplane that aims to achieve maximum separating margin with minimal number of classification errors, FSVM solves the following constrained optimization problem [275]:

$$\min \frac{1}{2}\|\mathbf{w}\|^2 + C \sum_{i=1}^{n} \lambda_i \xi_i$$

$$\text{s.t.} \quad y_i(\mathbf{w} \cdot \mathbf{u}_i + b) \geq 1 - \xi_i, \quad \xi_i \geq 0, \quad i = 1, \ldots, n \tag{11.33}$$

The parameter C is the regularization parameter that controls the trade-off between margin maximization and classification error. Larger value of C will put more emphasis on reduction of misclassification over margin maximization. ξ_i is called the slack variable that is related to classification errors in SVM. Misclassifications occur when $\xi_i > 1$. $\lambda_i = g(\mathbf{x}_P) \in [0, 1]$ is the fuzzy membership value associated with each training sample that indicates the relative importance of the data. The higher its value, the more reliable is its class label. In FSVM, the error term ξ_i is weighted by the membership value λ_i. This will ensure that the fidelity/reliability of the training samples is embedded into the soft penalty term during training. In this way, important samples with larger membership values or higher reliability will contribute more toward the FSVM training than those with smaller values.

The optimization problem of FSVM in Equation (11.33) can be transformed into the following equivalent dual problem using the Lagrangian framework:

$$\max \sum_{i=1}^{n} \alpha_i - \frac{1}{2} \sum_{i=1}^{n} \sum_{j=1}^{n} \alpha_i \alpha_j y_i y_j \mathbf{u}_i \cdot \mathbf{u}_j$$

$$\text{s.t.} \quad \sum_{i=1}^{n} y_i \alpha_i = 0, \quad 0 \leq \alpha_i \leq \lambda_i C, \quad i = 1, \ldots, n \qquad (11.34)$$

where α_i is the Lagrange multiplier. Solving the optimization problem in Equation (11.34), the decision function of the FSVM can be obtained as

$$f(\mathbf{x}) = \mathbf{w} \cdot \mathbf{u} + b = \sum_{i=1}^{n} \alpha_i y_i \varphi(\mathbf{x}_i) \cdot \varphi(\mathbf{x}) + b = \sum_{i=1}^{n} \alpha_i y_i K(\mathbf{x}_i, \mathbf{x}) + b \qquad (11.35)$$

where $K(\mathbf{x}_i, \mathbf{x})$ is the kernel function in the input space. This kernel function allows us to construct the optimal hyperplane in the feature space without having to know the mapping φ explicitly.

11.4.3 Experimental Results

The performance of the proposed PLFSVM method was evaluated using the same database shown in Figure 11.3. The same visual features listed in Table 11.1 were extracted and used in the experiments. The RBF kernel, $K(\mathbf{x}_i, \mathbf{x}) = \exp\left(-\frac{\|\mathbf{x}_i - \mathbf{x}_j\|^2}{2\sigma^2}\right)$ was used for the FSVM, with $\sigma = 3$ and $C = 100$. For each query, an initial set of m_0 images was shown to the user for labeling. The FSVM active learning was then used to generate m images for the user to label and feedback. In the experiment, m was set to be 20.

The performance of the PLFSVM method was compared with active learning using SVM [261]. The objective measure used is based on Corel's predefined ground truth. Eighty-five queries were selected for evaluation. Retrieval performance was evaluated by ranking the database images according to their directed distances from the SVM boundary after each iteration. Five iterations of feedback were recorded in total. The APR graphs after the first iteration of active learning for two different numbers of initially labeled images $m_0 = 5$ and $m_0 = 10$ are shown in Figure 11.8. It is observed that the PLFSVM method achieved better results than the standard SVM method in both cases. The PLFSVM method provided higher recall rate at the same precision level, and higher precision rate at the same recall level. For instance, the PLFSVM method achieved a precision rate of 54% at 10% recall rate for $m_0 = 10$. In comparison, the SVM method offered a lower precision rate of 39% at the same recall rate.

The retrieval accuracy of the system for $m_0 = 10$ over the first five iterations is given in Figure 11.9. From the figure, it is observed that the PLFSVM method outperformed the SVM method. For instance, the PLFSVM method achieved the retrieval accuracy of 60% (10 returned images) and 54% (20 returned

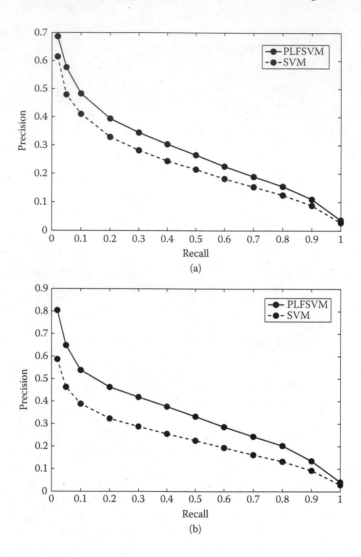

FIGURE 11.8
The APR graphs after the first active learning iteration: (a) APR for $m_0 = 5$; (b) APR for $m_0 = 10$.

images) compared to 43% (10 returned images) and 40% (20 returned images) offered by the SVM method after the first feedback iteration. Further, it is observed that the retrieval accuracy of the PLFSVM method increased quickly in the first few iterations for both the top 10 and 20 returned images. This is desirable as users often would like to obtain satisfactory results without going through too many feedback iterations. With the proper integration of labeled images and predicative-labeled images into training, the proposed PLFSVM method is able to provide satisfactory results.

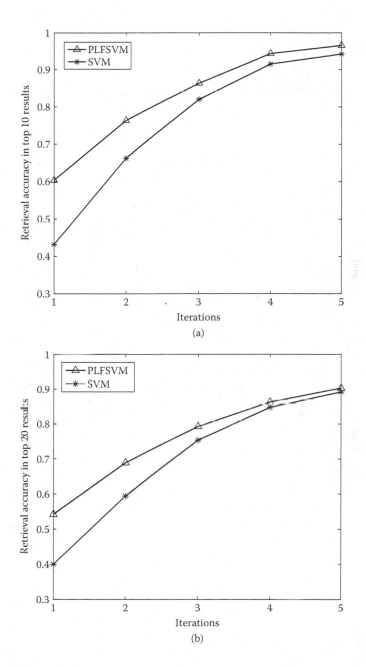

FIGURE 11.9
Retrieval accuracy of the PLFSVM and SVM methods for $m_0 = 10$: (a) retrieval accuracy in top 10 results; (b) retrieval accuracy in top 20 results.

11.5   Conclusion

This chapter demonstrates how computational intelligence can be used to address the challenges of fuzzy user perception and small sample problem in CBIR. The requirements for systematic signal identification, intelligent information integration, and robust optimization point to computational intelligence techniques. To address the issue of fuzzy user perception of feedback images in relevance feedback, the proposed method uses fuzzy logic to determine the relevance of uncertain images. These fuzzy images are then combined with the positive and negative images to train the RFRBFN classifier. The classifier possesses the following characteristics that make it a good choice in the context of CBIR: fast learning speed, simple network structure, multiple local modeling feature, and global generalization power. For the issue of small sample problem in CBIR, this chapter shows that fuzzy logic coupled with SVM are effective in handling the small sample problem. The proposed method uses fuzzy reasoning to perform predictive-labeling where selected images are assigned predictive labels, and their reliability is estimated. It then uses FSVM to incorporate these data in the training process. The results show the effectiveness of the proposed method. In summary, this chapter demonstrates the main aspects of computational intelligence in solving issues in CBIR, namely, identification of useful a priori information, integration of extracted knowledge into the schemes, and development of appropriate computational techniques to optimize the cost function.

References

1. A. K. Jain, 1981, Advances in mathematical models for image processing, *Proc. IEEE* 69(5): 502–528.
2. A. K. Jain, *Fundamentals of Digital Image Processing*. Englewood Cliffs, NJ: Prentice Hall, 1989.
3. A. Rosenfeld, ed., *Image Modeling*. New York: Academic Press, 1981.
4. R. C. Gonzalez and R. Woods, *Digital Image Processing*. Reading, MA: Addison-Wesley, 1992.
5. R. M. Haralick, 1984, Digital step edges from zero crossings of second directional derivatives, *IEEE Trans. Pattern Analysis and Machine Intelligence* 6(1): 58–68.
6. J. Canny, 1986, A computational approach to edge detection, *IEEE Trans. Pattern Analysis and Machine Intelligence* 8(6): 679–698.
7. J. J. Shen and S. S. Castan, 1992, An optimal linear operator for step edge detection, *CVGIP: Graphical Models and Image Processing* 54(2): 112–133.
8. D. Marr and E. Hildreth, 1980, Theory of edge detection, *Proc. Royal Soc.* B-207: 187–217.
9. P. Perona and J. Malik, 1990, Scale-space and edge detection using anisotropic diffusion, *IEEE Trans. Pattern Analysis and Machine Intelligence* 12(7): 629–639.
10. R. Chellappa and S. Chatterjee, 1985, Classification of textures using Gaussian Markov random fields, *IEEE Trans. Acoust., Speech, Signal Processing* 33(4): 959–963.
11. R. Chellappa and R. L. Kashyap, 1982, Digital image restoration using spatial interaction models, *IEEE Trans. Acoust., Speech, Signal Processing* 30(6): 461–472.
12. G. R. Cross and A. K. Jain, 1983, Markov random field texture models, *IEEE Trans. Pattern Analysis and Machine Intelligence* 5(1): 25–39.
13. R. L. Kashyap and R. Chellappa, 1983, Estimation and choice of neighbors in spatial interaction models of images, *IEEE Trans. Inform. Theory* 29(1): 60–72.
14. J. W. Woods, 1972, Two-dimensional discrete Markovian fields, *IEEE Trans. Inform. Theory* 18(2): 232–240.
15. H. Derin and H. Elliott, 1987, Modelling and segmentation of noisy and textured images using Gibbs random fields, *IEEE Trans. Pattern Analysis and Machine Intelligence* 9(1): 39–55.
16. S. Geman and D. Geman, 1984, Stochastic relaxation, Gibbs distribution, and the Bayesian restoration of images, *IEEE Trans. Pattern Analysis and Machine Intelligence* 6(6): 721–741.
17. M. Barnsley, *Fractals Everywhere*. New York: Academic Press, 1988.
18. B. Mandelbrot, *Fractals and a New Geometry of Nature*. San Francisco: Freeman, 1981.
19. G. R. Arce and R. Foster, 1989, Detail-preserving ranked-order based filters for image processing, *IEEE Trans. Acoust., Speech, Signal Processing* 37(1): 83–98.

20. A. Tekalp, H. Kaufman, and J. Woods, 1989, Edge-adaptive Kalman filtering for image restoration with ringing suppression, *IEEE Trans. Acoust., Speech, Signal Processing* 37(6): 892–899.

21. H. Andrews and B. Hunt, *Digital Image Restoration*. Englewood Cliffs, NJ: Prentice Hall, 1977.

22. M. R. Banham and A. K. Katsaggelos, 1997, Digital image restoration, *IEEE Signal Processing Magazine* 14(2): 24–41.

23. A. K. Katsaggelos, ed., *Digital Image Restoration*. Berlin: Springer Verlag, 1991.

24. M. Bertero, T. Poggio, and V. Torre, 1988, Ill-posed problems in early vision, *Proc. IEEE* 76(8): 869–889.

25. B. R. Hunt, 1973, The application of constrained least squares estimation to image restoration by digital computers, *IEEE Trans. Comput.* 22(9): 805–812.

26. T. Poggio, V. Torre, and C. Koch, 1985, Computational vision and regularization theory, *Nature* 317(6035): 314–319.

27. A. K. Katsaggelos and M. G. Kang, 1995, Spatially adaptive iterative algorithm for the restoration of astronomical images, *Int. J. Imaging Systems and Technology* 6(4): 305–313.

28. R. L. Lagendijk, J. Biemond, and D. E. Boekee, 1988, Regularized iterative image restoration with ringing reduction, *IEEE Trans. Acoust., Speech, Signal Processing* 36(12): 1874–1888.

29. H. S. Wong and L. Guan, 1997, Adaptive regularization in image restoration using a model-based neural network, *Opt. Eng.* 36(12): 3297–3308.

30. N. Ansari and E. Hou, *Computational Intelligence for Optimization*. Boston: Kluwer Academic, 1997.

31. W. Pedrycz, *Computational Intelligence: an Introduction*. Boca Raton, FL: CRC Press, 1997.

32. P. H. Winston, *Artificial Intelligence*. Reading, MA.: Addison-Wesley, 1984.

33. J. S. R. Jang, C. T. Sun, and E. Mizutani, *Neuro-Fuzzy and Soft Computing: A Computational Approach to Learning and Machine Intelligence*. Upper Saddle River, NJ: Prentice Hall, 1997.

34. S. Haykin, *Neural Networks: A Comprehensive Foundation*. NJ: Prentice Hall, 1999.

35. J. Hertz, A. Krogh, and R. G. Palmer, *Introduction to the Theory of Neural Computation*. Santa Fe Institute: Studies in the Science of Complexity. Redwood City, CA: Addison-Wesley, Pub. Co.

36. G. J. Klir and T. A. Folger, *Fuzzy Sets, Uncertainty and Information*. Englewood Cliffs, NJ: Prentice Hall, 1988.

37. B. Kosko, *Neural Networks and Fuzzy Systems*. Englewood Cliffs, NJ: Prentice Hall, 1992.

38. L. A. Zadeh, 1965, Fuzzy sets, *Information and Control* 8(3): 338–353.

39. T. Bäck, *Evolutionary Algorithms in Theory and Practice*. New York: Oxford University Press, 1996.

40. D. B. Fogel, *Evolutionary Computation: Toward a New Philosophy of Machine Intelligence*. Piscataway, NJ: IEEE Press, 1995.

41. E. Ardizzone, A. Chella, and R. Pirrone, A neural based approach to image segmentation, in *Proc. Int. Conf. on Artificial Neural Networks (ICANN'94)*: 1153–1156, 1994.

42. A. Ghosh, N. R. Pal, and S. K. Pal, 1991, Image segmentation using a neural network, *Biological Cybernetics* 66(2): 151–158.

43. C. L. Huang, 1992, Parallel image segmentation using modified Hopfield network, *Pattern Recogn. Lett.* 13(5): 345–353.

44. T. Kohonen, *Self-Organizing Maps*, 2d ed. Berlin: Springer-Verlag, 1997.
45. C. Peterson and B. Söderberg, 1989, A new method for mapping optimization problems onto neural networks, *Int. J. Neural Sys.* 1(1): 3–22.
46. Y. T. Zhou, R. Chellappa, A. Vaid, and B. K. Jenkins, 1988, Image restoration using a neural network, *IEEE Trans. Acoust., Speech, Signal Processing* 36(7): 1141–1151.
47. R. Anand, K. Mehrotra, C. K. Mohan, and S. Ranka, 1995, Efficient classification for multiclass problems using modular neural networks, *IEEE Trans. Neural Networks* 6(1): 117–124.
48. S. Y. Kung and J. S. Taur, 1995, Decision-based neural networks with signal/image classification applications, *IEEE Trans. Neural Networks* 6(1): 170–181.
49. E. R. Kandel and J. H. Schwartz, *Principles of Neural Science*. New York: Elsevier, 1985.
50. R. F. Thompson, *An Introduction to Neuroscience*. New York: W.H. Freeman & Company, 1985.
51. Y. S. Choi and R. Krishnapuram, 1997, A robust approach to image enhancement based on fuzzy logic, *IEEE Trans. Image Processing* 6(6): 808–825.
52. N. Fang and M. C. Cheng, 1993, An automatic crossover point selection technique for image enhancement using fuzzy sets, *Pattern Recogn. Lett.* 14(5): 397–406.
53. J. Hsieh, 1995, Image enhancement with a fuzzy logic approach, *Electron. Lett.* 31(9): 708–710.
54. R. Krishnapuram, J. M. Keller, and Y. Ma, 1993, Quantitative analysis of properties and spatial relations of fuzzy image regions, *IEEE Trans. Fuzzy Systems* 1(3): 222–233.
55. C. Y. Tyan and P. P. Wang, 1993, Image processing-enhancement, filtering and edge detection using the fuzzy logic approach, in *2nd IEEE Int. Conf. on Fuzzy Systems*: 600–605.
56. J. C. Bezdek, *Pattern Recognition with Fuzzy Objective Function Algorithms*. New York: Plenum Press, 1981.
57. R. J. Hathaway and J. C. Bezdek, 1986, Local convergence of the fuzzy c-means algorithms, *Pattern Recogn.* 19(6): 477–480.
58. N. R. Pal and J. C. Bezdek, 1995, On cluster validity for the fuzzy c-means model, *IEEE Trans. Fuzzy Systems* 3(3): 370–379.
59. M. J. Sabin, 1987, Convergence and consistency of fuzzy c-means/ISODATA algorithms, *IEEE Trans. Pattern Analysis and Machine Intelligence* 9(5): 661–668.
60. J. MacQueen, 1967, Some methods for classification and analysis of multivariate observations, in *Proc. 5th Berkeley Symp. on Math. Stat. and Prob.* 1: 281–296.
61. D. E. Goldberg, *Genetic Algorithms in Search, Optimization and Machine Learning*. Reading, MA: Addison-Wesley, 1989.
62. A. Katsaggelos, J. Biemond, R. Schafer, and R. Mersereau, 1991, A regularized iterative image restoration algorithm, *IEEE Trans. Signal Processing* 39(4): 914–929.
63. A. Tikhonov, A. Goncharsky, and V. Stepanov, Inverse problem in image processing, in *Ill-Posed Problems in the Natural Sciences*, 220–232. Moscow: Mir Publishers, 1987.
64. M. Sezan and A. Tekalp, 1990, Survey of recent developments in digital image restoration, *Opt. Eng.* 29(May): 393–404.
65. S. Geman and D. Geman, 1984, Stochastic relaxation, Gibbs distribution, and the Bayesian restoration of images, *IEEE Trans. Pattern Analysis and Machine Intelligence* 6(Nov.): 721–741.

66. G. Demoment, 1989, Image reconstruction and restoration: Overview of common estimation structures and problems, *IEEE Trans. Acoustics, Speech and Signal Processing* 37(Dec.): 2024–2073.

67. M. Sondhi, 1972, Image restoration: The removal of spatially invariant degradations, *Proc. IEEE* 60(Jul.): 842–853.

68. H. Andrews, 1974, Digital image restoration: A survey, *IEEE Computer* 7(May): 36–45.

69. B. Hunt, 1974, Digital image processing, *Proc. IEEE* 63(Apr.): 36–45.

70. B. Frieden, Image enhancement and restoration, in *Picture Processing and Digital Filtering*, ed., T. Huang. New York: Springer-Verlag, 1975.

71. S. Perry and L. Guan, 2000, Weight assignment for adaptive image restoration by neural networks, *IEEE Trans. Neural Networks* 11(Jan.): 156–170.

72. S. Perry, Adaptive image restoration: Perception based neural network models and algorithms. PhD thesis, School of Electrical and Information Engineering, University of Sydney, Apr. 1999.

73. T. Caelli, D. Squire, and T. Wild, 1993, Model-based neural networks, *Neural Networks* 6(5): 613–625.

74. J. P. Sutton, J. S. Beis, and L. E. H. Trainor, 1988, A hierarchical model of neuron-cortical synaptic organization, *Math. Comput. Modeling* 11: 346–350.

75. A. M. Turing, 1952, The chemical basis of morphogenesis, *Philos. Trans. R. Soc. B* 237: 5–72.

76. T. Kohonen, 1982, Self-organized formation of topologically correct feature maps, *Biological Cybernetics* 43: 59–69.

77. G. G. Yen and Z. Wu, 2005, Ranked centroid projection: A data visualization approach for self-organizing maps, in *Proc. Int. Joint Conf. Neural Networks*, (Montreal, Canada): 1587–1592.

78. A. Ultsch and H. P. Siemon, 1990, Kohonen's self organizing feature maps for exploratory data analysis, in *Proc. Int. Neural Network Conf.* (Dordrecht, Netherlands): 305–308.

79. R. Miikkulainen, 1990, Script recognition with hierarchical feature maps, *Conn. Sci.* 2: 83–101.

80. T. Martinetz and K. Schulten, 1991, A "neural-gas" network learns topologies, *Artificial Neural Network* 1: 397–402.

81. M. Dittenbach, D. Merkl, and A. Rauber, The growing hierarchical self-organizing map, in *Proc. Int. Joint Conf. Neural Networks* (Como, Italy): 15–19.

82. M. Dittenbach, D. Merkl, and A. Rauber, 2001, Hierarchical clustering of document archives with the growing hierarchical self-organizing map, in *Proc. Int. Conf. Artificial Neural Networks* (Vienna, Austria): 500–508.

83. B. Fritzke, 1995, A growing neural gas network learns topologies, in *Adv. Neural Inf. Proc. Sys. 7*: 625–632.

84. B. Fritzke, 1995, Growing grid: a self-organizing network with constant neighborhood range and adaption strength, *Neural Proc. Lett.* 2(5).

85. B. Fritzke, 1994, Growing cell structures: a self-organizing network for unsupervised and supervised learning, *Neural Networks* 7(9): 1441–1460.

86. I. Pitas and A. Venetsanopoulos, *Nonlinear Digital Filters: Principles and Applications*. Boston: Kluwer Academic Publishers, 1990.

87. H. Kong and L. Guan, 1995, Detection and removal of impulse noise by a neural network guided adaptive median filter, in *Proc. IEEE Int. Conf. Neural Networks* (Perth, Australia): 845–849.

88. F. Roddier, The effects of atmospheric turbulence in optical astronomy, in *Progress in Optics*, ed. E. Wolf, 281–376. Amsterdam: North-Holland, 1981.

89. J. Oakley and B. Satherley, 1998, Improving image quality in poor visibility conditions using a physical model for contrast degradation, *IEEE Trans. Image Processing* 17(Feb).

90. E. McCartney, *Optics of the Atmosphere: Scattering by Molecules and Particles*. New York: Wiley & Sons, 1976.

91. V. Zuev, *Propagation of Visible and Infrared Radiation in the Atmosphere*. New York: Wiley & Sons, 1974.

92. I. Dinstein, H. Zoabi, and N. Kopeika, 1998, Prediction of effects of weather on image quality: Preliminary results of model validation, *Appl. Opt.* 27: 2539–2545.

93. L. Scharf, *Statistical Signal Processing: Detection, Estimation and Time Series Analysis*. Reading, MA: Addison Wesley, 1991.

94. E. Chong and S. Zäk, *An Introduction to Optimization*. New York: John Wiley & Sons, 1996.

95. J. K. Paik and A. K. Katsaggelos, 1992, Image restoration using a modified Hopfield network, *IEEE Trans. Image Processing* 1(1): 49–63.

96. Y. Iiguni et al., 1992, A real-time learning algorithm for a multilayered neural network based on the extended Kalman filter, *IEEE Trans. Signal Processing* 40(Apr.): 959–966.

97. R. Steriti and M. A. Fiddy, 1993, Regularised image reconstruction using SVD and a neural network method for matrix inversion, *IEEE Trans. Signal Processing* 41(Oct.): 3074–3077.

98. S. Perry and L. Guan, 1995, Neural network restoration of images suffering space-variant distortion, *Electron. Lett.* 31(Aug.): 1358–1359.

99. S. Perry and L. Guan, 1995, Restoration of images degraded by space variant distortion using a neural network, in *Proc. IEEE Conf. Neural Networks* 4(Nov. 26–Dec. 2): 2067–2070, Perth, Australia.

100. S. Perry and L. Guan, 1995, Image restoration using a neural network with an adaptive constraint factor, in *Proc. IEEE Conf. Neural Networks in Signal Processing*, (Dec. 10–20): 1031–1034, Nanjing, P.R. China.

101. S. Perry and L. Guan, 1996, Image restoration using a neural network with an adaptive constraint factor (invited paper), in *Proc. Computational Engineering in Systems Applications* (Jul. 9–12): 854–859, Lille, France.

102. M. G. Kang and A. K. Katsaggelos, 1995, General choice of the regularization functional in regularized image restoration, *IEEE Trans. Image Processing* 4(5): 594–602.

103. J. Anderson and J. Sutton, 1995, A networks of networks: Computation and neurobiology, *World Congress of Neural Networks* 1.

104. J. Beck, B. Hope, and A. Rosenfeld, eds., *Human and Machine Vision*. Orlando, FL: Academic Press, 1983.

105. S. Perry and L. Guan, 1998, A statistics-based weight assignment in a Hopfield neural network for adaptive image restoration, in *Proc. IEEE International Joint Conference on Neural Networks* (May 5–9). 992–997, Anchorage, AK.

106. S. Perry and L. Guan, 1997, Adaptive constraint restoration and error analysis using a neural network, in *Proc. Australian Joint Conference on Artificial Intelligence* 87–95, Perth, Australia.

107. D. Levi and S. Klein, 1992, Weber law for position: The role of spatial frequency and contrast, *Vision Research* 32: 2235–2250.

108. R. Hess and I. Holliday, 1992, The coding of spatial position by the human visual system: Spatial scale and contrast, *Vis. Res.* 32: 1085–1097.
109. J. Beck, 1966, Effect of orientation and shape similarity on perceptual grouping, *Perception and Psychophysics* 1: 300–302.
110. A. Gagalowicz, 1981, A new method for texture field synthesis: Some applications to the study of human vision, *IEEE Trans. Pattern Analysis and Machine Intelligence* 3: 520–533.
111. B. Julesz, 1975, Experiments in the visual perception of texture, *Scientific American*, 232(4): 34–43.
112. M. Fahle and T. Poggio, 1981, Visual hyperacuity: Spatiotemporal interpolation in human stereo vision, in *Proc. R. Soc.* B 213: 451–477.
113. D. Marr, S. Ullman, and T. Poggio, 1979, Bandpass channels, zero-crossings and early visual information processing, *J. Opt. Soc. Am.* 69: 914–916.
114. F. Attneave, 1954, Some informational aspects of visual perception, *Psychological Review* 61: 183–193.
115. G. Johansson, 1975, Visual motion perception, *Scientific American* 232(6): 76–88.
116. R. Hess and D. Badcock, 1995, Metric for separation discrimination by the human visual system, *J. Opt. Soc. Am.* 12(Jan.): 3–16.
117. S. Perry and L. Guan, 1998, Peception based adaptive image restoration, in *Proc. IEEE International Conference on Acoustics, Speech and Signal Processing*, 5(May 12–15): 2893–2896, Seattle, WA.
118. S. Perry, L. Guan, and P. Varjavandi, 2006, Incorporating local statistics in image error measurement for adaptive image restoration, *Opt. Eng.* 45(Mar.): 037001-1–037001-14.
119. S. Perry, P. Varjavandi, and L. Guan, 2004, Adaptive image restoration using a perception based error measurement, in *Proc. IEEE Canadian. Conf. on Electrical and Computer Engineering* 3(May): 1585–1588.
120. B. L. M. Happel and J. M. J. Murre, 1994, Design and evolution of modular neural network architectures, *Neural Networks* 7(6–7): 985–1004.
121. R. A. Jacobs and M. I. Jordan, 1993, Learning piecewise control strategies in a modular neural network architecture, *IEEE Trans. Syst. Man Cybern.* 23(2): 337–345.
122. A. Kehagias and V. Petridis, 1997, Time-series segmentation using predictive modular neural networks, *Neural Computation* 9(8): 1691–1709.
123. S. Ozawa, K. Tsutsumi, and N. Baba, 1998, An artificial modular neural network and its basic dynamical characteristics, *Biological Cybernetics* 78(1): 19–36.
124. L. Wang, S. A. Rizvi, and N. M. Nasrabadi, 1998, A modular neural network vector predictor for predictive image coding, *IEEE Trans. Image Proc.* 7(8): 1198–1217.
125. L. Wang, S. Z. Der, and N. M. Nasrabadi, 1998, Automatic target recognition using a feature-decomposition and data-decomposition modular neural network, *IEEE Trans. Image Proc.* 7(8): 1113–1121.
126. H. Fu and Y. Y. Xu, 1998, Multilinguistic handwritten character recognition by Bayesian decision-based neural networks, *IEEE Trans. Signal Proc.* 46(10): 2781–2789.
127. S. Y. Kung, M. Fang, S. P. Liou, M. Y. Chiu, and J. S. Taur, 1995, Decision-based neural network for face recognition system, in *Proc. Int. Conf. on Image Processing*: 430–433.
128. S. H. Lin and S. Y. Kung, 1995, Probabilistic DBNN via expectation-maximization with multi-sensor classification applications, in *Proc. Int. Conf. on Image Processing*: 236–239.

129. S. H. Lin, S. Y. Kung, and L. J. Lin, 1997, Face recognition/detection by probabilistic decision-based neural networks, *IEEE Trans. Neural Networks* 8(1): 114–132.

130. A. C. Bovik, T. S. Huang, and D. C. Munson, 1983, A generalization of median filtering using linear combinations of order statistics, *IEEE Trans. Acoust., Speech, Signal Processing* 31(6): 1342–1350.

131. M. Ropert, F. M. de Saint-Martin, and D. Pele, 1996, A new representation of weighted order statistic filters, *Signal Processing* 54(2): 201–206.

132. J. T. Tou and R. C. Gonzalez, *Pattern Recognition Principles.* Reading, MA: Addison-Wesley, 1974.

133. L. A. Zadeh, 1968, Fuzzy algorithms, *Information and Control* 12(2): 94–102.

134. L. A. Zadeh, 1973, Outline of a new approach to the analysis of complex systems and decision processes, *IEEE Trans. Syst. Man Cybern.* 3(1): 28–44, 73.

135. L. A. Zadeh, 1978, Fuzzy sets as a basis for a theory of possibility, *Information and Control* 1(1): 3–28.

136. J. Hopfield and D. Tank, 1985, Neural computation of decisions in optimization problems, *Biol. Cybern.* 52(3): 141–152.

137. L. J. Fogel, A. J. Owens, and M. J. Walsh, *Artificial Intelligence Through Simulated Evolution.* New York: Wiley, 1966.

138. T. Bäck, D. B. Fogel, and Z. Michalewicz, eds., *Handbook of Evolutionary Computation.* New York: Oxford University Press and Institute of Physics, 1997.

139. T. Bäck, U. Hammel, and H. P. Schwefel, 1997, Evolutionary computation: Comments on the history and current state, *IEEE Trans. Evolutionary Comp.* 1(1): 3–17.

140. T. Bäck and H. P. Schwefel, 1996, Evolutionary computation: An overview, in *Proc. 3rd IEEE Conf. on Evolutionary Computation:* 20–29, Piscataway, NJ, IEEE Press.

141. J. H. Holland, *Adaptation in Natural and Artificial Systems.* Ann Arbor, MI: University of Michigan Press, 1975.

142. M. Mitchell, *An Introduction to Genetic Algorithms.* Cambridge MA: MIT Press, 1996.

143. H. P. Schwefel, *Numerical Optimization of Computer Models.* Chichester: Wiley, 1981.

144. H. P. Schwefel, *Evolution and Optimum Seeking.* New York: Wiley, 1995.

145. K. H. Yap and L. Guan, 2000, Adaptive image restoration based on hierarchical neural networks, *Opt. Eng.* 39(7): 1877–1890.

146. W. Pratt, *Digital Image Processing.* New York: Wiley & Sons, 1991.

147. M. Cannon, 1976, Blind deconvolution of spatially invariant image blur with phase, *IEEE Trans. Acoust. Speech, Signal Processing* 24: 58–63.

148. D. Kundur and D. Hatzinakos, 1996, Blind image deconvolution, *IEEE Signal Processing Magazine* 13(3): 43–64.

149. R. L. Lagendijk, J. Biemond, and D. E. Boekee, 1990, Identification and restoration of noisy blurred images using the expectation-maximization algorithm, *IEEE Trans. Acoust. Speech, Signal Processing* 38(Jul.): 1180–1191.

150. S. Reeves and R. Mersereau, 1992, Blur identification by the method of generalized cross-validation, *IEEE Trans. Image Processing* 1(Jul.): 301–311.

151. G. Ayers and J. Dainty, 1988, Iterative blind deconvolution method and its applications, *Opt. Lett.* 13: 547–549.

152. B. C. McCallum, 1990, Blind deconvolution by simulated annealing, *Opt. Commun.* 75(Feb.): 101–105.

153. D. Kundur and D. Hatzinakos, 1998, A novel blind image deconvolution scheme for image restoration using recursive filtering, *IEEE Trans. Signal Proc.* 46(Feb.): 375–390.

154. Y. You and M. Kaveh, 1996, A regularization approach to joint blur identification and image restoration, *IEEE Trans. Image Processing* 5(Mar.): 416–428.

155. T. F. Chan and C. K. Wong, 1998, Total variation blind deconvolution, *IEEE Trans. Image Processing* 7(Mar.): 370–375.

156. K. H. Yap and L. Guan, 2000, A recursive soft-decision PSF and neural network approach to adaptive blind image regularization, in *Proc. IEEE Int. Conf. Image Processing* 3(Sept.): 813–816.

157. D. B. Fogel, 1994, An introduction to simulated evolutionary optimization, *IEEE Trans. Neural Networks* 5(Jan.): 3–14.

158. K. S. Tang, K. F. Man, S. Kwong, and Q. He, 1996, Genetic algorithms and their applications, *IEEE Signal Processing Magazine* (Nov.): 22–37.

159. J. P. Sutton, Hierarchical organization and disordered neural systems. PhD thesis, University of Toronto, 1988.

160. L. Guan, J. Anderson, and J. Sutton, 1997, A network of networks model for image regularization, *IEEE Trans. on Neural Networks* 8(Jan.): 169–174.

161. A. A. Zhigljavsky, *Theory of Global Random Search. Mathematics and Its Applications.* Dordrecht: Kluwer, 1992.

162. P. Baldi and K. Hornik, 1989, Neural networks and principal component analysis: Learning from examples without local minima, *Neural Networks* 2(1): 53–58.

163. E. Oja, 1982, A simplified neuron model as a principal component analyzer, *J. Math. Biol.* 15(3): 267–273.

164. E. Oja, 1989, Neural networks, principal components and subspaces, *Int. J. Neural Systems* 1(1): 61–68.

165. T. Kohonen, 1990, The self-organizing map, *Proc. IEEE* 78(9): 1484–1490.

166. D. J. Willshaw and C. von der. Malsburg, 1979, A marker induction mechanism for the establishment of ordered neural mappings: Its application to the retino-tectal problem, in *Proc. R. Soc. London* B 287: 203–243.

167. J. A. Kangas, T. Kohonen, and J. T. Lassksonen, 1990, Variants of self-organizing maps, *IEEE Trans. Neural Networks* 1(1): 93–99.

168. G. A. Carpenter and S. Grossberg, Associative learning, adaptive pattern recognition and cooperative-competitive decision making by neural networks, in *Hybrid and Optical Computing*, ed. H. Szu. SPIE, 1986.

169. A. V. Skorokhod, *Studies in the Theory of Random Processes*. Reading, MA: Addison-Wesley, 1965.

170. S. Grossberg, 1969, On learning and energy-entropy dependence in recurrent and nonrecurrent signed networks, *J. Statist. Phys.* 1: 319–350.

171. B. Kosko, 1990, Unsupervised learning in noise, *IEEE Trans. Neural Networks* 1: 44–57.

172. D. E. Rumelhart and D. Zipser, 1985, Feature discovery by competitive learning, *Cognitive Sci.* 9: 75–112.

173. B. Kosko, 1991, Stochastic competitive learning, *IEEE Trans. Neural Networks* 2(5): 522–529.

174. W. K. Pratt, *Digital Image Processing*. New York: Wiley & Sons, 1991.

175. H. C. Andrews and B. R. Hunt, *Digital Image Restoration*. Englewood Cliffs, NJ: Prentice Hall, 1977.

176. R. R. Lawrence, R. S. Marvin, and E. S. Carolyn, 1975, Applications of a nonlinear smoothing algorithm to speech processing, *IEEE Trans. Acoustics, Speech, and Signal Processing* ASSP-23(6): 552–557.

177. Y. H. Lee and S. A. Kassam, 1985, Generalized median filtering and related nonlinear filtering techniques, *IEEE Trans. Acoustics, Speech, and Signal Processing* ASSP-33(3): 672–683.

178. N. Wiener, *Nonlinear Problems in Random Theory*. Cambridge, MA: The Technology Press, MIT and New York, John Wiley & Sons, Inc., 1958.

179. J. W. Tukey, *Exploratory Data Analysis*. Reading, MA: Addison-Wiley, 1977.

180. D. R. K. Brownrigg, 1984, The weighted media filter, *Comm. ACM* 27: 807–818.

181. S. J. Ko and Y. H. Lee, 1991, Center weighted median filters and their applications to image enhancement, *IEEE Trans. Circuits and Systems* CAS-38(9): 984–993.

182. O. Yli-Harja, J. Astola, and Y. Neuvo, 1991, Analysis of the properties of median and weighted median filters using threshold logic and stack filter representation, *IEEE Trans. Signal Processing* 39(2): 395–410.

183. G. R. Arce and M. P. McLoughlin, 1987, Theoretical analysis of max/median filters, *IEEE Trans. Acoustics, Speech, and Signal Processing* ASSP-35(1): 60–69.

184. A. Nieminen, P. Heinonen, and Y. Neuvo, 1987, A new class of detail-preserving filters for image processing, *IEEE Trans. Pattern Analysis and Machine Intelligence* PAMI-9(1): 74–90.

185. A. Kundu, S. K. Mitra, and P. P. Vaidyanathan, 1984, Application of two-dimensional generalized mean filtering for removal of impulse noises from images, *IEEE Trans. Acoustics, Speech, and Signal Processing* ASSP-32(3): 600–609.

186. I. Pitas and A. Venetsanopoulos, 1986, Nonlinear mean filters in image processing, *IEEE Trans. Acoustics, Speech, and Signal Processing* 34(3): 573–584.

187. H. Lin and A. N. Willson, 1988, Median filters with adaptive length, *IEEE Trans. Circuits and Systems* 35(6): 675–690.

188. R. Bernstein, 1987, Adaptive nonlinear filters for simultaneous removal of different kinds of noise in images, *IEEE Trans. Circuits and Systems* 34(11): 1275–1291.

189. J. E. Kenneth, 1995, Minimum mean squared error impulse noise estimation and cancellation, *IEEE Trans. Signal Processing* 43(7): 1651–1662.

190. E. Abreu, M. Lightstone, S. K. Mitra, and K. Arakawa, 1996, A new efficient approach for the removal of impulse noise from highly corrupted images, *IEEE Trans. Image Processing* 5(6): 1012–1025.

191. B. I. Justusson, Median filtering: Statistical properties, in *Two Dimensional Digital Signal Processing*, ed. T. S. Huang. Berlin: Springer Verlag, 1981.

192. H. Kong and L. Guan, 1998, Noise-exclusive adaptive filtering for removing impulsive noise in digital images, *IEEE Trans. Circuits and Systems* 45(3): 422–428.

193. H. Kong and L. Guan, 1998, Self-organizing tree map for eliminating impulse noise with random intensity distributions, *J. Electron. Imag.* 7(1): 36–44.

194. H. Kong and L. Guan, 1996, A neural network adaptive filter for the removal of impulse noise in digital images, *Neural Networks* 9(3): 373–378.

195. J. B. Bednar and T. L. Watt, 1984, Alpha-trimmed means and their relationship to median filters, *IEEE Trans. Acoust., Speech, Signal Processing* 32(1): 145–153.

196. J. Randall, L. Guan, X. Zhang, and W. Li, 1999, Investigations of the self-organizing tree map, in *Proc. of Int. Conf. on Neural Information Processing* 2: 724–728.

197. P. Munnesawang, Retrieval of image/video content by adaptive machine and user interaction. PhD thesis, University of Sydney, 2002.

198. Y. Rui, T. Huang, and S. Mehrotra, 1997, Content-based image retrieval with relevance feedback in MARS, in *Proc. IEEE Int. Conf. on Image Processing*: 815–818.

199. G. Salton and M. J. McGill, *Introduction to Modern Information Retrieval*. New York: McGraw-Hill, 1983.

200. J. Lay and L. Guan, 1999, Image retrieval based on energy histograms of the low frequency DCT coefficients, in *Proc. IEEE Int. Conf. on Acoustics, Speech and Signal Processing*: 3009–3012.

201. Z. Xiong and T. S. Huang, 2002, Subband-based, memory-efficient JPEG2000 images indexing in compressed-domain, in *IEEE Southwest Symposium on Image Analysis and Interpretation* (Apr.).

202. C. Lui and M. K. Mandal, 2001, Fast image indexing based on JPEG2000 packet header, in *Proc. Int. Workshop on Multimedia Information Retrieval* (Oct.).

203. M. N. Do and M. Vertterli, 2002, Wavelet-based texture retrieval using generalized Gaussian density and Kullback-Leibler distance, *IEEE Trans. Image Processing* 11(2): 146–158.

204. D. G. Sim, H. K. Kim, and R. H. Park, 2001, Fast texture description and retrieval of DCT-based compressed images, *IEEE Electron. Lett.* 37(1): 18–19.

205. J. Bhalod, G. F. Fahmy, and S. Panchanathan, 2001, Region based indexing in the JPEG2000 framework, in *Proc. Int. Workshop on Multimedia Information Retrieval*, (Sept.).

206. ISO/IEC, ISO/IEC 14496-2:1999: Information technology—coding of audio-visual objects—Part 1: Visual. Technical Report, International Organization for Standardization, Geneva, Switzerland, Dec. 1999.

207. ISO/IEC JTC 1/SC 29/WG 1, ISO/IEC FDIS 15444-1: information technology—JPEG 2000 image coding system: core coding system [WG 1 N 1890]. Technical Report, International Organization for Standardization, Geneva, Switzerland, Sept. 2000.

208. Z. Xiong and T. S. Huang, 2002, Fast, memory-efficient JPEG2000 images indexing in compressed domain, in *IEEE Southwest Symposium on Image Analysis and Interpretation* (Apr.).

209. M. K. Mandal, S. Panchanathan, and T. Aboulnasr, 1997, Image indexing using translation and scale-invariant moments and wavelets, *Storage and Retrieval for Image and Video Databases (SPIE)*: 380–389.

210. N. B. Karayiannis, P. I. Pai, and N. Zervos, 1998, Image compression based on fuzzy algorithms for learning vector quantization and wavelet image decomposition, *IEEE Trans. on Image processing* 7(8): 1223–1230.

211. M. Antonini, M. Barlaud, F. Mathieu, and I. Daubechies, 1992, Image coding using wavelet transform, *IEEE Trans. on Image Processing* 1(2): 205–220.

212. P. Muneesawang and L. Guan, 2000, Multiresolution-histogram indexing for wavelet-compressed images and relevant feedback learning for image retrieval, in *Proc. IEEE Int. Conf. on Image Processing*: 526–529.

213. B. S. Manjunath and W. Y. Ma, 1996, Texture features for browsing and retrieval of image data, *IEEE Trans. Pattern Analysis and Machine Intelligence* 18(8): 837–842.

214. B. S. Manjunath, J. Ohm, V. V. Vasudevan, and A. Yamada, 2001, Color and texture descriptors, *IEEE Trans. Circuit and Systems for Video Technology* 11(6): 703–715.

215. P. Salembier and J. R. Smith, 2001, MPEG-7 multimedia descriptor schemes, *IEEE Trans. on Circuit and Systems for Video Technology* 11(6): 748–759.

216. A. Jain and G. Healey, 1998, A multiscale representation including opponent color features for texture recognition, *IEEE Trans. Image Processing*, 7: 124–128.

217. W. Ma and B. Manjunath, 2000, EdgeFlow: A technique for boundary detection and image segmentation, *IEEE Trans. Image Processing* 9(8): 1375–1388.

218. W. B. Pennebaker and J. L. Mitchell, *JPEG Still Image Compression Standard*. New York: Van Nostrand Reinhold, 1993.

219. ISO/IEC JTC 1/SC 29/WG 1 N 751, *Coding of Still Pictures*. International Organization for Standardization, Geneva, Switzerland, 1998.

220. B. Furht, 1995, A survey of multimedia compression techniques and standards, *Real-Time Imaging* 1: 49–67.

221. A. W. M. Smeulders, M. Worring, S. Santini, A. Gupta, and R. Jain, 2000, Content-based image retrieval at the end of the early years, *IEEE Trans. Pattern Analysis and Machine Intelligence* 22(12): 1349–1380.

222. Y. Rui, T. S. Huang, and S. F. Chang, 1999, Image retrieval: Current techniques, promising directions and open issues, *J. Visual Comm. and Image Representation* 10(1): 39–62.

223. A. Gupta and R. Jain, 1997, Visual information retrieval, *Comm. ACM* 40(5): 71–79.

224. C. C. Hsu, W. W. Chu, and R. K. Taira, 1996, A knowledge-based approach for retrieving images by content, *IEEE Trans. on Knowledge and Data Engineering* 8(4): 522–532.

225. M. K. Mandal, F. M. Idris, and S. Panchanathan, 1999, A critical evaluation of image and video indexing techniques in the compressed domain, *Image Vision Computing* 17(7): 513–529.

226. E. Regentova, S. Latifi, and S. Deng, 2001, Images similarity estimation by processing compressed data in the compressed domain, *Image and Vision Computing*, 19(7): 485–500.

227. R. Chang, W. Kuo, and H. Tsai, 2000, Image retrieval on uncompressed and compressed domains DCT coefficients, in *Proc. Int. Conf. on Image Processing* 2: 546–549.

228. H. H. Yu, 1999, Visual image retrieval on compressed domain with Q-distance, in *Proc. ICCIMA '99, Proc. 3rd International Conference on Computational Intelligence and Multimedia Applications*: 285–289.

229. S. Climer and S. K. Bhatia, 2002, Image database indexing using JPEG coefficients, *Patt. Recog.* 35(11): 2479–2488.

230. N. Ahmed, T. Natarajan, and K. Rao, 1974, Discrete cosine transform, *IEEE Trans. Comput.* 23: 90–93.

231. M. Shneier and M. A. Mottaleb, 1996, Exploiting the JPEG compression scheme for image retrieval, *IEEE Trans. Pattern Analysis and Machine Intelligence* 18(8): 849–853.

232. J. Jian, A. J. Armstrong, and G. C. Feng, 2001, Direct content access and extraction from JPEG compressed images, *Patt. Recog.* 35(11): 2511–2519.

233. A. Vailaya, M. Figueiredo, A. K. Jain, and H. Zhang, 2001, Image classification for content-based indexing, *IEEE Trans. on Image Processing* 10(1).

234. A. Conci and E. Castro, 2002, Image mining by content, *Expert Systems with Applications* 23(4): 377–383.

235. J. M. Corridoni, A. D. Bimbo, and P. Pala, 1999, Image retrieval by color semantics, *Multimedia Systems* 7(3): 175–183.

236. J. Hafner, H. Sawhney, W. Equitz, M. Flickner, and W. Niblack, 1995, Efficient color histogram indexing for quadratic form distance functions, *IEEE Trans. Pattern Analysis and Machine Intelligence* 17(7): 729–736.

237. S. Cha and S. N. Srihari, 2002, On measuring the distance between histograms, *Patt. Recog.* 35(6): 1355–1370.

238. K. Sirlantzis and M. Fairhurst, 2001, Optimisation of multiple classifier systems using genetic algorithms, in *Proc. IEEE Int. Conf. on Image Processing*: 1094–1097.

239. K. Sirlantzis, M. C. Fairhurst, and R. M. Guest, 2002, An evolutionary algorithm for classifier and combination rule selection in multiple classifier systems, in *Proc. IEEE Int. Conf. on Pattern Recognition*: 771–774.

240. L. Xu, A. Krzyzak, and C. Suen, 1992, Methods of combining multiple classifiers and their applications to handwriting recognition, *IEEE Trans. System, Man and Cybernetics* 22(3): 418–435.

241. P. Smits, 2002, Multiple classifier systems for supervised remote sensing image classification based on dynamic classifier selection, *IEEE Trans. Geoscience and Remote Sensing* 40(4): 801–813.

242. J. Kittler and F. M. Alkoot, 2003, Sum versus vote fusion in multiple classifier systems, *IEEE Trans. Pattern Analysis and Machine Intelligence* 25(1): 110–115.

243. TelecomsEurope, 2008, Worldwide mobile phone user base hits record 3.25b. http://www.telecomseurope.net/article.php?id_article=4208.

244. InfoTrends, 2006, InfoTrends releases mobile imaging study results. http://www.infotrends-rgi.com/home/Press/itPress/2006/1.18.2006.html.

245. PRNewswire, 2008, Flickr adds video to its popular photo-sharing community. http://www.hispanicprwire.com/generarnews.php?l=in&id=11197&cha=0.

246. M. Flickher, H. Sawhney, W. Niblack, J. Ashley, Q. Huang, B. Dom, M. Gorkani, J. Hafner, D. Lee, D. Petkovic, et al., 1995, Query by image and video content: The QBIC system, *IEEE Computer* 28(Sept.): 23–32.

247. A. Pentland, R. Picard, and S. Sclaroff, 1997, Photobook: Content-based manipulation of image databases, *Commun. ACM* 40(Sept.): 70–79.

248. J. R. Smith and S. F. Chang, 1996, VisualSEEk: a fully automated content-based image query system, in *Proc. ACM Multimedia*: 87–98.

249. T. Gevers and A. W. M. Smeulders, 2000, PicToSeek: Combining color and shape invariant features for image retrieval, *IEEE Trans. Image Processing* 9(Jan.): 102–119.

250. I. J. Cox, M. L. Miller, T. P. Minka, T. V. Papathomas, and P. N. Yianilos, 2000, The Bayesian image retrieval system, PicHunter: Theory, implementation, and psychophysical experiments, *IEEE Trans. Image Processing* 9(Jan.): 20–37.

251. J. Huang, S. R. Kumar, and M. Metra, 1997, Combining supervised learning with color correlograms for content-based image retrieval, in *Proc. ACM Multimedia*: 325–334.

252. Y. Rui, T. S. Huang, M. Ortega, and S. Mehrotra, 1998, Relevance feedback: A power tool for interactive content-based image retrieval, *IEEE Trans. Circuits and Video Technology* 8(Sept.): 644–655.

253. Y. Rui and T. S. Huang, 2000, Optimizing learning in image retrieval, in *Proc. IEEE Int. Conf. Computer Vision and Pattern Recognition* 1(Jun.): 236–243.

254. N. Vasconcelos and A. Lippman, 1999, Learning from user feedback in image retrieval systems, in *Proc. Neural Information Processing Systems*: 977–986.

255. Z. Su, H. J. Zhang, S. Li, and S. P. Ma, 2003, Relevance feedback in content-based image retrieval: Bayesian framework, feature subspaces, and progressive learning, *IEEE Trans. Image Processing* 12(Aug.): 924–937.

256. H. K. Lee and S. I. Yoo, 2001, A neural network-based image retrieval using nonlinear combination of heterogeneous features, *Int. J. Computational Intelligence and Applications* 1(2): 137–149.

257. J. Laaksonen, M. Koskela, and E. Oja, 2002, PicSom: self-organizing image retrieval with MPEG-7 content descriptions, *IEEE Trans. Neural Networks* 13(Jul.): 841–853.

258. P. Muneesawang and L. Guan, 2002, Automatic machine interactions for content-based image retrieval using a self-organizing tree map architecture, *IEEE Trans. Neural Network* 13(Jul.): 821–834.

259. K. H. Yap and K. Wu, 2005, Fuzzy relevance feedback in content-based image retrieval systems using radial basis function network, in *Proc. IEEE Int. Conf. Multimedia and Expo*: 177–180.

260. K. H. Yap and K. Wu, 2005, A soft relevance framework in content-based image retrieval systems, *IEEE Trans. Circuits and Systems for Video Technology* 15(Dec.): 1557–1568.

261. S. Tong and E. Chang, 2001, Support vector machine active leaning for image retrieval, in *Proc. ACM Int. Conf. Multimedia*: 107–118.

262. Y. Chen, X. S. Zhou, and T. S. Huang, 2001, One-class SVM for learning in image retrieval, in *Proc. IEEE Int. Conf. Image Processing*: 815–818.

263. G. D. Guo, A. K. Jain, W. Y. Ma, and H. J. Zhang, 2002, Learning similarity measure for natural image retrieval with relevance feedback, *IEEE Trans. Neural Networks* 13(Jul.): 811–820.

264. L. Wang and K. L. Chan, 2003, Bootstrapping SVM active learning by incorporating unlabelled images for image retrieval, in *Proc. IEEE Int. Conf. Computer Vision and Pattern Recognition*: 629–634.

265. L. Wang and K. L. Chan, 2004, Incorporating prior knowledge into SVM for image retrieval, in *Proc. IEEE Int. Conf. Pattern Recognition*: 981–984.

266. Y. Wu, Q. Tian, and T. S. Huang, 2000, Discriminant-EM algorithm with application to image retrieval, in *Proc. IEEE Int. Conf. Computer Vision and Pattern Recognition*, 222–227.

267. S. Chiu, 1994, Fuzzy model identification based on cluster estimation, *J. Intelligent and Fuzzy Systems* 2(Sept.).

268. X. S. Zhou and T. S. Huang, 2003, Relevance feedback in image retrieval: A comprehensive review, *ACM Multimedia Systems J.* 8(6): 536–544.

269. J. Friedman, Regularized discriminant analysis, 1989, *J. Am. Stat. Assoc.* 84(405): 165–175.

270. M. Swain and D. Ballard, 1991, Color indexing, *Int. J. Comp. Vis.* 7(1): 11–32.

271. S. Markus and O. Markus, 1995, Similarity of color images, in *Proc. SPIE Storage and Retrieval for Image and Video Databases*: 381–392.

272. J. Huang, S. R. Kumar, M. Mitra, W. J. Zhu, and R. Zabih, 1997, Image indexing using color correlograms, in *Proc. IEEE Conf. Computer Vision and Pattern Recognition Conference*: 762–768, San Juan, Puerto Rico.

273. J. R. Smith and S. F. Chang, 1996, Automated binary texture feature sets for image retrieval, in *Proc. Int. Conf. Acoustics, Speech, and Signal Processing*: 2239–2242, Atlanta, GA.

274. X. F. He, O. King, W. Y. Ma, M. J. Li, and H. J. Zhang, 2003, Learning a semantic space from user's relevance feedback for image retrieval, *IEEE Trans. Circuits and Systems for Video Technology* 13(Jan.): 39–48.

275. C. F. Lin and S. D. Wang, Fuzzy support vector machines, 2002, *IEEE Trans. Neural Networks* 13(Mar.): 464–471.

276. V. N. Vapnik, *The Nature of Statistical Learning Theory*. New York: Springer-Verlag, 1995.

Index